ELECTRICAL
CONTROL
FOR MACHINES

7TH EDITION

DIANE **LOBSIGER**   PETER **GIULIANI**   KENNETH **REXFORD**

Australia • Brazil • Canada • Mexico • Singapore • United Kingdom • United States

Electrical Control for Machines, 7TH Edition

Diane Lobsiger, Peter Giuliani, Kenneth Rexford

Vice President, GM Skills & Product Planning: Dawn Gerrain

Product Team Manager: James DeVoe

Senior Director Development: Marah Bellegarde

Senior Product Development Manager: Larry Main

Senior Content Developer: John Fisher

Editorial Assistant: Andrew Ouimet

Vice President Marketing Services: Jennifer Baker

Market Manager: Linda Kuper

Senior Production Director: Wendy A. Troeger

Senior Production Manager: Andrew Crouth

Content Production Management and Art Direction: Lumina Datamatics, Inc.

Technology Project Manager: Joe Pliss

Cover Image(s): ©hüseyin harmandağlı/iStock; ©ildogesto/shutterstock

For product information and technology assistance, contact us at
Cengage Learning Customer & Sales Support, 1-800-354-9706

For permission to use material from this text or product, submit all requests online at **www.cengage.com/permissions.** Further permissions questions can be e-mailed to **permissionrequest@cengage.com**

Library of Congress Control Number: 2014939241

ISBN: 978-0-357-67116-0

Cengage Learning
20 Channel Center Street
Boston, MA 02210
USA

Cengage Learning is a leading provider of customized learning solutions with office locations around the globe, including Singapore, the United Kingdom, Australia, Mexico, Brazil, and Japan. Locate your local office at: **www.cengage.com/global**

To learn more about Delmar, visit **www.cengage.com/delmar**
Purchase any of our products at your local college store or at our preferred online store **www.cengagebrain.com**

*We dedicate this book to our families. Without their love
and support this book would not have been possible.*

CONTENTS

The seventh edition of *Electrical Controls for Machines* continues the tradition of this textbook as an introductory level to electrical machine controls. Significant changes have been made to this edition to update the text to reflect current trends in industry. As the electronics industry continues to change, it is imperative that the information obtained in this text accurately reflects these changes. The circuit diagrams and the descriptions of the circuits throughout the text have been updated to show a more modern representation of the control circuits.

This edition expands on the power and control circuitry required to operate electrical machinery. The bus system and the use of bus plugs are described as a means of power distribution throughout a factory. Transformers calculations and the sizing of transformers is included to assist the technician in understanding what values to expect (voltage and current) when troubleshooting equipment. Insulation classifications, conductor ampacity, and conductor color code is explained to assist the technician in identifying and selecting proper wiring for a given application. Finally, the trend toward 24 VDC as the standard control voltage for equipment is covered.

The introduction to PLCs has been modified to include updated information. The trend away from relay control to PLC control is discussed along with the advantages to using PLCs to control machinery. The need to still study relay circuits is explained as relays are still utilized on PLC equipment to switch loads. The study of relay circuits also provides a basic understanding of control circuits and knowledge that may be easily transferred to an understanding of PLC circuits. Clarification and expansion of important concepts are included such as PLC scan, addressing schemes, power supply functions, memory, I/O status, and peripheral devices. The use of Ethernet as a common protocol for control systems was also added in this edition.

Other updated technology included in this edition is the explanation of variable frequency drives. As the cost of electrical components continues to decline, the use of drives to control motors is continually becoming more affordable.

Several concepts have been explained in the new edition. One topic is the understanding of the normal state for switching devices. A thorough understanding of this concept is required to understand circuit diagrams and the proper operation of the devices. Another topic that is introduced is the concept of fail-safe designs. This practice is critical to the proper functionality of safety circuits, quality control systems, and processes by ensuring the integrity of switching devices.

The safety of workers and equipment is a number one priority in industrial settings. In this edition, an explanation of the risk assessment process and a safety relay circuit has been added. It is important to understand the need for safety rated devices and how they are integrated into a circuit so that unintentional changes to a safety circuit are not made. An explanation of Ground Fault Circuit Interrupters (GFCIs) is also included in this edition.

The globalization and integration of businesses has caused drastic changes to the way that industries function. It is very common for companies to purchase equipment from other regions of the world for use in their operations. It is also quite common for companies to relocate existing machinery to other regions of the world as production volumes fluctuate or as industry requirements change. This edition expands the content of the material to include international equipment specifications. Concerns related to relocating equipment to various regions of the world are also covered. In addition, the use of IEC617 symbols is included whenever a new device symbol is discussed. Questions have been added to the end of relevant chapters to include the conversion of circuits using JIC symbols to the equivalent circuits using IEC617 symbols.

It is imperative that someone troubleshooting equipment has a thorough understanding of the control circuit diagrams. To effectively work on the equipment and for the safety of workers and machinery, the troubleshooter must know how the circuit is supposed to function and what to expect when working on the system. Many modifications have been included in this edition to support the thorough understanding of the control drawings. Various cross-referencing schemes, parts lists, wiring methodology, and push-button layouts have been added to this edition. Some older material, although not readily utilized on new equipment, has been maintained in this edition. This decision was made to provide a comprehensive understanding of basic circuits and for the technician that will need to troubleshoot these devices on older equipment. Questions have been added at the end of each chapter to enhance the comprehension of the material covered in the text. In addition, the recommended web links at the end of every chapter have been updated to reflect accurate Web site addresses.

This text is designed for the student, maintenance technician, process engineer, and sales representative who need a clear and simple understanding of all elements of a complex manufacturing system. Therefore, this text can best be described as providing the reader with a practical understanding of electrical control principles. A prerequisite for the reader is knowledge of the basic theories of electricity and electrical circuits. To meet this prerequisite, a student should have taken a course in basic electricity. A practitioner (technician or engineer) should review a book on the principles of electrical theory and circuits.

The text is designed to be functional in either an educational or industrial environment. In education, no matter the level (vocational high school, two-year technical school, or four-year college), the entire contents should be examined and understood throughout the course. If the book is used as a reference, then appropriate chapters can be read as needed.

We need to teach the fundamentals of process and equipment control. Those mainly affected are the thousands of small builders and users, including the large manufacturer. Most selling organizations, both large and small, include many thousands of manufacturers' agents, distributors, sales engineers, and maintenance personnel. These are the people charged with the responsibility for selling machines and keeping the machines operating. The success or failure of installations may depend on the ability of these people to maintain and troubleshoot equipment properly. The personnel involved need to understand electrical components and their symbols. With this knowledge, they are in a better position to read and understand elementary circuit diagrams.

Six specific areas in which education is needed are (1) electrical and electronic components, (2) control techniques and circuits, (3) troubleshooting, (4) maintenance, (5) electrical standards, and (6) keeping current with changing technology.

1. *Components* are the building blocks of all systems. The core knowledge of the principles and application of each component in a system sets the groundwork for a complete understanding of all facets of a control system.
2. *Techniques and circuits* is the process of building the system from the components. Like building a house, without a concept and plan the structure will fail over time. Techniques and circuits are the lead to a successful plan.
3. *Troubleshooting* machine control circuits involves locating and properly identifying the nature and magnitude of a fault or error. This fault may be in the circuit design, physical wiring, or components and equipment used. The time required and the technique or system used to locate and identify the error are important. Of similar importance are the time and expense involved to put the machine back into normal operating condition.
4. *Preventive maintenance* would eliminate the need for most troubleshooting. Many machines are allowed to operate until they literally fall apart.
5. *Applicable standards* should be followed. If the intended result is the improvement of design and application to reduce downtime and promote safety, electrical standards can be extremely helpful. Where should the education start? The answer is at the beginning, and keep it simple. Even a basic concept, such as the relation between a component and its symbol, can be of benefit to the user.
6. *Keeping current with changing technology* means implementing a strategy for life-long learning. Today the Internet has become a medium for presenting information and expanding knowledge. To expand your knowledge, this text notes important Web sites that should be reviewed occasionally to keep up with latest technological changes.

In addition to the Web sites for specific technologies listed at the end of each chapter, the following Web sites are broad-based sites devoted to technical and training issues.

Standards Organizations

Electrical Inspectors Information: www. joetedesco.com

Institute of Electrical and Electronic Engineers (IEEE): www.ieee.org

International Brotherhood of Electrical Workers (IBEW): www.ibew.org

National Fire Protection Association (NFPA): www.nfpa.org

National Electrical Safety Foundation (NESF): www.nesf.org

National Joint Apprenticeship Training Committee (NJATC): www.njatc.org

National Electrical Contractors Association (NECA): www.necanet.org

National Electrical Manufacturers Association (NEMA): www.nema.org

Underwriters Laboratories Inc. (UL): www.ul.com

Council for the Harmonization of Electrotechnical Standardization of Nations of the Americas (CANENA): www.canena.org

Canadian Standards Association (CSA): www.csa.ca

Indexes of Technical Products

Process index: www.processindex.com

Norm's Industrial Electronics: www. compusmart.ab.ca/ndyrvik/

Process Mart: www.iprocessmart.com

Graybar Electric: www.graybar.com

Electronic Engineer's Master (EEM): www. eem.com

Omega Engineering: www.omega.com

On-Line Technical Publications

Control Engineering: www.controleng.com/

Motion: www.motion.org

Fluid Power Society—*Fluid Power Journal*: www.fluidpowerjournal.com

Allen Bradley—*View* Magazine: www.ab.com/ viewanyware/the_view

Allen Bradley—*AB Journal*: www.ab.com/ abjournal

ControNews and LogixNews: www.ab.com/ controlnews/

Current Issues and News about Manufacturing: www.manufacturing.net

National Electrical code: www.nfpa.org/NEC/ NEChome.org

Independent Web Sites Devoted to Automation Issues

PLC Tutor: www.plcs.net

NEC Information and Training: www.mikeholt.com

Brief Overview of the Chapters

Significant changes have been made to this edition to:

- Explain crucial components of industrial control systems in more depth
- Provide an introduction and general explanation of topics that are important to modern industrial machine control
- Delete obsolete information. Note that some material, although not readily utilized on new equipment, has been maintained. This decision was made to provide a comprehensive understanding of basic circuits and for the technician that will need to troubleshoot these devices.
- Update text to current trends in industry
- Update circuit diagrams to show a more modern representation of control circuits. The description of each circuit has also been updated.
- Questions have been added to Achievement Review section
- Recommended Web Links have been updated to reflect accurate Web site addresses
- Expand text content to international equipment specifications

Chapter 1 ("Transformers and Power Supplies") provides an overview of the power systems utilized on industrial equipment. New topics covered in this edition include:

- Bus system and bus plugs for power distribution

- Transformer calculations to determine turns ratio, voltage, and current values
- Sizing of transformers (moved from Appendix) with example calculations provided
- Concerns related to relocating equipment in various regions of the world
- Trend toward 24 VDC control
- Circuit diagrams
- Insulation classifications, conductor ampacity, and conductor color code
- Use of IEC617 symbols

Chapter 2 ("Fuses, Disconnect Switches, and Circuit Breakers") provides an overview of the means for disconnecting power to machinery. The construction and characteristics of different fuses are also covered. The use and operation of Ground Fault Circuit Interrupters (GFCIs) has been included in this edition.

Chapter 3 ("Control Units for Switching and Communication") introduces operator interface devices. Emphasis is given to pushbuttons, selector switches, and pilot lights. Significant changes have been made to this edition to clarify the normal state of switching devices to further comprehend the device symbols represented on electrical drawings. The operation of the devices has been expanded upon along with the different types of operating heads and switching configurations that are available. Series and parallel circuits have been explained and examples of control circuits have been included.

Chapter 4 ("Relays") describes the construction and operation of electrically operated relays. The relay is a functional device that can be used in many different control system operations, such as logical sequencing and control of motions. Some of the changes in this edition include the explanation of the normal states for relay contacts, the clarification and inclusion of timing diagrams for pneumatic timing circuits, the progression to PLC systems, and circuit modifications.

Chapter 5 ("Solenoids") is an introduction to the general operation of solenoids and solenoid-operated control valves. Some of the changes in this edition include the clarification and detailed explanation of solenoid/valve operations along with circuit modifications.

Chapter 6 ("Types of Control") covers the different types of control theories applied to controlling a process actuator. Each control method has unique characteristics that provide a reference for determining the expected controllability, which ultimately will affect the quality and productivity of the system. In this edition, the discussions involving different types of control are expanded. Additional examples are also provided.

Chapter 7 ("Motion Control Devices") is a comprehensive review of the control devices such as limit switches, proximity switches, and photoelectric transducers used in the control of moving actuators. The normal states for switching devices have been explained and circuit modifications have been incorporated into this edition.

Chapter 8 ("Pressure Control") is devoted to achieving an understanding of systems that require the precise control of pressures exerted by an actuator onto a process. The normal states for pressure switches and the implementation of pressure transducers have been explained in this edition. Circuit modifications have been also been made.

Chapter 9 ("Temperature Control") analyzes circuits and controllers used in industrial systems in which precise temperature control must be maintained within the process. In this edition, some clarifications to temperature control are provided along with modifications to circuits.

Chapter 10 ("Time Control") describes the operation and application of timers to timed, sequentially controlled events. Some changes to this edition include clarifications in regards to position sensing, the progression to timing control in PLCs, and circuit modifications.

Chapter 11 ("Count Control") covers counters and their applications in control sequences that depend on counted events. Some changes to this edition include clarifications in regards to counter operations and circuit modifications.

Chapter 12 ("Control Circuits") incorporates information learned in previous chapters to practical applications of electrical control circuits using ladder logic diagrams. Significant changes have been made in this edition to the control circuits and the supporting descriptions of the circuits throughout this chapter.

Chapter 13 ("Motors") provides insights into the theory and operation of AC and DC motors. Changes have been made in this edition to the order of the material and the content in this chapter. Some of the modifications include the analysis of a single loop DC motor and the addition of the Variable Frequency Drives section.

Chapter 14 ("Motor Starters") explains how motor starters are used to protect and control motors. Full-voltage and reduced-voltage magnetic types as well as solid state types are covered. Significant changes have been made in this edition to the control circuits and the supporting descriptions of the circuits throughout this chapter.

Chapter 15 ("Introduction to Programmable Control") is an introduction to the concepts associated with Programmable Logic Controllers. In this edition, numerous modifications have been made to this chapter. These changes include:

• Clarification of concepts from relay control that apply to PLCs
• Clarification of how the classification of devices change from a relay circuit to a PLC design
• Discussion in regards to the advantages of PLCs over relay circuits included
• Basic ladder logic fundamentals provided
• Explanation of different addressing schemes provided
• Clarification of PLC scan
• Clarification of PLC power supply functions
• Detailed explanation of PLC memory contents and I/O status as PLC scan is executed
• Updated information for peripheral and support devices
• Significant modifications to PLC circuits and their associated descriptions
• Introduction to fail-safe design practices

Chapter 16 ("Industrial Data Communications") explores the terminology, configuration, and issues of data communication within an industrial environment. A section on Ethernet is included in this edition.

Chapter 17 ("Quality Control") is a review of the devices and control concepts used to monitor and control product quality in a production process. In this edition, a section explaining the costs associated with the lack of a good quality control system and the need for fail-safe design practices has been added.

Chapter 18 ("Safety") presents the issues and technology that affect worker and equipment safety. In this edition, an explanation of the risk assessment process and a safety relay circuit has been added. Also, circuit modifications have been made throughout this chapter.

Chapter 19 ("Troubleshooting") provides the principles and techniques needed to isolate a problem associated with a control circuit. Changes have been made in this edition to the control circuits and the supporting descriptions of the circuits throughout this chapter.

Chapter 20 ("Designing Control Systems for Easy Maintenance") is devoted to the general requirements to be considered when designing and maintaining control circuits. Various cross-referencing schemes, parts lists, wiring methodology, and push-button layouts have been added to this edition. Also, changes have been made to the control circuits and the supporting descriptions of the circuits in this chapter.

Supplements

Lab Manual: A lab manual to accompany this text is also available. ISBN 9781285169057.

An online Instructor Companion Web site contains an Instructor Guide with answers to end of chapter review questions, testbanks, and Chapter presentations done in PowerPoint.

Accessing an Instructor Companion Web site from Single Sign On Front Door

1. Go to: http://login.cengage.com and login using the Instructor email address and password.
2. Enter author, title or ISBN in the Add a title to your bookshelf search box, click on Search button
3. Click Add to My Bookshelf to add Instructor Resources
4. At the Product page click on the Instructor Companion site link

New Users
If you are new to Cengage.com and do not have a password, contact your sales representative.

ACKNOWLEDGMENTS

The authors and Cengage Learning gratefully acknowledge the review panel for their suggestions in the development of this edition. Our thanks to:

Brett McCandless
Vincennes University
Vincennes, IN

Russ Davis
Delta College
University Center, MI

Denny Owen
Kennebec Valley Technical College
Fairfield, ME

Murry D. Stocking,
Ferris State University,
Big Rapids, MI

Bill Welborn,
Alamance CC,
Graham, NC

Mickey Giorgano,
Mississippi State University,
Mississippi State, MS

Technical guidance and illustrations were provided by the following companies. Appreciation is expressed to them for their cooperation and assistance.

Acme Electric—Acme Transformer Division, Lumberton, NC 28358 (www.acme-electric. com)

Allen-Bradley, a Rockwell International Co., Milwaukee, WI 53204 (www.ab.com)

Allen-Bradley, a Rockwell International Co., Highland Heights, OH 44143 (www.ab.com)

Automatic Timing and Controls Co., Inc., King of Prussia, PA 19406 (www.automatictiming. com)

Banner Engineering Corporation, Minneapolis, MN 55441

Barber-Colman Co., Loves Park, IL 61132-2940 (www.barber-colman.com)

Barksdale, Division of Crane Company, Los Angeles, CA 90058-0843 (www.barksdale. com)

Chromalox, Pittsburg, PA 15238 (www. chromalox.com)

Columbus Controls, Columbus, OH 43081 (www. columbuscontrols.com)

Detroit Coil Co., Ferndale, MI 48270

Divelbiss Corporation, Fredericktown, OH 43019 (www.divelbiss.com)

Eagle Signal Controls, Austin, TX

Fenwal Inc., Ashland, MA 01721-2150 (www. fenwalcontrols.com)

Ferraz Shawmut, Newbury, MA 01950 (www. gouldshawmut.com)

Fluke Corporation, Everett, WA 98206 (www. fluke.com)

General Electric Co., Motor Sales Division, Fort Wayne, IN 46801 (www.geindustrial.com)

General Electric Co., GE Electrical Distributor of Controls, Plainville, CT 06062 (www. geindustrial.com)

Hoffman, Division of Pentair Company, Anoka, MN 55303 (www.hoffmanonline.com)

HPM Division of Taylor Industrial Services, Mount Gilead, OH 43338 (www.hpmcorp.com)

ifm efector inc., Exton, PA 19341 (www.ifmefector.com)

Larry Flanery, Kendall Electric, Saginaw, MI

Liebert Corporation, Columbus, OH 43229 (www.liebert.com)

Logex, Inc., Columbus, OH 43085

McNaughton-McKay Electric Company (www.mc-mc.com)

Mercury Displacement Industries, Inc., Ewardsburg, MI 49112 (www.mdius.com)

MTS Sensors Division, Cary, NC 27513 (www.mtssensors.com)

National Fire Protection Association, Quincy, MA 02269-9101 (www.nfpa.org)

POWERTEC Industrial Motors, Rock Hill, SC 29732 (www.powertecmotors.com)

Ronan Engineering Co., Woodland Hills, CA 91367 (www.ronan.com)

Siemens Energy and Automation Inc., Programmable Controls Division, Peabody, MA 01960 (www.sea.siemens.com)

Solid Controls, Inc., Hopkins, MN 55343 (www.solidcontrols.com)

Square D/Schneider Electric, Milwaukee, WI 53201 (www.squared.com)

Standish Industries, Lake Mills, WI 53551 (www.hitekelec.com)

Superior Electric Co., Bristol, CT 06010 (www.superiorelectric.com)

Temposonics Inc., Division of MTS Systems Corporation, Plainsville, NY 11803 (www.temposonics.com)

Vickers Inc., Division of Eaton Aeroquip, Troy, MI 48084 (www.eatonhydraulics.com)

TECO-Westinghouse Motor Company, Pittsburgh, PA 15222 (www.teco-wmc.com)

XYMOX Technologies, Inc., Milwaukee, WI 53201 (www.xymoxtech.com

Yellow Springs Instrument Co., Inc., Industrial Division, Yellow Springs, OH 45387 (www.ysi.com)

Revising author Diane Lobsiger is an Assistant Professor at Delta College in University Center, Michigan. She is also the coordinator of the electrical department for the Technical Trades and Manufacturing Division. Diane enjoys working with the students and the faculty at Delta College.

Diane received her Bachelor of Science degree in Electrical Engineering from Auburn University, Alabama. While attending graduate school, she was employed as a teachers' assistant and received an Honorable Mention—Outstanding Student-Teacher Award from Michigan State University. During this time, she was inducted into several honoraries including Eta Kappa Nu (EE Honorary), Tau Beta Pi (Engineering Honorary), Pi Mu Epsilon (Math Honorary), and Sigma Pi Sigma (Physics Honorary). Diane received her Master of Science degree in Electrical Engineering from Michigan State University, Michigan. She also obtained her Secondary Teaching Certification in Mathematics and Physics from Saginaw Valley State University located in University Center, Michigan.

Diane worked as a Senior Manufacturing Controls Engineer for 24 years at a manufacturing facility in Michigan. During this time, she gained a wealth of knowledge and extensive experience in the design, troubleshooting, specification writing, and the approval of machine controls systems. Her expertise includes relay control systems, programmable logic controllers (PLCs), human-machine interfaces (HMIs), servo systems, adjustable frequency drives, and fluid power systems.

Diane is married and has three children. She would like to thank her family for their continued love and support throughout her endeavors. Her life has been filled with opportunities, challenges, rewards, and blessings. The most fulfilling experiences have come from the opportunity to raise children, mentor college students, and work with young people throughout the community. She enjoys a wide variety of interests from sports to scrapbooking. In her free time, she can usually be found fishing at her cabin.

INTRODUCTION to Electrical Control—Development of Circuits

To understand electrical control circuits, it is necessary to examine three basic steps in developing a circuit.

The FIRST step is to know what work or function is to be performed. For example, a simple problem may be to light a lamp. The solution can be achieved by completing a path for electrical energy from a source such as a battery to a load such as a lamp. For convenience, a switch is used to open or close the path. When the switch is open, electrical energy is removed from the lamp, which is said to be *de-energized*. When the path of electrical energy is closed, the lamp is said to be *energized* and performs a function of illumination. See Figures 1

and 2, in which a battery is used as the source of electrical energy. These drawings are known as *pictorial* drawings because they show a picture of the actual components—battery, switch, and lamp.

In industrial electrical control circuits, symbols are used to represent the components. Figures 3 and 4 show the use of symbols for the components (battery, switch, and lamp). These diagrams are referred to as *schematic* or *control circuit diagrams*.

Figures 3 and 4 can be redrawn in a slightly different form as shown in figure 5. The circuit performs exactly the same function since the lamp is energized, when the switch is closed. This type of drawing is called a *ladder* diagram. (The ladder-type diagram is used throughout the rest of the book.) In this drawing, the battery is shown as the source of electrical energy feeding the circuit, or the loads. In a ladder diagram, the voltage source is always depicted by two vertical lines (or sides of

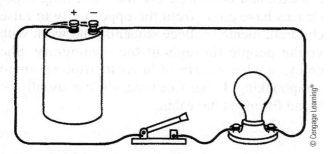

Figure 1 With the switch open, the path is open and the lamp is de-energized.

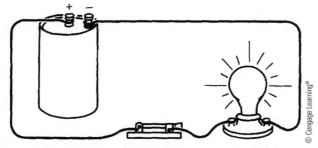

Figure 2 With the switch closed, the path is closed and the lamp is energized.

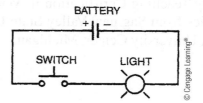

Figure 3 Schematic showing the switch and path open and the light de-energized.

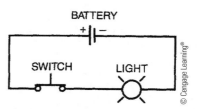

Figure 4 Schematic showing the switch and path closed and the light energized.

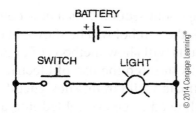

Figure 5 **Ladder diagram showing the source along the vertical lines and the load connected in a circuit drawn horizontally.**

the ladder). The circuit is then drawn horizontally with all switching devices shown in their normal or "at rest" condition being connected to the load. In this case, the switch is drawn as normally open and is connected to the light. Since the switch is open in this condition, there is no path for electricity to flow to the light. However, when the switch is closed, a path will be established between the light and the battery. Electricity will flow in the circuit and the light will illuminate. The vertical lines extend downward past the light and switch circuit indicating the potential to connect additional loads to the existing circuit.

Notice in Figure 5 that an important symbol has been introduced. The symbol is "conductors connected" and is shown separately in Figure 6. A similar symbol, which is not used here but appears many times in later diagrams, is "conductors not connected" and is shown in Figure 7.

The SECOND step is to know the operating conditions under which the starting, stopping, and controlling of the process is to take place. Practically all conditions fall into one or more general groups, as affected by:

• Position
• Time
• Pressure
• Temperature

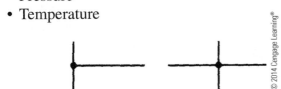

Figure 6 **Symbols for conductors connected.**

Figure 7 **Symbols for conductors not connected.**

In the chapters on components that follow, each of these conditions can generally be associated with certain components. In many cases, although the actual initiating of a cycle may be through the manually operated push-button switch, certain conditions must be met before the circuit can be closed. This cycle initiation could be one or any combination of the conditions previously listed.

The THIRD step in the development of a circuit is selecting the desired control conditions. There are many times that a circuit must be capable of operating under certain sets of conditions to produce the desired results. For example, a circuit may be required to operate a machine under manual, semiautomatic (single-cycle) or fully automatic (continuous-cycle) operation.

After a decision is made on which of these types of operation is to be used, a selection is made. For reasons of safety to both the machine and operating personnel, the machine must operate in the selected manner.

Review the three basic steps in developing the control circuit diagram.

1. Know what work or function is to be performed.
2. Know the conditions for starting, controlling, and stopping the process.
3. Arrange for selecting the desired control conditions: manual, semiautomatic, or automatic.

As the student progresses through the study of components, the symbol for each component will be prominently displayed. It is very important that the symbol becomes closely associated with the component. In Appendix A, all of the symbols used will be shown for review.

In understanding electrical control circuit diagrams, there are fundamental problems that should be recognized and overcome if progress is to be made. Some of these problems are:

• Starting with a circuit that is too large or too complicated.
• Failing to carry through a mental picture of the component into the electrical circuit.
• Failing to relate physical, mechanical, or environmental actions into devices that convert these actions into electrical signals.

• Failing to understand that an electrical circuit must perform the correct functions and not perform those actions that will result in damaged components, danger to the operator or machine, or a faulty product.

In addition, one of the biggest problems in reading circuits is gaining a clear understanding of a switch or contact condition. The condition must be properly presented in the ladder diagram, and the user must properly interpret its use in a process or on a machine. Diagrams will be shown in each of the component sections. As a new component is introduced, the symbol will be shown in the circuit. These circuits will show methods of obtaining specific actions through the use of electrical components. Ultimately all circuits designed for specific actions will need to be assembled into a complete circuit for the overall operation of a machine.

The important point here is to become acquainted with components and their use in the small circuits, and work with only one or a few components at a time.

Transformers and Power Supplies

OBJECTIVES

After studying this chapter, you should be able to:

- Give two reasons for energizing machine control systems at 24 volts DC (VDC).

- Explain how to obtain 120 volts from a higher line voltage through the use of a transformer.

- Define *turns ratio* in a transformer.

- Identify the symbol for a dual-primary, single-secondary control transformer.

- Draw a connection diagram for a dual-primary, single-secondary control transformer to a higher voltage line and to a 120-volt control circuit.

- Define *regulation* in a transformer.

- Explain the method for calculating regulation in a transformer.

- Calculate the size of a transformer for a given load.

- Explain what causes temperature rise in a transformer.

- Explain the considerations to be taken into account when converting between 50 Hz and 60 Hz systems.

- Explain the basic operation of different types of power supplies.

- Explain the basic function of the uninterruptible power system.

- List the uses of uninterruptible power systems during undesirable power disturbances.

1.1 Control Transformers

In the electrical control circuits shown in "Introduction to Electrical Control—The Development of Circuits" (Figures 1 through 5), a battery is the source of electrical energy. It supplies a form of electrical energy known as *direct current* (DC). Most control circuits in industry today utilize DC to provide electrical energy for devices that control equipment. The DC voltage source on machinery is a power supply instead of a battery. The power supply is connected to a form of electrical energy called *alternating current* (AC). The power supply converts the incoming AC voltage to DC for use in the control circuit.

The main power that is supplied to industries in the United States is three phase (3ϕ), 480 volts alternating current (VAC), at a frequency of 60 Hertz (Hz). Some companies can run their facility from a 240 VAC system. However, the use of 480 VAC is most prominent.

The 480 VAC supply is typically distributed throughout a plant by the use of a bus system (see Figure 1-1). A bus system consists of copper bars that are capable of handling the large amount of current that is required to run many machines simultaneously. Bus plugs may be attached to the bus to provide a connection point from the power distribution system to the equipment (see Figure 1-2).

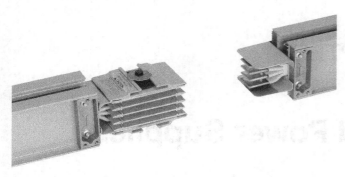

Figure 1-1 **Bus system.** *(Siemens Industry, Inc. provided the image. All rights reserved.)*

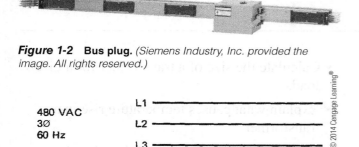

Figure 1-2 **Bus plug.** *(Siemens Industry, Inc. provided the image. All rights reserved.)*

480 VAC
3Ø
60 Hz

L1 ——————————————
L2 ——————————————
L3 ——————————————

Figure 1-3 **Three-phase power line.**

These can be plug-in or bolt-on type. The plugs may be fused with a disconnect or they may contain a circuit breaker. The bus plugs must be locked out any time power must be disconnected to the incoming lines supplying the equipment. The power wiring is usually run from the bus plug to the electrical cabinet of the machine inside of the metallic conduit.

The electrical drawings for a piece of equipment will contain the wiring schematic for every electrical device connected to the equipment. The three-phase power source will be identified at the beginning of the drawings in the upper left-hand corner. The incoming wires are labeled L1, L2, and L3 and are shown as three parallel horizontal lines on the drawings (Figure 1-3).

Motors, heaters, and other devices that require 480 VAC for proper operation are connected as loads to the incoming power wiring circuit. Devices that operate at other voltage levels require the use of a transformer to convert the incoming 480 VAC to the proper voltage level.

Traditionally, most electrical control systems on machines were energized at 120 VAC. There were at least two good reasons for this: safety and the use of standard-design components. The

transformer that converts the 480 VAC to 120 VAC for use in the control circuit is called the *control transformer.* Over the past decade, industry has moved away from the traditional 120 VAC design. New equipment generally uses 24 VDC as the primary control circuit voltage. This conversion to low level DC has made the equipment even safer for the operators who run the machines and for the personnel who services the equipment. As this trend continues, 120 VAC devices are becoming obsolete and low voltage DC devices are becoming more commonplace.

Even when the control devices on a machine are powered by DC circuits, the need for a 120 VAC control transformer still exists. The standard voltage for use in households throughout the United States is 120 VAC. Therefore, convenience receptacles included on the equipment will require this voltage level. These receptacles provide the power necessary to run peripheral devices for troubleshooting the equipment. They also can be used to operate devices such as lights, fans, tools, or computers.

In addition, some components still require the 120 VAC for proper operation. The power supply that supplies the 24 VDC control voltage for the machine is typically powered from the 120 VAC circuit. For these reasons, the 120 VAC control transformer will usually still be included on new equipment that is built for use in the United States.

The control transformer is single phase, requiring a connection to any two of the three power lines. The transformer used in industrial machine control consists of at least two separate coils wound on a laminated steel core. The line voltage is connected to one coil, called the *primary.* The control load is connected to the other coil, called the *secondary.* Where the voltage is reduced from the primary to the secondary, the transformer is called a *step-down transformer.* In a *step-up transformer*, the voltage is increased from the primary to the secondary.

The simplest arrangement uses only two coils. One coil is used for the primary, the other for the secondary. The voltage is directly proportional to the number of turns in each coil. This relationship may be shown by the equation:

$$\frac{V_p}{V_s} = \frac{N_p}{N_s} \qquad \textbf{Eq 1.1}$$

where: V_p = primary voltage
V_s = secondary voltage
N_p = number of turns of wire on the primary coil
N_s = number of turns of wire on the secondary coil

The quantity N_p/N_s is referred to as the *turns ratio* for the transformer. This value is expressed as a ratio and is not simplified to a single value.

Example 1

Find the turns ratio for a transformer with a primary voltage of 240 VAC and a secondary voltage of 120 VAC.

Solution

Using Equation 1.1 $V_p/V_s = N_p/N_s$

Substituting V_p = 240 VAC and V_s = 120 VAC we obtain N_p/N_s = 240 VAC/120 VAC

Reducing this equation by dividing the numerator and the denominator by a common factor of 120, results in a final solution of N_p/N_s = 2/1

Note that the answer was left as a ratio of 2 / 1 (stated as 2 to 1) instead of an answer of 2.

The turns ratio is commonly written as a ratio $N_p : N_s$. For this problem, we would write the answer as 2:1 and state the answer as 2 to 1.

Example 2

A transformer has a turns ratio of 4:1. If the transformer is connected to a primary voltage of 480 VAC, what is the secondary voltage?

Solution

Using Equation 1.1 $V_p/V_s = N_p/N_s$
Substituting the given values of V_p = 480 VAC
$$N_p/N_s = 4/1$$

we obtain the equation
480 VAC/V_s = 4/1

Using cross multiplication to solve ratio equations results in
480 VAC $\times$ 1 = 4 $\times$ V_s

Solving for V_s by dividing both sides of the equation by 4 shows that
V_s = 480 VAC/4 = 120 VAC

It is possible for the line voltage in a plant to vary from approximately 460 VAC to 500 VAC. As the primary voltage varies, the secondary voltage will also vary accordingly. For example, if the line voltage in a plant were to drop from its normal 480 VAC to 460 VAC, the transformer in Example 2 would deliver 115 VAC to the secondary. Similarly, if a plant's power source voltage were to drop from its normal 240 VAC to 230 VAC, a transformer having a 2:1 ratio would deliver 115 VAC to the secondary.

Another useful equation for transformers involves the current in the primary and secondary circuits. The current is inversely proportional to the number of turns in each coil. This relationship may be shown by the equation:

$$\frac{N_p}{N_s} = \frac{I_s}{I_p} \qquad \textbf{Eq 1.2}$$

where: I_s = secondary current
I_p = primary current
N_p = number of turns of wire on the primary coil
N_s = number of turns of wire on the secondary coil

Combining Equations 1.1 and 1.2 results in the equation:

$$\frac{V_p}{V_s} = \frac{N_p}{N_s} = \frac{I_s}{I_p} \qquad \textbf{Eq 1.3}$$

Example 3

A transformer is connected to a 480 VAC line. The secondary voltage is 240 VAC. The load connected to the secondary draws 10 Amps.

a. What is the turns ratio of the transformer?
b. How much current flows in the primary circuit?

Solution

a. Using Equation 1.1 $V_p/V_s = N_p/N_s$
Substituting the given values of $V_p = 480$ VAC
$$V_s = 240 \text{ VAC}$$
we obtain the equation
480 VAC/240 VAC $= N_p/N_s$

Reducing this equation by dividing the numerator and the denominator by a common factor of 240, results in a final solution of $N_p/N_s = 2/1$

b. Using Equation 1.3 $V_p/V_s = I_s/I_p$
Substituting the given values of $V_p = 480$ VAC
$$V_s = 240 \text{ VAC}$$
$$I_s = 10 \text{ A}$$
we obtain the equation
480 VAC/240 VAC $= 10$ A/I_p

Using cross multiplication to solve ratio equations results in
480 VAC $\times I_p = 240$ VAC $\times 10$ A

Dividing both sides by 480 VAC results in
$I_p = (240$ VAC $\times 10$ A$)/480$ VAC $= 5$ A

Returning to Equation 1.3 $V_p/V_s = I_s/I_p$ and cross multiplying results in the relationship:

$$\mathbf{V_p \times I_p = V_s \times I_s} \qquad \textbf{Eq 1.4}$$

The power delivered by a source or the power consumed by a load is equal to the voltage multiplied by the current. Therefore, Equation 1.4 shows that the power in the primary is equal to the power in the secondary circuit. This relationship will be true for ideal transformers where it can be assumed that there are no losses incurred in the system. Transformers are rated by this power relationship, which is expressed in volt-amps (VA). Larger transformers are rated in kVA where 1 kVA is equal to 1000 VA. Transformers are available in sizes from 50 VA to 10 kVA.

Multiple coil windings may be used on the primary and secondary sides of a transformer. The user may connect the coil windings in a manner that allows the transformer to provide the proper

secondary voltage depending on the line voltage that is available. Multiple coil windings are usually provided on the primary side of the transformer to make it easier to connect to different power voltages. Special primary windings with multiple taps for 200 - 208 - 240 - 480 - 575 V and a 120 V secondary are available.

The most widely used control transformers have a dual-voltage primary of 240/480 VAC. They have an isolated secondary winding to provide 120 V for the load.

The symbol for the control transformer is shown in Figures 1-4 and 1-5. Two primary coils and one secondary coil are used. When the control transformer primary coils are connected to a 240-V power source, the two coils are connected in *parallel*. When the primary coils are connected to a 480-V power source, the coils are connected in *series*.

When two coils with the same number of turns (same voltage rating) are connected in parallel, the effective number of turns for determining the turns ratio remains the same as if only one coil were used. When two coils with the same number of turns are connected in series, the numbers of turns on each coil are added together.

When two separate coils are used on the primary side, the arrangement is called a *dual primary*. When two separate coils are used on the secondary side, it is called a *dual secondary*.

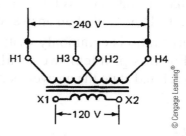

Figure 1-4 Primary coils connected in parallel.

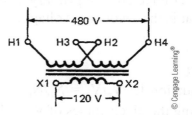

Figure 1-5 Primary coils connected in series.

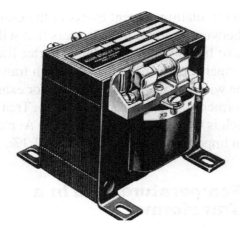

Figure 1-6 Control-circuit transformer with built-in fuse block.
(Courtesy of Rockwell Automation, Inc.)

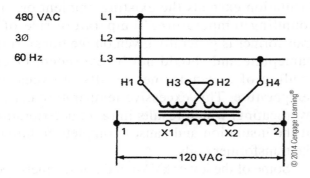

Figure 1-7 Transformer connected to three-phase power line.

The transformer is available either in an open type or in its own enclosure. At one time, the general practice was to use the open type and panel mount it in the control cabinet. With panel-mounted transformers, however, high allowable temperature rise may create unwanted high temperatures in the control cabinet. With this condition, the transformer in its own enclosure should be used, mounted on the outside of the cabinet.

The transformer either has screw-type terminals on the coil, or the leads are brought out to a terminal block. In some cases, fuses or circuit breakers are offered as an integral part of the transformer installation.

A typical transformer is shown in Figure 1-6. A diagram of a three-phase power supply and control transformer is shown in Figure 1-7. In that figure, the primary side of the transformer is connected to one phase (L1 to L3) of the three phase power source. The secondary is supplying 120 VAC to the vertical sides of the ladder-type control circuit diagram.

1.2 Transformer Regulation

Voltage *regulation* in transformers is the difference between the no-load voltage and the full-load voltage. This value is usually expressed in terms of a percent. It can be calculated as follows:

$$\frac{\text{No-load voltage} - \text{Full-load voltage}}{\text{Full-load voltage}} \times 100$$

Using an example in which the transformer delivers 100 V at no load and the voltage drops to 95 V at full load, the regulation would be calculated as follows:

$$\frac{100 - 95}{95} = \frac{5}{95} \times 100 = 5.26\%$$

It is generally desirable to have the regulation for a control transformer in the range of 2 to 3%.

1.3 Sizing a Transformer

Transformers are available in many different sizes. A transformer will be selected based on the primary voltage and frequency available, the secondary voltage required, and the capacity required in volt-amps (or kVA).

The transformer must be sized to handle the full load current for all of the devices that may be in operation at any given time. This process involves determining the current requirement for each load in the circuit individually. The overall current draw for the circuit may then be calculated by adding together the individual loads. Two calculations should be performed to determine the size of the transformer. Consideration should be given to the continuous or sealed current for all loads in the system. In addition, the inrush current characteristics of the coils on relays, contactors, motor starters, solenoids, etc. must be taken into account. This information is available from the manufacturer of the component. The procedure is as follows:

- Calculate the total maximum continuous or sealed current by adding the continuous current drawn by all the loads that will be in an energized condition at the same time. Multiply this figure by ¾ (1.25).

- Calculate the total maximum inrush current by adding the inrush current of all the coils that will be energized together at any one time. Multiply this figure by ¼ (.25).

Using the larger of the two figures you have just calculated, multiply this current by the control voltage. This product is the VA required by the transformer load. If the resulting figure drops below a commercially available transformer size, use the next larger size. For example, you may have calculated that the total VA required is 698. Then use a 750 VA transformer, which is commercially available.

Example 4

Calculate the kVA requirement for a transformer whose secondary will be connected to a 120 VAC circuit with the following loads (assume inrush calculation is not needed):

20 relays with a full load current (FLA) of .5 A each

10 pilot lights with a FLA of .2 A each

4 solenoids with a FLA of 1 A each

Solution

Calculate the total current required for all of the devices to operate simultaneously.

Device	FLA (A)
Relays	10
Pilot Lights	2
Solenoids	4
Total	16 A

Total current rating = 16 × 1.25 = 20 A

kVA rating = 120 VAC × 20 A = 2400 VA

Select the next highest available size. You would order a 2.5 kVA transformer.

1.4 Operating Transformers in Parallel

Single-phase transformers can be used in parallel only when their voltages are equal and impedances are approximately equal. If unequal voltages are used, a circulating current exists in the closed network between the two transformers that will cause excessive heating and result in a shorter life of the transformer. Impedance values of each transformer must be within 7.5% of one another. For example, if Transformer A has an impedance of 4%; Transformer B, which is to be connected parallel to A, must then have an impedance between 3.7% and 4.3%.

1.5 Temperature Rise in a Transformer

Temperature rise in a transformer is the amount by which the temperature of the windings and insulation exceeds the existing ambient or surrounding temperature. Temperature rise of a transformer is generally given on the transformer nameplate and should not be exceeded. Overloading of a transformer results in excessive temperature. This excessive temperature causes overheating, which results in rapid deterioration of the insulation and causes complete failure of the transformer coils.

Some of the lower kVA-rated transformers (below 1 kVA) that are rated at 60 Hz can be operated satisfactorily at 50 Hz. However, at higher kVA ratings 60-Hz rated transformers operating at 50 Hz will produce a greater heat rise. Therefore, it is not advisable to use higher kVA-rated transformers in 50-Hz power circuits.

1.6 50-Hz vs 60-Hz Operation

Globalization has had a significant impact on industry. It is now very common to purchase equipment produced in other countries or to build equipment for use in other countries. In addition, companies frequently transport existing equipment to other regions of the world as the production volumes or economic situations vary within their organizations. It is important to understand the governing codes and regulations, standards, and mains power supply available at the destination location for the machine to ensure the proper and safe operation of the equipment. Worldwide, the power distribution systems vary by the use of different voltages, frequencies, and grounding systems. Even the types of plugs and receptacles used for common household items vary in different regions of the world.

50 Hz is the standard frequency that is utilized in Europe, Asia, and many countries throughout the world. North America has standardized on the use of 60 Hz for its power distribution systems. When moving equipment, the ratings of each transformer must be evaluated. A transformer that is designed for use in a 60-Hz system may not be capable of withstanding the heat generated by use in a 50-Hz circuit. A transformer designed for 50-Hz operation will actually run cooler when placed in a 60-Hz system and would not be a concern in regards to heat generation. However, the nominal voltage of the primary circuit and the amount of variation of the voltage from the nominal value must also be taken into consideration. Remember that the turns ratio of the transformer dictates the amount of voltage available on the secondary of the transformer as the primary voltage is changed. Therefore, the correct voltage may not be available on the secondary to operate the loads on the equipment at a safe operating level. Also, if the voltage level varies significantly around the nominal value, special considerations may need to be implemented to maintain the voltage at a safe operating range for the devices. For these reasons, it is a common practice to replace all of the transformers on equipment when the machinery is relocated to different regions of the world.

Another item that is typically replaced on relocated equipment is the induction motors. Induction motors vary in speed based on the frequency of their supply voltage. A motor designed for use with a 50-Hz system will run faster when connected to a 60-Hz supply. Motors and the relationship between frequency and speed will be covered in more detail in Chapter 13.

1.7 Constant Voltage Regulators*

The constant voltage regulator (CVR) consists of a leakage reactance, a ferroresonant transformer

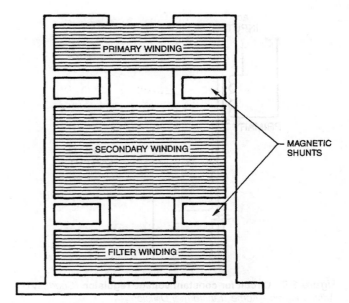

Figure 1-8 **Constant voltage regulator.** *(Courtesy of Acme Electric—Acme Transformer Division.)*

with an additional pair of magnetic shunts, and a filtering winding. Together they develop a regulated, low-distortion sinusoidal output. The circuit is designed so that the segment of the core under the secondary winding (Figure 1-8) will saturate and ferroresonate with the AC capacitor once each half-cycle, limiting the output voltage to a fixed value. The primary-to-secondary leakage reactance and AC capacitor are tuned to achieve ferroresonant regulation of the output over a broad range of input voltage.

The second pair of magnetic shunts and filter winding is incorporated to soften the secondary core saturation effect, cancelling the harmonic voltages that are present in conventional CVRs. The filtering winding is connected in series with the AC capacitor (Figure 1-9). This forms an LC (inductance/capacitance) trap to filter out the low order of harmonics generated by ferroresonant action. A cutaway view of the Acme transformer coil is shown in Figure 1-10.

Ferroresonance (transformer) is a phenomenon usually characterized by overvoltages and very irregular wave shapes. It is associated with the excitation of one or more saturable inductors through capacitance in series with the inductor.

---

*Information courtesy of Acme Electric—Acme Transformer Division.

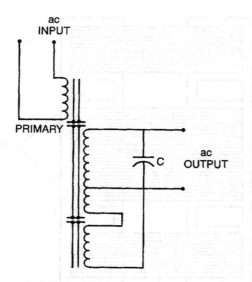

Figure 1-9 Circuit for constant voltage regulator. *(Courtesy of Acme Electric—Acme Transformer Division.)*

Figure 1-10 Cutaway view of a CVR transformer coil. *(Courtesy of Acme Electric—Acme Transformer Division.)*

1.8 Power Supplies for Control Voltage

The primary purpose of a power supply is to convert the incoming AC voltage to a DC voltage for use by the control circuit. The power supply is usually connected to the 120 VAC circuit on equipment used in the United States.

The simplest method to convert from AC to DC is by using an unregulated power supply. This unit will consist of a transformer, a rectifier, and a filter circuit. The transformer will change the incoming 120 VAC to a lower voltage level. The rectifier will then convert this lower level AC signal to a pulsating DC signal. The filtering circuit will smooth out (filter) the pulsating waveform to provide the DC voltage signal output. However, the output voltage of an unregulated power supply will vary as the load changes or as the incoming AC line voltage varies. If the system voltage drops to an unusually low level, components may begin to malfunction and drop out (de-energize).

A linear regulated power supply will generate a consistent output voltage that does not vary with the incoming voltage level or with the load applied to the system. A somewhat complex regulator circuit is used to maintain the voltage at the proper level. These power supplies provide excellent regulation with very little ripple on the output signal. However, they are inefficient and have a limited input range.

Switching regulated power supplies use complex circuitry to obtain a constant output voltage that is isolated from the incoming voltage supply. These power supplies convert the incoming AC voltage to a DC signal by rectifying the source voltage and filtering the signal. This DC signal is then switched at a high frequency to obtain a pulse width modulation (PWM) signal. The PWM signal is then sent through a transformer and rectified again to produce the DC output signal. This conversion from AC to DC to AC and finally back to DC essentially isolates the output power from variations in the utility supply. In addition, a sensing circuit can monitor the output voltage and adjust the PWM switching to maintain a consistent output voltage.

An uninterruptible power supply (UPS) will provide an emergency backup source of electricity in the event of a power outage.

All machine tool electrical control circuits shown in this text through Chapter 13 use the basic ladder-type diagram. The voltage source between the two vertical sides in these diagrams is 24 VDC. Therefore, the complete three-phase power circuit, transformer, and power supply symbols will not always be shown, though they will generally be shown on industrial schematics.

1.9 Uninterruptible Power Systems (UPS)*

Certain types of electrical equipment, such as programmable logic controllers and computers, are very sensitive when it comes to the quality of their power supply. Small voltage fluctuations or variations in frequency can cause serious malfunctions, and a total power outage can result in the loss of data stored in the memory.

When electric power is generated, it is both clean and stable but during transmission and distribution, it is subjected to a variety of detrimental influences. Electrical storms, noisy and largely varying loads, and accidents all lead to a less than perfect supply emerging from the utility power supply.

These supply problems can be overcome by connecting a UPS between the utility supply and sensitive load equipment. It will not only clean up any supply aberrations but will also maintain the critical load during a complete outage. The Liebert Corporation Uninterruptible Power System provides both power conditioning and supply backup (Figure 1-11). It takes the raw utility power and, using state-of-the-art solid-state power electronic technology, converts it into a DC form. A microprocessor-controlled inverter then reconverts the DC power into controlled AC that can be used to supply equipment. Due to this double conversion technique, from AC to DC to AC, the output power is essentially isolated from the utility supply so that the equipment will be oblivious to any utility supply variations.

Backup is provided by an internal battery that is automatically connected to the DC portion of the UPS when the input power fails. The standby battery can maintain the unit's fully rated load for 10 minutes, but this period will be longer if lighter loads are used. If the supply break exceeds the battery backup time, the UPS will shut down once the battery charge has been exhausted. However, the unit will sound an alarm to warn that this is about to occur to give adequate time to shut down the load in an orderly fashion.

Figure 1-11 **Uninterruptible power system.** *(Courtesy of Liebert Corporation.)*

This UPS will automatically restart when the utility supply returns, and the battery will quickly recharge to prepare for further use. A block diagram shows the arrangement of the various elements within this UPS (Figure 1-12).

1.10 Circuit Diagram

To troubleshoot or maintain a piece of equipment, it is crucial to understand the symbols that are shown on the electrical drawings.

A typical drawing is provided (Figure 1-13) to summarize many of the devices that were discussed in this chapter. Notice that line numbers are typically given on the drawing set to provide a means for cross-referencing devices to other places within the drawing set. The first number of

*Information courtesy of Liebert Corp.

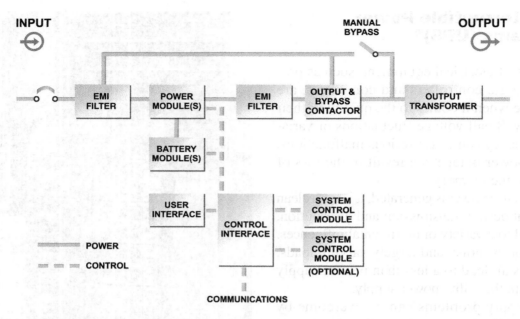

Figure 1-12 Block diagram of various elements of the UPS. *(Courtesy of Liebert Corporation.)*

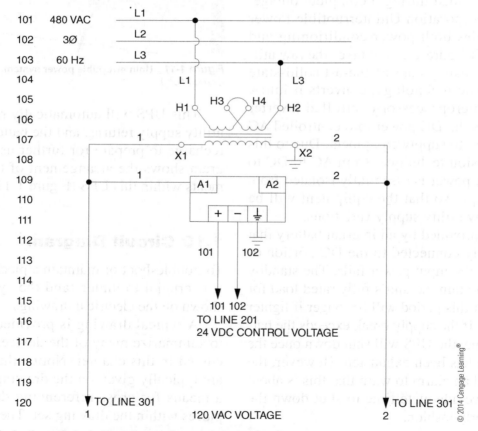

Figure 1-13 Electrical schematic including wire numbers and line numbers.

the line number is typically representative of the sheet number for the given drawing.

Notice also that wire numbers have been included in the drawing. All wires on a piece of equipment will be assigned a number. A label will be physically attached to both ends of the wire with the wire number printed on the labels. The wire numbers will agree with the wire number identified on the electrical drawing. The rule for assigning wire numbers is that the wire number will change every time the wire goes through a device. Therefore, several wires may have the same wire number if they are connected together to form a common node. This situation applies to wire numbers 1, 2, L1, and L3 in Figure 1-13.

1.11 Insulation Classifications

The Underwriters Laboratory (UL) is a global consumer safety company. The UL mark on a product signifies that the device has undergone extensive testing to ensure that it complies with its rigorous standards. This organization also publishes standards for safety and compliance. One such standard is UL83, which specifies the insulation requirements for 600 V wires used in electrical circuits. The National Electrical Manufacturers Association (NEMA) also classifies insulation systems according to the maximum temperature that is allowable for proper and safe operation. Other sources for requirements include the International Electrotechnical Commission (IEC) 60085 standard and the Japanese Industrial Standards (JIS) C4003. The important point to remember is the applicable standards for the location where a machine will operate must be consulted and complied with. The maximum allowable temperature for various types of insulation is consistent between these standards for the classes given in Figure 1-14.

It is imperative that the maximum temperature for a given type of wire is not exceeded or the insulation may melt or burn. Note also that the insulation classes listed in Figure 1-14 also apply to other devices such as motors and transformers.

Insulation Class	Maximum Temperature (Degree C)
A	105
B	130
F	155
H	180

Figure 1-14 Insulation Classifications.

1.12 Conductor Ampacity

The amount of current flowing through a wire also impacts the amount of heat generated in the wire. Therefore, ampacity tables are provided in electrical standards. The ampacity tables indicate the maximum amount of current allowed to flow through a given size of wire, for a given ambient temperature and insulation class of wire. The information presented in the tables must be derated to allow for different ambient temperatures or numbers of wires run together in a conduit or raceway. An example of an ampacity table for use on equipment in the United States is provided by the National Fire Protection Association (NFPA). Table 6 from NFPA79: Electrical Standard for Industrial Machinery is included in Appendix E.

1.13 Conductor Color Code

A consistent color coding scheme for conductors used on industrial equipment provides an additional level of safety for individuals working on the equipment. Compliance to these standards allows an electrician to quickly identify the amount and type of voltage that will be present when troubleshooting the equipment. Field devices may be wired using multi-conductor cable that will not comply with the color coding scheme. However, the color of conductors wired inside the main control cabinet should comply with the applicable standards for the region in which the machine resides. NFPA79: Electrical Standard for Industrial Machinery provides the color coding scheme for machinery in operation within the United States. The example wiring schematic shown in

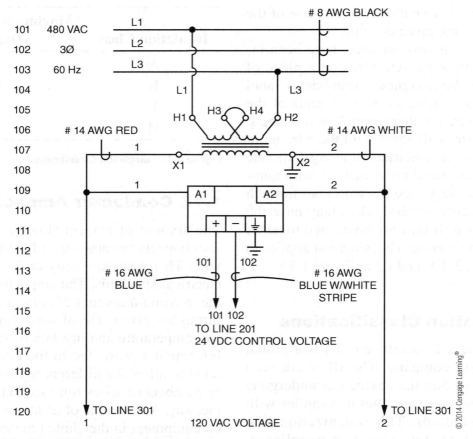

Figure 1-15 **Electrical schematic including wire size and color.**

Figure 1-13 may now be finalized as shown in Figure 1-15 to include the wire size and color for various conductors.

1.14 Electrical Symbols

The Joint International Committee (JIC) developed a set of symbols to be used in machine control drawings to standardize electrical drawings. The JIC symbols are shown in this text and are used in the electrical circuits shown throughout the chapters. European countries utilize symbols from the IEC 617 standard for use in electrical schematics. These symbols have been provided throughout the text whenever a new symbol is introduced. However, to avoid confusion, only the JIC symbols are used in circuit diagrams.

The European countries utilize a different format for their drawings along with a different standard sheet size. Drawings for North America utilize vertical lines to show power and common wires. Devices are then drawn horizontally between the power rails. Devices in European drawings are shown vertically in the drawings.

Recommended Web Links

Students are encouraged to view the following Web sites as a supplement to the concepts presented in this textbook. Review and analyze the array of products that are available for electrical control applications. Many of these sites offer technical information that can help in converting the principles to practical applications. To view catalogs, your PC may require Adobe Acrobat Reader software.

Transformers

1. The Osborne Transformer Co.
   www.osbornetransformer.com
   Review: Products and Applications

2. Sola/Hevi-Duty
   www.solahevidutysales.com
   Review: Products and Power Supplies

3. Acme Electric Corp.
   www.acmepowerdist.com
   Review: Technical Services and Transformers

Power Supplies and UPS

1. Majorpower.com
   www.majorpower.com
   Review: Products

2. APC
   www.apc.com
   Review: Products and Services

3. Liebert Corp.
   www.liebert.com
   Review: Products—Surge Suppressors, Power Conditioners, and UPS

Standards

1. Underwriters Laboratory
   www.ul.com

Achievement Review

1. What type of electrical energy is normally supplied to industries?

2. What are the two important reasons for using 24 VDC in machine control systems?

3. Why is 120 VAC still available on machine systems in the United States?

4. There is 480 VAC available in a given plant. To obtain 120 V, what turns ratio is required between the primary and secondary of the control transformer?

5. There is 460 VAC available in a given plant. To obtain 230 VAC, what turns ratio is required between the primary and secondary of the transformer?

6. A transformer has a turns ratio of 4:1. If the transformer is connected to a primary voltage of 360 VAC, what is the secondary voltage?

7. A transformer is connected to a 460 VAC line. The secondary voltage is 230 VAC. The load connected to the secondary draws 30 A.

   a. What is the turns ratio of the transformer?
   b. How much current flows in the primary circuit?

8. You find that under unusually heavy loads, the voltage in your plant drops to 456 V. What will be the resulting secondary voltage if you use a transformer with a 4:1 primary-to-secondary turns ratio?

9. Draw the symbol for a dual-primary, single-secondary control transformer. Show all lead designations.

10. Draw a complete circuit showing the primary of a dual-primary control transformer connected to a three-phase, 480-V power line, and the single secondary connected to a 120-V control system.

11. In a given transformer the voltage is reduced from the primary to the secondary. What is this transformer called?

    a. Current transformer
    b. Step-up transformer
    c. Step-down transformer

12. A transformer has a no-load voltage of 120 and a full-load voltage of 110. What is the percent regulation for this transformer?

13. Calculate the kVA requirement for a transformer whose secondary will be connected to a 230 VAC circuit with the following loads:

    a. 16 relays with FLA of .2A each (inrush 2A)
    b. 10 pilot lights with a FLA of .1A each (inrush 2A)
    c. 6 solenoids with a FLA of .5A each (inrush 5A)

14. What parts of a transformer generally contribute to temperature rise in the transformer?

    a. The enclosure
    b. Copper winding
    c. Iron core

15. What will generally result if you operate a transformer (1 kVA or larger) designed for 60 Hz on a 50-Hz supply?

16. What may happen to components that are energized in a control system if, owing to poor transformer regulation, the voltage drops to an unusually low level?

17. Draw a block diagram showing all the standard components that make up a UPS.

18. When using a UPS, what happens if the power is lost?

19. What factors will generally lead to a less than perfect supply emerging from the utility power system?

20. What problems can exist within a utility power supply that will make the use of an uninterruptible power supply advisable?

2

CHAPTER

Fuses, Disconnect Switches, and Circuit Breakers

OBJECTIVES

After studying this chapter, you should be able to:

- Describe basic fuse construction.
- List three different types of fuses and some of their uses.
- Identify four different types of circuit breakers and uses for each.
- Describe the steps to take when first setting up electrical control and power circuits.
- Explain why time-delay fuses are used with motor starter circuits.
- List the voltage and current ratings available for fuses and circuit breakers.
- Discuss the important factors to consider when selecting protective devices.

- Draw the symbols for important protective and disconnecting devices.
- Know what is meant by *interrupting capacity*.
- Understand the use of rejection-type fuses.
- Explain the two measures of the degree of current limitations provided by a fuse.
- Explain voltage and frequency surges caused by lightning or switching.
- Explain the operation of a GFCI circuit.

2.1 Protective Factors

Once the appropriate electrical power is determined for the control circuits, methods to protect the circuit components from current and temperature surges must be considered. There are two factors to be considered when providing control circuit protection:

1. A means of disconnecting electrical energy from the circuits.
2. Protection against sustained overloads and short circuits.

The power circuit can be disconnected by using a disconnect switch or nonautomatic circuit breaker (circuit interrupter). Protection is provided by adding adequate fusing to the disconnect switch, and thermal and/or magnetic trip units to the circuit interrupter.

The 120 VAC circuit is normally protected by a single fuse or circuit breaker. In some cases, two or more fuses may be used. This arrangement is covered in more detail in Chapter 19 "Troubleshooting."

Figure 2-1 shows an example of a commercially available three-pole, fusible disconnect switch ganged with a handle.

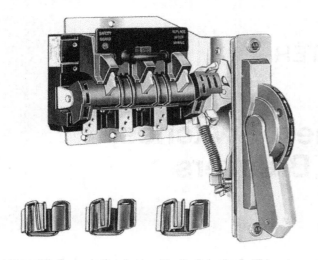

Figure 2-1 **Fusible disconnect switch.** *(Courtesy of Rockwell Automation, Inc.)*

2.2 Fuse Construction and Operation*

The typical fuse consists of an element surrounded by a filler and enclosed by the fuse body. The element is welded or soldered to the fuse contacts, blades, or ferrules (Figure 2-2).

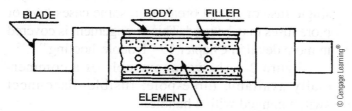

Figure 2-2 **Cross section through a typical fuse.**

The element is a calibrated conductor. Its configuration, mass, and the materials employed are varied to achieve the desired electrical and thermal characteristics. The element provides the current path through the fuse. It generates heat at a rate that depends on its resistance and the load current.

The heat generated by the element is absorbed by the filler and passed through the fuse body to the surrounding air. A filler such as quartz sand provides effective heat transfer and allows for the small-element cross section typical in modern fuses. The effective heat transfer allows the fuse to carry harmless overloads. The small-element cross section melts quickly under short-circuit conditions. The filler also aids fuse performance by absorbing arc energy when the fuse clears an overload or short circuit.

When a sustained overload occurs, the element generates heat at a faster rate than the heat can be passed to the filler. If the overload persists, the element will reach its melting point and open. Increasing the applied current heats the element faster and causes the fuse to open sooner. Thus, fuses have an inverse time-current characteristic; that is, the greater the overcurrent, the less time required for the fuse to open the circuit.

2.3 Fuse Types*

Fuses are available in numerous types; here are three common ones:

1. Standard one-time fuse (Figure 2-3)
2. Time-delay fuse (Figure 2-4)
3. Current-limiting non-time-delay fuse (Figure 2-5)

Standard voltage ratings for fuses are 125 V, 250 V, 300 V, 480 V, and 600 V. Higher-voltage fuses are available. Current ratings range from a fraction of an ampere (A) to 6000 A, in all voltage ratings.

The ability of a protective device (fuses or circuit breakers) to interrupt excessive current in an electrical circuit is important. All protective devices have a published *interrupting capacity*, which is defined as the highest current at rated voltage that a device can safely interrupt.

*Information courtesy of Ferraz Shawmut.

Figure 2-3 One-time
fuse—class K-5. *(Courtesy of
Ferraz Shawmut.)*

Figure 2-4 Time-delay
fuse—class RK-5. *(Courtesy
of Ferraz Shawmut.)*

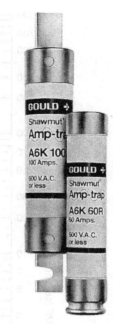

Figure 2-5 Current-limiting
fuse—class RK-1. *(Courtesy of
Ferraz Shawmut.)*

It is therefore important when installing or replacing a protective device that at least three items be considered:

1. Voltage rating
2. Current rating
3. Interrupting capacity

Correct fuses must be used as replacements for continued safety in the protection of equipment. To help in this area, manufacturers provide a "rejection"-type fuse. It is called Class R (R for rejection). The ferrule sizes (0–60 A) have an annular groove in one ferrule. The blade sizes (61–600 A) have a slot in one blade. Replacement of this fuse with a fuse of lower voltage or lower interrupting rating is not possible provided that this fuse is used with rejection fuse blocks. The rejection fuse block is similar to the standard fuse block except physical changes are made in the block to accommodate the annular ring in the ferrule-type fuse and the slot in blade type.

The physical configuration of the rejection-type fuse is shown in Figures 2-4 and 2-5. The rejection-type fuse has a 200,000-A interrupting rating as contrasted to class K-5 with the standard fuse

configuration, ferrule or blade, shown in Figure 2-3, with an interrupting capacity of 50,000 A.

To further understand the operation of fuses, a melting time–current data curve is shown in Figure 2-6. This curve is for a typical 100-A, 250-V, time-delay fuse. It shows an inverse time relationship between current and melting time; that is, the higher the current, the faster the melting time. For example, referring to this curve, it can be seen that at 1300 A, the melting time is 0.2 seconds. At 200 A, the melting time is 300 seconds. This characteristic is desirable because it parallels the characteristic of conductors, motors, transformers, and other electrical apparatus. This equipment can carry low-level overloads for relatively long times without damage. However, under high current conditions caused by short circuits, damage can occur quickly. Because of the inverse time characteristic, a properly applied fuse can provide effective protection over a broad current range from overloads to short circuits.

The standard one-time fuse (class K-5) link consists of a low-temperature-melting metal strip with several reduced area sections. On overloads that exceed the rating of the fuse, the narrow center section will melt, thus opening the circuit. The heat to

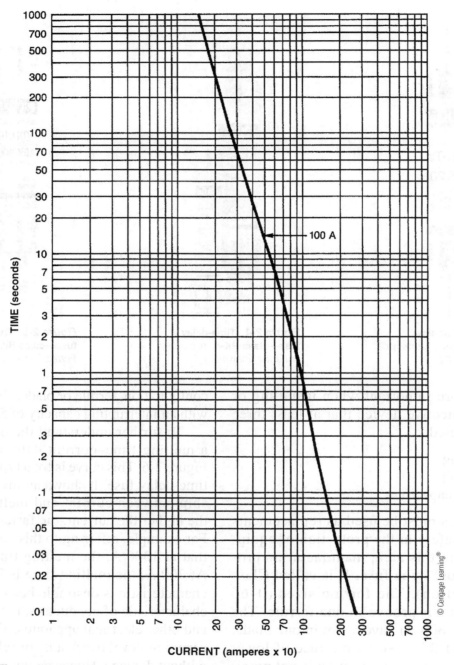

Figure 2-6 Melting time–current data for a typical time-delay fuse—100 amperes.

melt the strip comes from excessive current passed through the fuse. In case of short circuits, all the reduced cross-sectional areas will melt and vaporize simultaneously. The tube holding the strip (fuse link) must be of sufficient strength to withstand the internal pressure developed during this time.

The class K-5 fuses are suitable for the protection of mains, feeders, and branch circuits serving lighting, heating, and other nonmotor loads, provided that the interrupt capacity is adequate.

The time-delay fuses (class RK-5 and RK-1) were developed for use where a heavy overloading might exist for a short time and the fuse is not expected to open unless the overload persists. An example of a short overload is normal motor starting. These time-delay delay fuses are also suitable for general-purpose protection of transformers, service entrance equipment, feeder circuits, and branch circuits. The time-delay characteristic of an efficient fuse allows it to withstand normal surge conditions without compromising short-circuit protection.

Current-limiting time-delay fuses are used today in almost all fuse applications because in most electrical power systems, a fault can produce currents so great that much damage may result. The current-limiting fuse will limit both the magnitude and duration of the current flow under short-circuit conditions. This type of fuse will clear available fault currents in less than one half cycle, thus limiting the actual magnitude of the current flow. It is generally supplied as a rejection-type fuse. The replacement of these fuses with a fuse of another UL class is not possible.

The current-limiting fuse (class RK-1) is available as a non-time-delay fuse or a time-delay fuse with interrupting capacity of 200,000 A and higher. The non-time-delay fuse is suitable for the protection of capacitors, circuit breakers, load centers, panel boards, switch boards, and bus ducts in which high available short-circuit currents may exist. They are also available in time delay versions for motor loads with a high degree of current limitation for minimizing short-circuit current damage.

With the increased use of solid-state power devices, fuse designs have changed to match solid-state protection demands.

Solid-state devices operate at high current densities. Cooling is a prime consideration. Cycling conditions must be considered. The ability of solid-state devices to switch high currents at high speed subjects fuses to thermal and mechanical stresses. Solid-state devices have relatively short thermal time constants. A short-circuit current that may not harm an electromechanical device can cause catastrophic failure of a solid-state device in a very short time.

Most programmable controllers (covered in Chapter 15) use a semiconductor fuse with a blown-fuse indicator. There also may be add-on switches that can be used to energize an indicator light. This light can be used for remote indication. The trigger actuator may be a part of the fuse or a field-mounted blown-fuse indicator that would be wired in parallel with the fuse being monitored.

2.4 Peak Let-Thru Current (I_p) and Ampere Squared Seconds (I^2t)

Current limitation is one of the important benefits provided by modern fuses. Current-limiting fuses are capable of isolating a faulted circuit before the fault current has sufficient time to accelerate to its maximum value. This current-limiting action provides several benefits:

- It limits thermal and mechanical stresses created by the fault currents.
- The magnitude and duration of the voltage drop caused by the fault currents is reduced, improving overall power quality.
- Current-limiting fuses can be precisely coordinated to minimize unnecessary service interruption.

Peak let-thru current (I_p) and ampere squared seconds (I^2t) are two measures for the degree of current limitation provided by a fuse. Maximum allowable I_p and I^2t values are specified in UL standards for all UL-listed current-limiting fuses and by the fuse manufacturer for semiconductor and special purpose fuses.

Let-thru current is that current which passes by a fuse while the fuse is interrupting a fault within the fuse's current-limiting range (Figure 2-7). It is

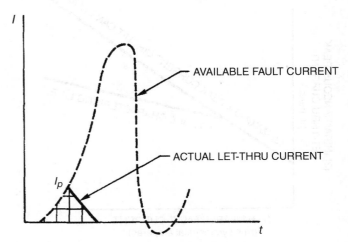

Figure 2-7 Let-thru current is expressed as a peak instantaneous value (I_p). *(Courtesy of Ferraz Shawmut.)*

expressed as a peak instantaneous value (I_p) in amperes. I^2t is a measure of the heat generated with current flow over time.

The I_p data is generally presented in the form of a graph. Key information is provided in the peak let-thru graph in Figure 2-8. As shown on the figure, the important components are:

1. The x-axis, labeled "Available Fault Current" in rms symmetrical amperes.
2. The y-axis, labeled "Instantaneous Peak Let-Thru Current" in amperes.
3. The line labeled "Maximum Peak Current Circuit Can Produce," which gives the worst-case peak current possible with no fuse in the circuit.
4. The "Fuse Characteristic Line," which is a plot of the peak let-thru currents that are passed by a given fuse at various available fault currents.

Figure 2-9 illustrates the use of the peak let-thru current graph. Assume that a 200 A, class J fuse is to be applied where the available fault current is 35,000 A rms. The graph shows that with 35,000 A rms available, the peak available current is 80,500 A (35,000 × 2.3) and that the fuse will limit the peak let-thru current to 12,000 A.

You may wonder why the peak available current is 2.3 times greater than the rms available current. In theory, the peak available fault current can be anywhere from 1.414 × (rms available) to

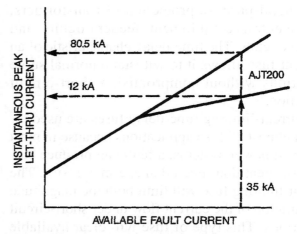

Figure 2-9 **Use of a peak let-thru graph.** *(Courtesy of Ferraz Shawmut.)*

2.828 × (rms available) in a circuit in which the impedance is all reactance with no resistance. In reality all circuits include some resistance, and the multiplier has been chosen as a practical unit.

The I_p value is relatively useful in determining the magnitude of the peak current available downstream of a fuse for general purpose circuits. If the fuse limits the current to a value less than the available fault current, then all the equipment downstream of the fuse may be selected to withstand this lower value, rather than the full available fault current. The level of protection for the equipment provided by the current-limiting fuse is not complete without considering the I_p and I^2t. I_p provides the peak current value, and I^2t provides a measurement of the amount of heat energy that the fuse passes during opening.

Two fuses can have the same I_p but different total clearing times (Figure 2-10). The fuse that clears by time A will provide greater component protection than the fuse that clears in time B.

Fuse I^2t depends on both I_p and total clearing time. The I^2t passed by a given fuse depends on the characteristics of the fuse and on the applied voltage. The I^2t passed by a given fuse will decrease as the application voltage decreases. Unless otherwise stated, published I^2t values are based on AC testing. The I^2t passed by a fuse in a DC application may be higher or lower than in an AC application. The voltage available fault current and time constant of the DC circuit are the determining factors.

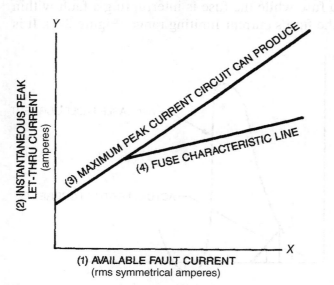

Figure 2-8 I_p **data—peak let-thru graph.** *(Courtesy of Ferraz Shawmut.)*

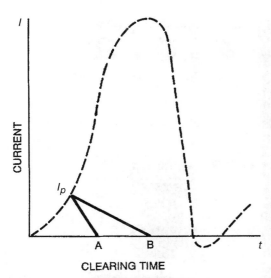

Figure 2-10   Clearing time versus current.
(Courtesy of Ferraz Shawmut.)

Fuse I^2t and I_p values can be used to determine the degree of protection provided for all circuit components under fault current conditions.

Manufacturers of diodes, thyristors, triacs, and cable publish I^2t withstand ratings for their products. The fuse chosen to protect these products should have a clearing I^2t that is lower than the withstand I^2t of the device being protected.

2.5 Voltage and Frequency Surges

Protecting electrical equipment against voltage surges caused by switching or lightning is very important. Even low surges that exceed the equipment insulation rating can cause the insulation to eventually weaken to the point of failure. For example, the insulation in motor windings can be stressed to the point that failure will result between windings and frame or between stator coils.

In case of a surge caused by lightning, the most severe damage occurs if there is a direct hit. Surge voltages from the lightning can also be caused by induction on the line. The voltage from a direct hit will be twenty to thirty times that caused by induction. In these cases it can be considered a high-voltage and high-frequency pulse.

A lightning arrester will limit the crest of the surge by breaking down and conducting to ground. After grounding, the arrester then returns to its initial condition of nonconducting.

Switching voltage surges are not as significant as lightning voltage surges. In most cases they will not exceed three times normal voltage. The highest overvoltage will be present when there is a ground fault in the system.

For switching voltage surges, the current-limiting fuse is used. The operating characteristic of this type of fuse is such as to interrupt in the first quarter of the cycle.

2.6 Circuit Breaker Types

The circuit breaker is available in four types:

1. Nonautomatic (circuit interrupter)
2. Thermal
3. Magnetic
4. Thermal magnetic

The *nonautomatic circuit breaker* (circuit interrupter) is used for load switching and isolation. Adding a *thermal trip unit* to the nonautomatic circuit breaker, by using a bimetallic element in each pole of the breaker, provides automatic tripping. This unit then carries the load current. When the conductors carry a current in excess of the normal load, the breaker thermal element increases in temperature as it carries the same current. This temperature increase deflects the thermal element and trips the breaker. Since the tripping action depends on temperature rise, a time lag is present. Therefore, the tripping action is not affected by momentary overloads.

To clear a circuit in case of a short circuit, a more rapid opening system is required. To achieve such a system, a *magnetic trip unit* is added either to the nonautomatic breaker or to the breaker with the thermal trip unit. The magnetic trip unit operates through a magnet that trips the breaker instantaneously on short-circuit current.

A combination of thermal and magnetic trip units is desirable. The thermal element provides inverse time tripping on overloads. The magnetic trip provides instantaneous trip on short circuits. *Molded-case* circuit breakers are available from 100 A to 2500 A. The voltage ratings are 240 V, 480 V, and 600 V with the interrupting capacity in some circuit breakers to 100,000 A. Figure 2-11 is a cutaway view of a molded-case circuit breaker.

Figure 2-11 Cutaway view of a molded-case circuit breaker. *(Courtesy of Square D/Schneider Electric.)*

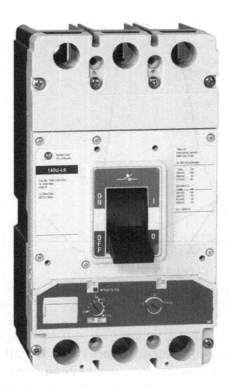

Figure 2-12 Molded-case circuit breaker. *(Courtesy of Rockwell Automation, Inc.)*

The fused disconnect switch and the circuit breaker are available in many types of enclosures. They can also be obtained as open types for panel mounting. Remote operators can be mounted on the door of the control cabinet and interlock mechanically with the door. The operator may also be mounted in a "dead" or stationary portion of the cabinet. The door is mechanically interlocked with the breaker trip switch. This allows the breaker to be locked open or closed with the cabinet door open.

Figure 2-12 shows a typical molded-case circuit breaker design. It incorporates, in frame ratings 150 A to 1600 A, interrupting capacities as high as 100 kA at 480 V AC (200 kA at 240 V AC) in physical sizes normally associated with standard interrupting capacity breakers.

The L-frame circuit breaker provides higher interrupting capacities and improved current-limiting capabilities compared to previous standard line circuit breakers.

Thermal magnetic and electronic trip unit designs are available. This circuit breaker is available with a fixed, thermal-adjustable magnetic trip unit that provides inverse time and instantaneous tripping. Also available is the electronic trip unit, including current-sensing circuits that provide an inverse time-delay tripping action for overload condition and either short delay or instantaneous tripping for protection against short-circuit conditions.

A push-to-trip button located on each trip unit provides a local means of manually exercising the trip mechanism.

High-strength glass-polyester bases and covers have excellent dielectric qualities and are inherently fungus proof. Cover design reduces the possibility of accidental contact with live terminals.

2.7 Programmable Motor Protection

The unit shown in Figure 2-13 is a multifunction, motor-protective relay that monitors three-phase AC current and makes separate trip and alarm decisions based on preprogrammed motor current and temperature conditions. This unit's motor protection algorithm is based on proven positive

Figure 2-13 **Motor-protective relay.** *(Courtesy of TECO-Westinghouse Electric Corporation.)*

and negative (unbalance) sequence current sampling and true rms calculations. (An algorithm is a special set of rules or instructions that perform a specific operation.)

By programming this unit with the motor's electrical characteristics (such as full-load current and locked-rotor current), the unit's algorithm will automatically tailor the optimal protection curve to the motor being monitored. No guesswork or approximation is needed in selecting a given protection curve because this unit matches the protection from an infinite family of curves to each specific motor.

Application-related motor-load problems are further addressed through the use of such functions as jam, underload, and ground fault protection.

A few of the many protective features of this unit are:

- Instantaneous overcurrent trip level and start delay
- Instantaneous overcurrent disable setting
- Locked-rotor current
- Maximum allowable stall time
- Ultimate trip current level
- I^2t alarm level
- Zero-sequence ground fault trip level with start and run delays

2.8 Electrical Metering and Voltage Protection

The unit shown in Figure 2-14 is a microprocessor-based monitoring and protective device. It provides complete electrical metering and affords system voltage protection. In one compact, standard package this unit provides an alternative to individually mounted and wired ammeters, voltmeters, ammeter and voltmeter switches, wattmeters, watthour meters, and more. Its protection features include:

- *Phase loss.* Voltage phase loss occurs if less than 50% of the nominal line voltage is detected. Current phase loss occurs if the smallest phase current is less than 1/16 of the largest phase current. (It updates itself twice per second and all other protection functions once per second.)
- *Phase unbalance.* It occurs if the maximum deviation between any two phases exceeds the amount of unbalance as a percent of nominal line voltage preset by dual in-line package (DIP) switches. Range: 5% to 40% (5% increments).
- *Phase reversal.* It occurs if any two phases become reversed for more than 1 second.
- *Overvoltage.* DIP switch setting as a percent of nominal line voltage. Range: 105% to 140% (5% increments).

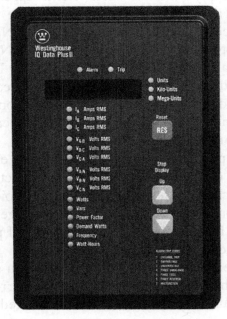

Figure 2-14 **Electric metering and voltage protection.** *(Courtesy of TECO-Westinghouse Electric Corporation.)*

- *Undervoltage.* DIP switch setting as a percent of nominal line voltage. Range: 95% to 60% (5% increments).
- *Delay.* It allows existence of overvoltage, undervoltage, or voltage unbalance before an alarm or trip occurs. Range: 0 to 8 seconds (1 second increments).

2.9 Selecting Protective Devices

The important factors to consider when selecting protective devices are:

- Size (current rating)
- Whether a time lag is required
- Interrupting capacity
- Ambient temperature where the device is to be located
- Voltage rating
- Number of poles
- Mounting requirements
- Type of operator
- Enclosure, if required

If the line voltage and load in a circuit are known, the size and voltage rating of the protective device can be determined. The conductor size to feed the load must be properly determined. Tables are available to supply this information (refer to Appendix E).

The actual sizing of the protective device should come from sources such as the National Fire Protection Association (NFPA), the National Electrical Code (NEC), and the NFPA-79 Electrical Standard for Industrial Machinery. Local electrical code requirements should also be consulted. The manufacturers of protective devices can also be helpful in obtaining this information.

The problem in providing a time-lag element depends on the nature of the load. If inductive devices such as motors, motor starters, solenoids, and contactors are involved, some time lag is generally needed. Here again, the manufacturers of protective devices can help.

In selecting a protective device, you should know the available short-circuit current in your plant. The first information needed is the impedance of the transformer supplying power. This information can be read from the unit's nameplate or obtained from the manufacturer of the unit. *Impedance* is the current-limiting characteristic of a transformer and is expressed as a percent.

The impedance is used to determine the interrupting capacity of a circuit breaker or fuse employed to protect the primary of a transformer.

Example 1

Determine the minimum circuit breaker trip rating and interrupting capacity for a 10-kVA, single-phase transformer with 4% impedance. It is to be operated from a 480-VAC, 60-Hz source.

Solution:

Normal full-load current =

$$\frac{\text{Nameplate volt-amperes}}{\text{Line voltage}} = \frac{10,000}{480} = 20.8 \text{ A}$$

Maximum short-circuit amperes =

$$\frac{\text{Full-load amperes}}{4\%} = \frac{20.8}{0.04} = 520 \text{ A}$$

The breaker or fuse would have a minimum interrupting rating of 520 A at 480 VAC.

Example 2

Determine the interrupting capacity in amperes required of a circuit breaker or fuse for a 75-kVA, three-phase transformer with a primary of 480 VAC delta and secondary of 208Y/120 VAC. The transformer impedance (Z) is 5%.

Solution

If the secondary circuit is shorted, the following capacities are required:

Normal full-load current =

$$\frac{\text{Volt-amperes}}{3 \times \text{Line voltage}} = \frac{75,000 \text{ VA}}{3 \times 480} = 52 \text{ A}$$

Maximum short-circuit line current $=$

$$\frac{\text{Full-load amperes}}{5\%} = \frac{52}{0.05} = 1040 \text{ A}$$

The breaker or fuse would have a minimum interrupting rating of 1040 A at 480 VAC.

The ambient temperature in the location where the protective device is to be located is important. Overload protection devices can be thermally operated. Therefore, an increase in the ambient temperature can affect the trip setting of the device. In some cases, ambient temperature compensation is provided by the manufacturer in the design of the device.

Standard pole arrangements for most disconnecting devices and protective devices are two poles for single-phase lines and three poles for three-phase lines. One- and four-pole devices are also available. The multiple-pole (multipole) devices are generally ganged together with a single common operator.

Symbols for various protective and disconnecting devices are shown in Figure 2-15. By using symbols, these components can be added to power source and control source diagrams.

Figure 2-16 illustrates a complete diagram: a three-phase power source, disconnect and protection, and control transformer with protection added in the secondary circuit.

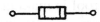

A. SINGLE-FUSE ELEMENT

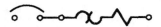

D. ONE-POLE, THERMAL-MAGNETIC CIRCUIT BREAKER

B. THREE-POLE DISCONNECT SWITCH, GANGED WITH HANDLE

E. THREE-POLE CIRCUIT INTERRUPTER, GANGED WITH HANDLE

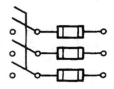

C. THREE-POLE FUSED DISCONNECT SWITCH, GANGED WITH HANDLE

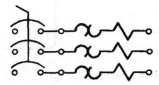

F. THREE-POLE THERMAL-MAGNETIC CIRCUIT BREAKER, GANGED WITH HANDLE

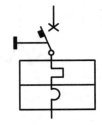

G. IEC 617 CIRCUIT BREAKER

Figure 2-15 **Symbols for protective devices.**

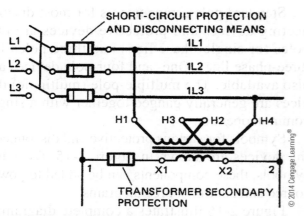

Figure 2-16 Power and control protection. Conductors carrying load current at line voltage are denoted by heavy lines.

2.10 Ground Fault Circuit Interrupter*

The main purpose of a fuse is to protect wiring from overheating when excessive current flows through the wire. The fuse will open to disconnect the flow of current in the circuit. A ground fault circuit interrupter (GFCI) will also safely open an electrical circuit. However, the main purpose of a GFCI is to protect a person from electrical shock.

In residential dwellings, GFCIs are required in potentially wet locations such as a bathroom, kitchen, garage, and basement. For industrial applications, GFCIs are used for receptacles that are installed external to the main control enclosure where the potential exists for any exposure to liquids such as water, oil, coolant, etc. In addition, any receptacles or cord sets used for portable equipment should have GFCI circuit protection.

GFCI circuits monitor the current flow to and from the electrical devices that are connected to the circuit. For normal operation, the incoming current is equal to the return current. If the amount of current differs by more than 5 mA (.005 A), the circuit will be interrupted quickly (25 msec) to avoid electrocution. For example, suppose a hot wire inside a tool is making contact with the metal casing. The case is now energized. Touching the case would then cause current to flow through you to ground (Figure 2-17). The additional current path causes the incoming current to no longer equal the return current. The GFCI circuit would quickly detect the difference in current and disconnect the circuit.

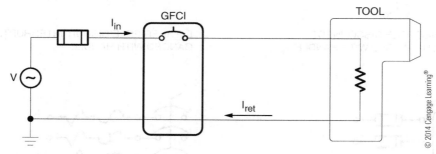

Figure 2-17A GFCI Circuit Normal Operation ($I_{in} = I_{ret}$)

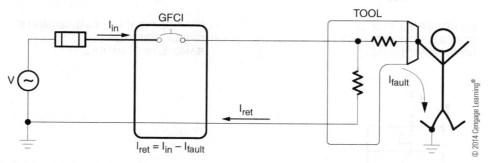

Figure 2-17B GFCI Circuit – Fault Detected ($I_{in} \neq I_{ret}$)

*Courtesy of the U.S. Department of Labor, http://www.dol.gov

Recommended Web Links

Students are encouraged to view the following Web sites as a supplement to the concepts presented in this textbook. Review and analyze the array of products that are available for electrical control applications. Many of these sites offer technical information that will help in converting the principles to practical applications. To view catalogs, your PC may require Adobe Acrobat Reader software.

Fuses

1. Littelfuse
   www.littelfuse.com
   Review: Education, Technical Resources, Technical Articles

Disconnect Switches

1. Hubbell Ltd.
   www.hubbell.co.uk/fuse.htm
   Review: Fused and Nonfused Disconnect Switches

2. Advance Controls, Inc.
   www.acicontrols.com
   Review: Disconnect Switches

Circuit Breakers

1. IDEC Corp.
   www.idec.com
   Review: Industrial Components, Circuit Breakers

Achievement Review

1. What two factors must be provided for in any electrical power or control system?

2. Under what conditions would you use the time-delay fuse? Give an example.

3. What are the four types of circuit breakers?

4. Under what conditions would you want to use the magnetic trip feature in a circuit breaker?

5. List at least five important factors to consider when selecting a protective device.

6. Are more than two poles available on disconnecting devices? If so, how many?

7. Draw the symbol for each of the following:
   a. Single-fuse element
   b. Three-pole fused disconnect switch, ganged with handle
   c. Three-pole circuit interrupter, ganged with handle
   d. Three-pole, thermal-magnetic circuit breaker, ganged with handle

8. What is meant by the interrupting capacity of a fuse?

9. What does quartz sand filler in a fuse provide?
   a. Increased voltage rating
   b. Effective heat transfer
   c. An inexpensive filler

10. Why is the inverse time characteristic of a fuse important?
    a. The fuse cost is reduced.
    b. It increases the interrupting capacity.
    c. It parallels the characteristics of conductors, motors, transformers, and other electrical apparatus.

11. What are two measures of the degree of current limitations provided by a fuse?

12. How are I^2t values of a fuse derived?

13. Under what condition will the highest overvoltage occur from a surge caused by switching?

14. Explain how a lightning arrester protects electrical equipment for a surge current caused by lightning.

15. Determine the interrupting capacity in amperes required of a circuit breaker or fuse for a 15-kVA, single-phase transformer with 4% impedance, to be operated from a 480-V, 60-Hz source.

3

Control Units for Switching and Communication

OBJECTIVES

After studying this chapter, you should be able to:

- Describe two methods of mounting push-button switch units.

- List four types of operators for the push-button switch.

- Explain why different colors are used for push-button switch operators.

- Draw the symbols for push-button switch units with (1) flush or extended head, (2) mushroom head, and (3) maintained contact attachment.

- List several arrangements available for selector switches.

- Draw the symbol for the selector switch.

- Draw the symbol for the foot switch.

- Discuss the advantages of push-to-test pilot lights.

- Draw the symbol and detailed circuit for the push-to-test pilot light.

- Explain the meaning of the letter in the pilot light.

- Draw basic circuits using selector switches, push-buttons, and pilot lights.

- Discuss the use of annunciators to obtain process information.

- Explain the use of the LED.

- Discuss how the environment may affect the design of a membrane switch.

3.1 Oil-Tight Units

Almost all push-button switches, selector switches, and pilot lights offered to industry today are oil-tight units. Since this text is concerned with industrial electrical control, only oil-tight types of units are discussed.

3.2 Push-Button Switches

Switches generally consist of two parts: the contact unit and the operator. This composition allows for many combinations that cover almost every application required.

The *normal* state of a switch is an important concept to understand. The symbol used to identify a device on an electrical drawing will show the component in its normal state. Therefore, to comprehend how a circuit functions, a thorough understanding of this concept must be mastered.

The normal state of a switch shows the contact configuration when the device is NOT being actuated (being acted upon by an external force). A discrete device can only have two states—*on* or *off*. Since a switch is a discrete device, each individual contact will either be *normally open* (NO) or *normally closed* (NC).

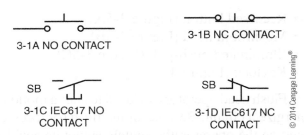

Figure 3-1 **Electrical symbols for push buttons.**

© 2014 Cengage Learning®

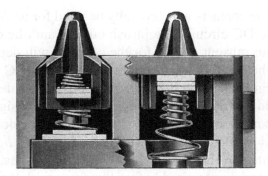

Figure 3-2 **Standard double-circuit contact block.** *(Courtesy of Eaton Corporation.)*

When the push-button is pressed (actuated) every contact associated with the device will transition states. The NO contact will close and the NC contact will open. When the push-button is released the contacts will return to their normal state. It is important to distinguish between a contact transitioning and the normal state of the contact. For example, an NO contact will close when the device is *actuated* but the contact will not become an NC contact. Likewise, an NC contact will open when the device is actuated but the contact will not become an NO contact.

The circuit symbols for a push-button device are shown in Figure 3-1. Remember that for electricity to flow in a circuit there must be a complete path throughout the circuit (commonly referred to as having continuity). Therefore, when the switch is not being actuated (or pressed), Figure 3-1A shows the NO contact as being an open in the circuit. Electricity cannot pass through the push-button device. When the push-button is pressed the contact will close and provide a path for electricity to flow.

Figure 3-1B shows the circuit symbol for a NC contact. This contact configuration allows a path for continuity in its normal state. However, when the push-button is pressed this contact will open and stop the flow of electricity through the push-button device.

Figure 3-2 shows a typical contact block. In the deactivated condition, the spring is released and the contacts are in their normal state of operation. When the push-button is pressed the spring compresses and the contacts transition states. When the push-button is released, the spring will push on the actuating bar and return the contacts to their normal state.

Push-buttons may be purchased with different contact configurations. The number and type of contacts will be determined by the part number of the device. Additional contact blocks may typically be purchased to add on to existing contact assemblies. One contact block will consist of one NO contact, one NC contact, or a combination of both contacts. Figure 3-3 shows typical contact blocks. The maximum number of contact blocks that may exist on a device will vary by manufacturer and the actual component. Product specifications should be verified when modifying existing assemblies.

Figure 3-3 **Contact blocks.** *(Courtesy of Square D/Schneider Electric.)*

The contacts will generally be rated for use in an AC or DC circuit. The inrush current and the continuous current ratings for the contacts will depend on whether the device is connected to an AC or a DC circuit. The product specifications for the device should be consulted when installing a new design or when replacing a component with another device.

In some cases, the contact block is *base-mounted* in the push-button enclosure. Thus, the units can be prewired before the cover with the operators is put in place. This method also eliminates cabling conductors to the cover.

The typical method is *panel mounting*. The base of the operator, with the contact block attached, is mounted through an opening in the cover or door of an enclosure. The push-button is then secured in its place by a threaded ring installed from the front of the panel. The ring is part of the operator assembly. Later in the text it will be shown that this arrangement has the advantage of providing a space for a terminal block installation in the base of the enclosure. Thus, all connecting circuits can be terminated at an easily accessible checkpoint.

There are slight differences in the way manufacturers machine mounting holes. Also, push-button devices are available in different sizes. The standard sizes are 30 mm for NEMA devices and 16 mm or 22 mm for IEC components. Therefore, modifications may be required when substituting one unit for another.

Figure 3-4 shows a cutaway section of a typical operator. Many types of operators are available to suit almost any application. They include:

- Recessed button (Figure 3-5A)
- Mushroom head (Figure 3-5B)
- Illuminated push-pull (Figure 3-5C)
- Keylock (Figure 3-5D)

Flush head operators are used to start a motion or energize a circuit. Because of the flush (or recessed) head on the operator, the switch cannot be accidentally pressed; thus requiring an intentional action to initiate any motion.

Figure 3-5A Recessed button push-button unit. *(Courtesy of General Electric Company, General Purpose Control, Bloomington, IL.)*

Figure 3-5B Mushroom head push-button unit. *(Courtesy of General Electric Company, General Purpose Control, Bloomington, IL.)*

Figure 3-5C Illuminated push-pull unit. *(Courtesy of General Electric Company, General Purpose Control, Bloomington, IL.)*

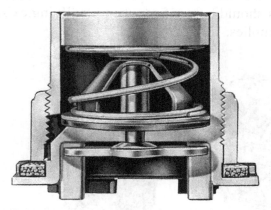

Figure 3-4 Standard-duty, flush button operator. *(Courtesy of Eaton Corporation.)*

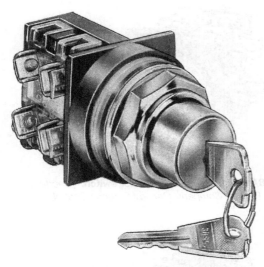

Figure 3-5D Cylinder lock push-button unit. *(Courtesy of General Electric Company, General Purpose Control, Bloomington, IL.)*

Extended head operators are used to stop a motion or de-energize a circuit. The stop button on a piece of equipment should be readily available to stop the machine. Therefore, mushroom type operators may be used to perform emergency stop functions. The placement of extended or mushroom head push-buttons should be taken into consideration to avoid nuisance or unintentional stopping of the equipment. However, this placement should not jeopardize the ability of the person running the machine to have easy access to the device in the event of an emergency.

Illuminated push-buttons will turn on a light when the button is pressed to provide a clear indication that the button is pressed or the circuit has been activated. The push-pull push-button does not utilize a spring to return the contacts to their normal state when the push-button is released. Instead, a maintained or detented type of operation is utilized. When this type of push-button is pressed the operator will remain in the pressed position until the push-button is physical pulled back out to its normal operating position. A twisting motion instead of a pulling force may be used to return the operating head to its normal state. This style of operator is referred to as a push-pull twist release.

Figure 3-6 shows the symbol used to identify the different types of operating heads. All contact

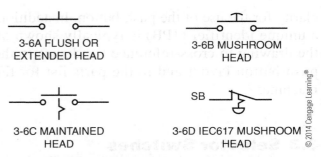

Figure 3-6 Symbols for push-button units with various operator heads (shown NO).

configurations are shown with one NO contact to avoid any confusion.

A keylock push-button may be used to prevent the operation of the device by untrained personnel. The key must be inserted into the device to operate the push-button.

Color designation is an important factor in push-button switch operators. It not only provides an attractive panel but, more importantly, it lends itself to safety. Quick identification is important. Certain functions become associated with a specific color. Standards have been developed that specify certain colors for particular functions. For example, in the machine tool industry, the colors red, yellow, and black are assigned the following functions:

Red: Stop, Emergency Stop

Yellow: Return, Emergency Return

Black: Start Motors, Cycle Start, Initiate Motion

Figure 3-7 shows the symbol for a push-button with multiple contact blocks. In this example, one mushroom head operator is used to switch three contacts (two NC and one NO). The contacts are shown as connected to the same device through the use of a dashed line. Also, note the use of wording added to the drawing to provide

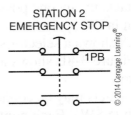

Figure 3-7 Push-button with multiple contacts.

clarity for the use of the push-button. In addition, a unique identifier (1PB) is typically shown on the drawing to cross-reference the device to the push-button layout and to the parts list for the machine.

3.3 Selector Switches

Selector switches function using the same general principles as push-button devices. The selector switch device consists of two parts: the contact unit and the operator.

A push-button requires a pressing force to actuate the contacts. The selector switch requires a rotary motion to operate the device. The contacts are actuated (switched) through the use of a rotary cam device. The operator head can have two or more positions. Therefore, a two position selector switch can be rotated to the left or to the right. A three position selector switch can be rotated to the left position, the center position, or the right position. Selector switches may be obtained with up to 24 different positions.

A maintained selector switch will remain in the current position when the operator head is released. The contacts will also maintain their state in the given position. However, a three position selector switch can be purchased with a spring return feature. The spring return function will return the selector switch to the center position when the operator is released. The contacts will also switch back to the center position when released. Three position spring return selector switches may be purchased to spring return from the right, from the left, or from both right and left.

The operators are available in standard knob, knob lever, or wing lever types. A cylinder lock can be used and the switch locked in any one position or all positions. The arrangement for opening or closing contacts in any one position or more depends on a cam in the operator.

Figure 3-8A shows a selector switch with a lever operator. Figure 3-8B shows a selector switch keylock operator.

Multiple contacts may be connected to a single selector switch (SSW). Each contact can exist in

Figure 3-8A **Lever-operated selector switch.** *(Courtesy of General Electric Company, General Purpose Control, Bloomington, IL.)*

Figure 3-8B **Cylinder lock selector switch.** *(Courtesy of General Electric Company, General Purpose Control, Bloomington, IL.)*

only one of two possible states; open or closed. However, the contact may or may not change states when the operator is rotated. The state of each contact depends on the cam configuration purchased for the switch and on the current location of the operator head. Therefore, it is very important to draw the symbol correctly so that the state of each contact is clearly identified on the drawing.

Figure 3-9 shows the basic drawing symbol for a three position selector switch. The filled in arrow indicates the normal (center) position of the

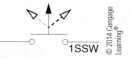

© 2014 Cengage Learning®

Figure 3-9 **Electrical symbol for 3 position SSW.**

operator. The left and right positions are indicated by dashed lines and open (not filled in) arrow heads.

There are two different methods that are typically used in industry to identify the state of the contacts for any given position of the operator. Both methods will be discussed for clarity purposes in deciphering a control circuit.

Method A shows the switch contacts as they exist in the state when the operator is in the position indicated by the filled in arrow. For example, Figure 3-10 shows a three position selector switch. When the operator is in the center position:

- Contact 1 is open
- Contact 2 is closed
- Contact 3 is open
- Contact 4 is closed

To show the contact closures for the right and left positions of the operator, an X is placed on the wire if the given contact is closed when the operator is located in that position. For example, Figure 3-11A shows the complete drawings for a three position selector switch with the following contact configurations:

Switch	Operator Positions		
Contact	Left	Center	Right
1	closed	open	open
2	closed	closed	open
3	closed	open	closed
4	open	closed	closed

Method B shows the three positions of the operator head to indicate the number of positions available for the operator. All contacts are shown on the drawing as NO contacts. A key is then provided for each contact to indicate the condition of

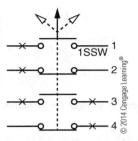

Figure 3-11A   3 Position SSW with multiple contacts (Method A).

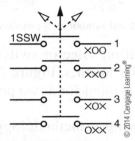

Figure 3-11B   3 Position SSW with multiple contacts (Method B).

the contact in any given position. This method for drawing a selector switch may easily be modified for any number of positions. Since a three position selector switch is being discussed, there will be three positions identified on each contact. The first position of the identifier code indicates the condition of the contact when the operator is rotated to the left. An X will be used to indicate the condition for the contact being closed. An O will be used to indicate the condition for the contact being opened. The second identifier will correspond to the center position of the operator. The third identifier will correspond to the right position of the operator. Figure 3-11B shows this drawing method being used for the same contact configuration illustrated using method A in Figure 3-11A. Either method may be used to identify the contact configurations for a selector switch. However, a combination of the two different methods should not be used for the same selector switch.

Figure 3-12 shows the inclusion of the spring return symbol to a three position selector switch. Contact 1 will close when the operator head is rotated and held to the left position. Contact 2 will close when the operator head is rotated and held to the right position. When the operator head is released, the operator will spring back to the center position and both contacts will be in the opened condition.

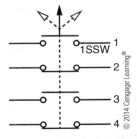

Figure 3-10   3 Position SSW with multiple contacts (Method A).

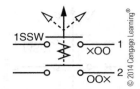

Figure 3-12   Electrical symbol for 3 Position spring return SSW.

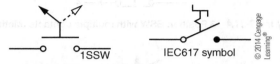

Figure 3-13   Electrical symbol for 2 position SSW.

A two position selector switch is identified in a very similar manner. Figure 3-13 shows the basic drawing symbol for a two position selector switch.

Using drawing A methodology, the contacts are drawn in the correct state when the operator is in the position indicated by the filled in arrow. An X is not needed on the drawing for a two position selector switch. The contacts will simply transition to the other state when the operator head is rotated.

Using drawing B methodology, the contacts will always be shown in their NO state. The methodology is identical to a three position selector switch except that only two positions are possible for the operator head (left and right). Figure 3-14 shows the proper design symbols for a two position selector switch with the following contact arrangement:

Switch Contact	Operator Positions	
	Left	Right
1	open	closed
2	closed	open
3	open	closed
4	closed	open

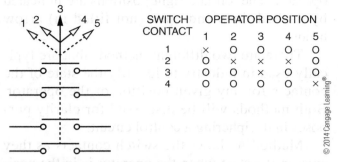

Figure 3-15   Electrical symbol for 5 Position SSW with multiple contacts.

SWITCH CONTACT	OPERATOR POSITION				
	1	2	3	4	5
1	×	O	O	O	O
2	O	×	×	×	O
3	O	O	×	O	O
4	O	O	O	O	×

As the number of positions increases to four and the number of contacts (poles) increases to eight, the manufacturer's reference to a specific operator is generally coded through a symbol chart or function table. Such charts or tables display which contacts are closed or open in the different positions of the selector switch operator.

Figure 3-15 shows a possible configuration for a five position selector switch with four contacts. For selector switches with more than three positions, a chart is usually included on the drawings to indicate the state of each contact when the operator head is placed in any of the possible positions.

One final item worth mentioning in regard to selector switches is the use of annotation (words) to describe the function of the selector switch at each operator position. The unique identifier for the device is also shown near the selector switch symbol (1SSW).

Examples are provided in Figure 3-16 for several different selector switch configurations.

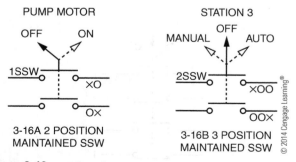

3-14A (METHOD A)   3-14B (METHOD B)

Figure 3-14   Electrical symbol for 2 Position SSW with multiple contacts.

PUMP MOTOR

OFF   ON

1SSW

3-16A 2 POSITION MAINTAINED SSW

STATION 3

OFF
MANUAL   AUTO

2SSW

3-16B 3 POSITION MAINTAINED SSW

Figure 3-16   *continued*

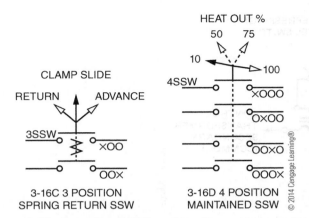

Figure 3-16 Examples of SSW configurations.

3.4 Heavy-Duty Switches

Heavy-duty switches and special application switches are oil tight units in which the switch is made an integral part of the operator and enclosures. A typical example of such a switch is the foot- operated unit shown in Figure 3-17A. The symbols are shown in Figure 3-17B.

3.5 Indicating Lights

The indicating or pilot light is available in different types and in different sizes. The full voltage type is typically rated for use in 120 VAC, 24 VAC, or 24 VDC circuits. Figure 3-18 shows the symbol for a standard pilot light.

Figure 3-17A Foot-operated switch. (Courtesy of Square D/ Schneider Electric.)

NC CONTACT-OPENS BY FOOT PRESSURE   NO CONTACT-CLOSES BY FOOT PRESSURE

Figure 3-17B Symbols for foot-operated switch.

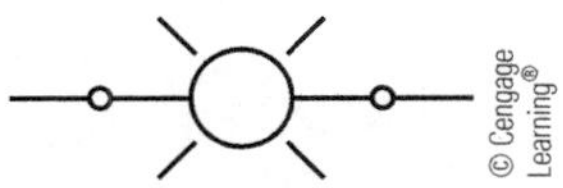

Figure 3-18 Electrical symbol for pilot light.

The pilot light will illuminate when continuity is provided to the light bulb. The lens cap is available in either plastic or glass, and in a variety of colors. The color designation allows for quick identification and lends itself to safety similar to the colors used for push-buttons. For example, in the machine tool industry, the following assignment for pilot light colors is typical:

Red—Danger, Abnormal Conditions

Amber—Attention

Green—Safe Condition

White or Clear—Normal Condition

Lens caps may be purchased separately and installed on existing units in the event the lens is broken or the function of the pilot light changes. Figure 3-19 shows the complete symbol for a pilot light. Note that the color code for the lens cap has been included as a part of the symbol.

The transformer type pilot light utilizes a low voltage bulb that operates at 6 to 8 V. The low voltage level is obtained through the use of a transformer. For 120 VAC control circuits, a low voltage bulb is usually preferred when vibration is present on a machine. In addition, the lower voltage bulb also provides a safer voltage level for the person operating the machine.

The push-to-test indicating light provides an additional feature. Consider, for example, an indicating lamp that will not illuminate. It may be that the bulb is not energized, or the bulb may be burned out. Depressing the lens unit will connect the bulb directly across the control voltage source. This provides a check on the condition of the bulb.

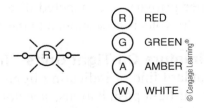

(R) RED

(G) GREEN

(A) AMBER

(W) WHITE

Figure 3-19 Complete electrical symbol for pilot light.

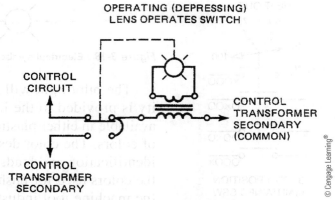

Figure 3-20A Circuit for push-to-test pilot light.

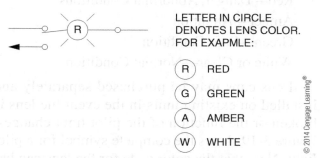

LETTER IN CIRCLE
DENOTES LENS COLOR.
FOR EXAPMLE:

- (R) RED
- (G) GREEN
- (A) AMBER
- (W) WHITE

Figure 3-20B Symbol for push-to-test pilot light.

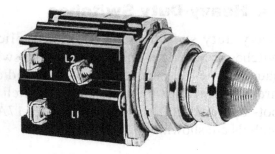

Figure 3-20C Standard transformer-type push-to-test indicating light. *(Courtesy of Eaton Corporation.)*

A detail circuit of the transformer-type, push-to-test indicating light is illustrated in Figure 3-20A. Figure 3-20B shows the symbol for the push-to-test indicating light. Figure 3-20C shows one type of standard transformer-type push-to-test indicating light.

Note that the push-to-test pilot light consists of a 120:6-V transformer, a 6-V bulb covered by a lens, and a standard contact block. When the pilot light lens is depressed (operated), it mechanically operates the contact block. The normally closed contacts open and the normally open contacts close. The normally closed contacts are connected to the control circuit. The normally open contacts are connected directly to a source of 120 VAC. Thus, when the lens is depressed (operated), the pilot light transformer primary is connected directly across 120 VAC. Six volts is now applied to the bulb.

3.5.1 Miniature Oil-Tight Units

In addition to the standard line of indicating lights, a line of miniature oil-tight push buttons, selector switches, and pilot lights are also available. Miniature units can often be used to an advantage where space is at a premium, such as where a great deal of indicating is required. They also have the same feature selection as the standard line. Figure 3-21 shows a group of miniature oil-tight push buttons.

3.6 General Information on Oil-Tight Units

All oil-tight operators mount on the panel of an enclosure with some type of sealing ring to give them an oil-tight feature.

A nameplate is provided to properly identify the unit. Nameplates are available in different sizes. The size generally depends on the amount of information required on the nameplate to properly describe the function of the unit. Typical nameplates for oil-tight units are shown in Figure 3-22.

3.7 Circuit Applications

We can now show a complete circuit, using the symbols illustrated previously. The electrical

Figure 3-21   **New CR204 miniature oil-tight push buttons.** *(Courtesy of General Electric Company, General Purpose Control, Bloomington, IL.)*

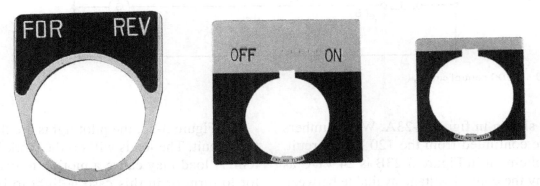

Figure 3-22   **Nameplates for oil-tight units.** *(Courtesy of Eaton Corporation.)*

power source, electrical control source disconnecting device, protection, and an output or load that can be controlled are shown in Figures 3-23.

Figure 3-23A shows a 120 VAC control circuit where the control voltage is available between the two vertical lines (sides of the ladder). The control voltage is present between wire numbers 4 and 2 if the control selector switch is rotated to the right (On position). The red light will illuminate when the "Light On" push-button is pressed. The red light remains energized as long as the push-button is actuated. Figure 3-23B shows a continuation of

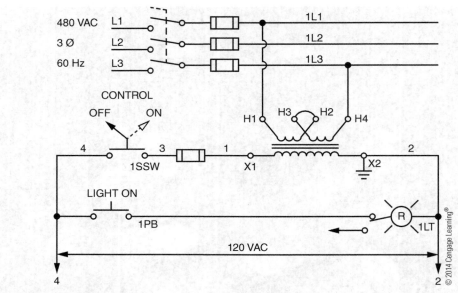

Figure 3-23A **120VAC control circuit.**

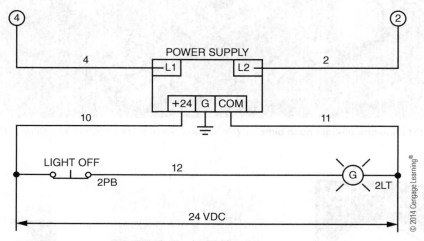

Figure 3-23B **24VDC control circuit.**

the circuit shown in figure 3-23A. Wire numbers 4 and 2 are continued from the 120 VAC circuit. The control circuit in Figure 3-23B is 24 VDC as indicated by the control voltage available between the sides of the ladder (wire numbers 10 and 11). In this circuit, the green light will be illuminated until the "Light Off" push-button is pressed, opening the circuit.

After Figure 3-23, the primary power source and the control power source will not be shown in circuit diagrams. The control voltage will be assumed to be available between the two vertical lines (sides of the ladder).

In Figure 3-23, the pilot lights are the loads in the circuit. The loads will do the work in the system. A load may cause a motion to occur, a motor to turn, or in this case, a light to illuminate. The push-button devices function as the switching devices to control the load. When the switching device is in a closed position, a path will be completed in the circuit, and electricity will flow, allowing the load to be energized.

Multiple switching devices may be placed in series with one another to control a load. For example, Figure 3-24 shows one selector switch and two push-buttons placed in series to control a pilot light.

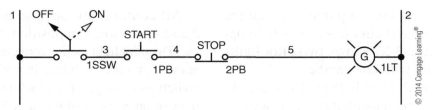

Figure 3-24 **Series circuit.**

To turn on the green pilot light, the switches must be in the following state:

- 1SSW must be placed in the right (On) position
- 1PB must be pressed (actuated)
- 2PB must be released (de-actuated)

The series circuit is typically referred to as an AND circuit since ALL devices must be in the proper state for the circuit to function.

Multiple switching devices may also be placed in parallel to control a load. For example, Figure 3-25 shows two push-buttons placed in parallel to control a pilot light.

To turn on the amber pilot light, continuity must be established in one of the paths. For this circuit, either 1PB or 2PB must be pressed (actuated) for the light to illuminate.

The parallel circuit is typically referred to as an OR circuit since at least one device must be in the proper state for the circuit to function. If both push-buttons were to be pressed, the light will still illuminate—there will just be two paths for electricity to flow. Although both paths will operate the circuit, only one path for continuity is required.

A thorough understanding of series and parallel circuits will be necessary to understand the function of more complicated circuits. Figure 3-26 shows a combination of series and parallel connected devices to control the pilot light.

To turn on the red pilot light, the switches must be in the following state:

- 1PB must be pressed (actuated) OR
- 2PB must be pressed (actuated) AND
- 3PB must be released (de-actuated) AND
- 1SSW must be in the right (On) position

Loads may also be combined in control circuits if the required conditions of the switching devices are identical for both loads. However, load devices must be connected in parallel instead of in

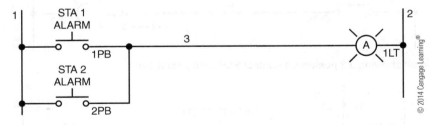

Figure 3-25 **Parallel circuit.**

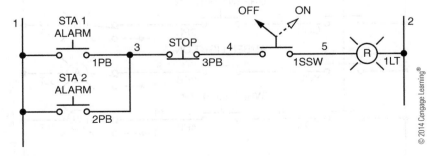

Figure 3-26 **Combination circuit.**

series. Loads are connected in parallel so that the full voltage of the control circuit is available to operate the load. Figure 3-27 shows two pilot lights connected in parallel with one another.

In this circuit, both pilot lights (1LT and 2LT) will illuminate when the push-button is pressed.

In the circuit in Figure 3-28, a three-position, four-contact selector switch is used. The symbol shows that contacts 1 and 3 are closed in position 1. All contacts are open in position 2. Contacts 2 and 4 are closed in position 3.

The pilot light load is connected so that the red and amber lights are energized in position 1 of the selector switch, and the green and white lights are energized in position 3. None of the pilot lights are energized in position 2, as all selector switch contacts are open.

Figure 3-29 shows additional control added to the circuit in Figure 3-28. Momentary contact

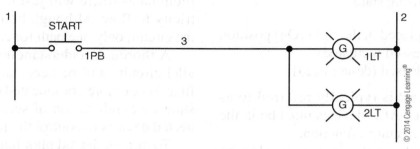

Figure 3-27 **Parallel loads.**

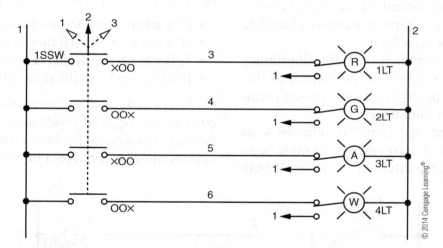

Figure 3-28 **Control circuit showing a 3 position, 4 contact SSW, and pilot lights.**

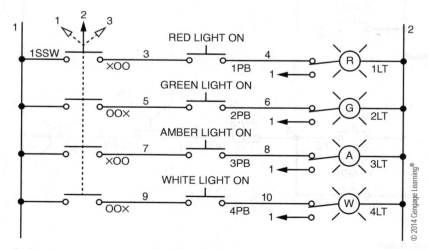

Figure 3-29 **Control circuit showing additional control added to Figure 3-28.**

push-button switches are added to each pilot light circuit. Now, in addition to selecting either position 1 or position 3 of the selector switch, a push-button switch must be operated to energize a pilot light.

3.8 Annunciators

The visual annunciator system as produced by Ronan displays the plant or process status by lighting individual windows identifying process functions. Input to the annunciator is derived from dry or live field contacts or analog signals indicating the process conditions or status. Integral alarm logic modules operate the window lights as directed by any Instrument Society Association (ISA) or custom sequence.

The visual annunciator system architecture, configured so that the electronic modules are remote from the window display, is ideally suited to large systems in power plants and process plant control rooms. The modular logic is housed in a relay rack-mount or surface-mount chassis, connected via multiconductor wiring to the lamp display. Connection to the display and logic chassis may be made by means of individual wires to terminals or via prefabricated cable assemblies with multipin connectors. The split architecture provides ease of maintenance with access to the electronics without interfering with the control room operation.

The lamp displays are available in a large variety of construction types, overall dimensions, and window sizes. The lamp cabinets can be configured any number of windows high or wide, with window sizes ranging from 0.9 inches high by 1.6 inches wide to 3 inches high by 5 inches wide.

The semigraphic series displays the custom process or plant layout with light indicators providing operating status (or alarm) of pumps, valves, hoppers, tank levels, etc. The unique construction of the Ronan graphic allows infinite resolution in positioning the status/alarm light on the back plane. The graphic layout, contained between a clear front plexiglas and a rear diffuser, can easily be modified in case of a process change or addition of equipment. The indicator lights can be driven by either solid-state or relay-type remote logic alarm system. See Figures 3-30 and 3-31 for typical annunciators.

3.9 Light-Emitting Diodes

In many digital systems the output unit is in the form of a readout display, which may call for a readout of several digits. Multidigit displays are

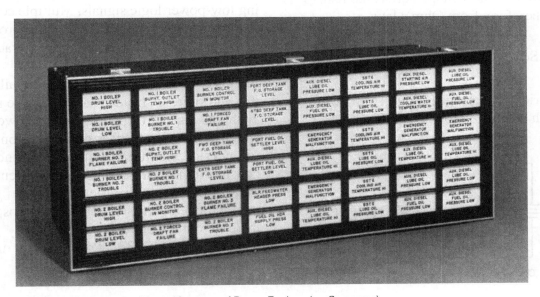

Figure 3-30   6 × 8 annunciator-amp cabinet. *(Courtesy of Ronan Engineering Company.)*

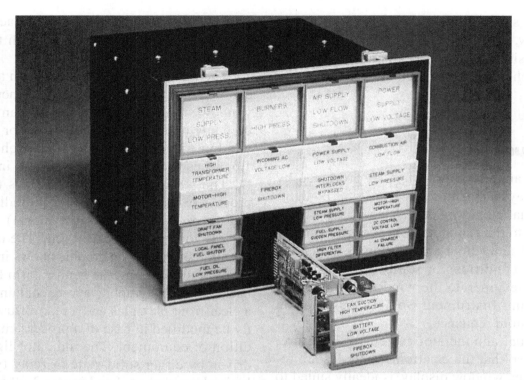

Figure 3-31   Solid state annunciator with on board alarm logic for multiple alarm points. *(Courtesy of Ronan Engineering Company.)*

available using light-emitting diodes (LEDs). They are also used as indicators of both outgoing and incoming signals. The symbol for the LED is shown in Figure 3-32, which shows a resistor (R) that prevents the current from exceeding the maximum current rating of the diode. The resistor has a voltage on the left side of V_B and a voltage on the right side of V_C. The voltage across the resistor is $V_B - V_D$. Therefore, from Ohm's law:

$$\text{Series current} = \frac{V_B \pm V_D}{R}$$

The series current will range from 10 mA to 50 mA with a voltage drop of 1.5 V to 2.5 V.

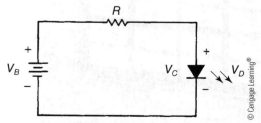

Figure 3-32   Circuit for an LED.

3.10 Membrane Switches

The membrane switch as produced by Xymox Technologies is a normally open, momentary contact, push-to-operate-type device. Layers of insulating materials, conductive coatings, decorative graphics, and adhesives are combined to form a completely sealed switch ideally suited to switching low-power logic signals. Multiple contacts and interconnections can be employed to provide complex switching networks that are attractive, reliable, and cost-effective.

There are two types of switches available: flexible and rigid. Outwardly, most flexible membrane switches appear to be the same. However, their construction must differ depending on the environment in which the switch must function. The construction must be chosen with attention to all factors in its operating environment, including maximum temperature, quality of atmosphere, and method of interconnection.

See Figure 3-33 for a typical membrane switch.

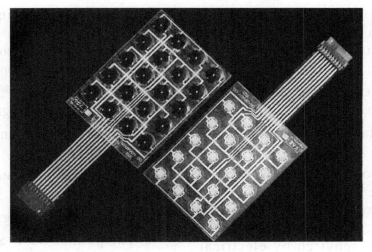

Figure 3-33 **Membrane switch circuit board.** *(Courtesy of Xymox Technologies Inc.)*

Transparent LED areas are provided in switch strip and keypad component units. They can be used with custom face plates having LED windows in the keys.

Various forms of feedback are compatible with Xymox stock membrane switches. A device that operates upon the depression of each switch will provide audible feedback and is readily available from a number of sources. Visual feedback can be provided in the form of backlighted areas.

Recommended Web Links

Students are encouraged to view the following Web sites as a supplement to the concepts presented in this textbook. Review and analyze the array of products that are available for electrical control applications. Many of these sites offer technical information that will help in converting the principles to practical applications. To view catalogs, your PC may require Adobe Acrobat Reader software.

Control Devices

1. Allen-Bradley
www.rockwellautomation.com
Review: Products/Product Catalogs/Industrial Controls/Push-buttons and Pilot Lights

2. General Electric
www.geindustrial.com
Review: Products/Push-buttons & Pilot Devices

3. Signaworks
www.signaworks.com
Review: Industrial Push-buttons Pilot Lights

4. Square D
www.schneider-electric.us
Review: Products and Services/Products/Push-buttons

Annunciators

1. Ronan Engineering
www.ronan.com
Review: Alarm Display Products, Annunciator

2. Rochester Instruments
www.rochester.com
Review: Annunciators and Alarm Management

Achievement Review

1. Describe how you would panel mount an oil-tight unit.

2. Why is the low-voltage bulb in the pilot light preferred?

a. It is much smaller in size.
b. It will withstand vibration.
c. It will not consume much electrical energy.

3. Draw the symbol for a mushroom head push-button switch (one contact block, two contacts; one normally open, one normally closed).

4. Are lenses for pilot lights available in more than one color? If so, why?

5. What is the advantage of panel mounting a contact block?

   a. The mounting provides space for terminal block installation.

   b. The mounting allows for more space on the panel.

   c. The mounting is required by electrical standards.

6. Draw the symbol for a red push-to-test light.

7. Draw the symbol for a three-position, four-contact selector switch. Contacts 1 and 3 are closed in position 1, all contacts are open in position 2, and contacts 2 and 4 are closed in position 3.

8. Explain how the push-to-test pilot light functions. Draw the complete circuit diagram.

9. Draw a circuit showing a two-position selector switch and a green push-to-test pilot light. The pilot light is to be energized when the selector switch is placed in the ON position.

10. What uses are made of the LEDs in a programmable controller?

11. What are some of the uses for an annunciator?

12. Draw a circuit showing:

   a. NO push-button (Start)

   b. NC push-button (Stop)

   c. Three position SSW (On – Off – Manual)

   d. Green pilot light

The green light will illuminate when the following conditions are met:

   a. The start push-button is pressed

   b. The stop push-button is not pressed

   c. The SSW is in the manual position

13. Draw a circuit showing:

   a. Two NO push-buttons

   b. Red pilot light

The red light will illuminate when either push-button is pressed.

14. Draw a circuit showing:

   a. Two position SSW (Sta. 1 – Sta. 2)

   b. Green pilot light

   c. Amber pilot light

The green light will illuminate when the SSW is in position for Station 1.

The amber light will illuminate when the SSW is in position for Station 2.

15. Draw a circuit showing:

   a. Two position SSW (Off – On)

   b. Green pilot light

   c. White pilot light

The green and amber lights will both illuminate when the SSW is placed in the On position.

16. Redraw the circuit diagram shown in Figure 3-24 using IEC 617 symbols.

17. Redraw the circuit diagram shown in Figure 3-25 using IEC 617 symbols.

18. Redraw the circuit diagram shown in Figure 3-26 using IEC 617 symbols.

19. Redraw the circuit diagram shown in Figure 3-27 using IEC 617 symbols.

4

CHAPTER

Relays

OBJECTIVES

After studying this chapter, you should be able to:

- Identify two main parts of a control relay.

- Explain the operation of a control relay.

- List the three published ratings for relays.

- Explain why silver is used in relay contacts.

- Discuss several factors involved with relay operation, such as contact bounce, overlap contacts, contact wipe, and split or bifurcated contacts.

- Describe the seal-in circuit and explain why it is used.

- Draw the symbols for the relay coil, relay contacts, and time-delay relay contacts.

- Explain the difference between inrush and holding current in a relay coil.

- Describe how the latching relay operates.

- List the basic uses for the contactor.

- Compare and contrast the use and operation of a relay and a contactor.

4.1 Control Relays and Their Uses

The relay is the basic building block in a control circuit. It is used to create memory circuits that allow a machine to methodically step through the sequences that are required for proper operation. The contacts from the relay may be used to:

- Control motions
- Make logical decisions for energizing additional memories
- Operate loads
- Turn off (de-energize) existing memories
- Communicate information

Relays may be used to control an entire machine or process. This was typically the case until the invention of the programmable logic controller (PLC). The PLC is an industrial computer that was designed to replace relay control on equipment. Therefore, the ladder logic code used to program a PLC functions as and looks very similar to a relay circuit. As the price of PLCs continue to drop, the use of relays for general machine control has declined. Even in a PLC design, however, a relay must still be utilized for switching higher current loads, interfacing to various voltage levels, or in certain hardware control functions. Therefore, the understanding of relay control circuits is important. The concepts learned throughout this study may easily be applied to PLC systems. The operation and symbols used to perform various functions in the PLC will vary drastically among different PLC systems. However, relay circuits will use the same symbols and will logically function similarly regardless of the manufacturer. Therefore, the understanding of relay logic control circuits is still important as an introductory study for the basics of machine control.

The relay is an electromechanical device that consists of two parts: a coil and the contacts. The major parts of the relay are shown in Figures 4-1A and 4-1B. The contact block assembly consists of both stationary contacts and a pole piece that is attached to the plunger. Contacts are referred to as *normally open* (NO) and *normally closed* (NC). The normal condition of the contact refers to the state of the contact when the relay coil is off (de-energized). At this time, the NO contacts are open (blocking current flow) and the NC contacts are closed (allowing current flow). When the coil of the relay is energized, the plunger moves through the coil, closing the normally open contacts and opening the normally closed contacts. Figure 4-1A shows the contacts in the normal or de-energized condition of the coil, as well as the symbols for the relay coil and contacts. The IEC symbols are shown in Figure 4-1C.

When the coil of the relay is connected to a source of electrical energy, the coil is energized. When the coil is energized, the plunger moves up, as shown in Figure 4-1B, because of a force produced by a magnetic field. The distance that the plunger moves is generally short—about 1/4 inch or less.

There is a difference in the current in the relay coil from the time the coil is first energized and when the contacts are completely operated. When the coil is initially energized, the plunger is in an out position. Because of the open gap in the magnetic path (circuit), the initial current in the coil is high. The current level at this time is known as *inrush current*. As the plunger moves into the coil, closing the gap, the current level drops to a lower value called the *sealed current*. The inrush current is approximately six to eight times greater than the sealed current.

Coils are available to cover most standard voltages from 24 V to 600 V (AC or DC). Relay coils are typically made of a molded construction to reduce moisture absorption and increase mechanical strength.

The level of voltage at which the relay coil is energized to move the contacts from their normal, unoperated position to their operated position is called *pickup voltage*. After the relay is energized, the level of voltage on the relay coil at which the contacts return to their unoperated condition is called *drop-out voltage*.

Coils on electromechanical devices such as relays, contactors, and motor starters are designed so as not to drop out (de-energize) until the voltage drops to a minimum of 85% of the rated voltage. The relay coils also will not pick up (energize) until the voltage rises to 85% of the rated voltage. This voltage level is set by the National Electrical Manufacturers Association (NEMA). As a rule, 85% is found to be a conservative figure. Most electromechanical devices will not drop out until a lower voltage level is reached. Also, most electromechanical devices will pick up at a lower rising voltage level. Generally, coils on electromechanical devices will operate continuously at 110% of the rated voltage without damage to the coil.

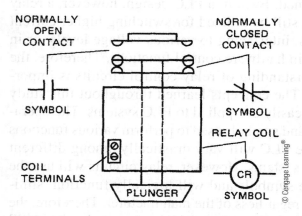

Figure 4-1A De-energized condition of relay showing contact position.

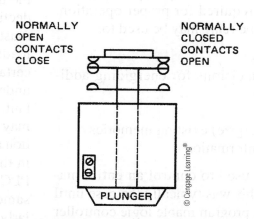

Figure 4-1B Energized condition of relay. With coil energized, magnetic force pulls the plunger up, operating the contacts.

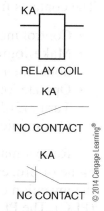

Figure 4-1C IEC617 Symbols

The two important parts in a relay are the coil and contacts. Of these, the contacts generally require greater consideration in practical circuit design.

There are some single-break contacts used in industrial relays. However, most of the relays used in machine tool control have double-break contacts. The rating of contacts can be misleading. The three ratings generally published are:

1. Inrush or "make contact" capacity
2. Normal or continuous carrying capacity
3. The opening or break capacity

For example, a typical 120 VAC industrial relay has the following contact ratings:

- 10-A noninductive continuous load (AC)
- 6-A inductive load at 120 VAC
- 60-A make and 60-A break inductive load at 120 VAC

A resistance is an example of a noninductive load, such as would occur in a resistance unit used as a heating element. An inductive load is basically a coil, such as a solenoid, contactor coil, or motor starter coil.

The point to remember is that in determining the contact rating, it must be clear what rating is given.

Relay contacts are usually silver or a silver-cadmium alloy. This material is used because of the excellent conductivity of silver. Silver oxide, which forms on the contacts, is also a good conductor. Adding a small amount of cadmium to the silver increases the interrupting capacity of the contacts. Some manufacturers offer gold-plated contacts as optional features. Gold plating improves shelf life and provides improved contact reliability in low-energy, low-duty-cycle applications. The circuit designer should consider the following points in the use of control relays:

- *Changing contacts normally open to normally closed, or vice versa.* Most machine tool relays have some means of making this change, with either a simple flip-over contact, to replacement contacts, or spring loaded switch.
- *Universal contacts.* The term *universal* refers to relays in which the contacts may be NO or NC. Only one type may be selected in the use of each contact, but not both types.
- *Split or bifurcated contacts.* As the name implies, a split or bifurcated contact is divided into two parts. This division provides for double the contact make points, improving reliability of contact and reducing contact bounce.
- *Contact bounce.* All contacts bounce (spring apart or vibrate) on closing. The problem is to reduce the bounce to a minimum. In rapid-operating relays, contact bounce can be a source of trouble. For example, a seal in a transfer circuit can be lost by a contact opening momentarily. The use of split contacts, overlapping contacts, other type of antibounce contacts or a circuit change may help.
- *Overlap contacts.* To overlap contacts, one contact can be arranged to operate at a different time relative to another contact on the same relay. For example, the NO contact can be arranged to close before the NC contact opens. This arrangement is called "make before break."
- *Contact wipe.* This factor results from the relative motion of the two contact surfaces after they make contact, that is, one contact wipes against the other. One advantage to be gained from this motion is that it induces self-cleaning of the contacts.

Relays shown in this book are of the 600-V design. Figure 4-2 shows a typical four-pole, 600-V control relay, with plug-in contact cartridges. The reference to 300-V and 600-V relays applies to the voltage being carried by the contacts. The voltage level used to operate the relay coil may be different than the voltage rating of the contacts. Both ratings must be considered when purchasing a relay or when connecting the coil or the contacts into a circuit.

These relays are generally available with up to 12 contacts in combinations of NO and NC contacts. They may be either convertible or fixed. In most cases, the relays can be combined with a pneumatic timing head or mechanical latch. However, when adding the pneumatic timing head, the relay is restricted to four instantaneous contacts.

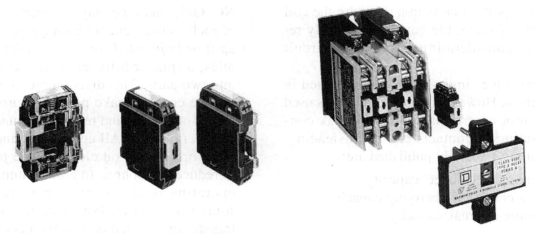

Figure 4-2 **Industrial control relays.** *(Courtesy of Square D/Schneider Electric.)*

Figure 4-3 shows a circuit diagram using a relay with two NO contacts. The two NO contacts are shown as 1CR-1 and 1CR-2 (referring to the physical contact positions on the relay). Contact 1CR-1 is the first contact on relay 1CR and contact 1CR-2 is the second contact on relay 1CR. This designation for contacts is common for build drawings at the original equipment manufacturer (OEM) facility. This designation also provides an easy method to differentiate between contacts when explaining circuits in this text. However, typical machine control drawings will not include the specific location of each contact on the relay. Therefore, both contacts will usually be designated as 1CR on the electrical drawings. The OEM build drawing designation will be utilized throughout this text to simplify discussion of the circuit.

As shown in Figure 4-3, one contact, 1CR-1, is used as a seal-in around the START push button. Thus, a *seal-in circuit* is a path provided for electrical energy to the load after the initiating path has been opened. The second relay contact, 1CR-2, is used to energize a light. Remember that when a relay coil is energized, the NO contacts close. The circuit can be de-energized by pressing the STOP push-button.

Several points are worth reiterating in regards to Figure 4-3. Initially, coil 1CR is energized when the START push-button is pressed. Contact 1CR-1 seals-in the circuit after the START push-button is released. Relay coil 1CR will serve as a memory circuit to keep the circuit energized until the STOP push-button is pressed. Devices that are utilized to initiate the circuit will be located inside of the seal. In this case, the START push-button is required to initiate the circuit. Devices that are utilized to stop (or open) the circuit will be located outside of the seal. In this case, the STOP push-button is required to de-energize (open) the circuit. Because coil 1CR will remain energized, the contact 1CR-2

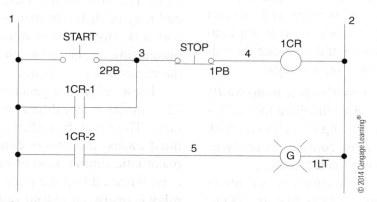

Figure 4-3 **Control circuit using a relay with two NO contacts.**

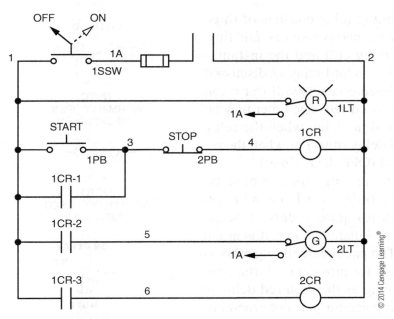

Figure 4-4 Control circuit from Figure 4-3 expanded.

may be used to control the load (1LT) to indicate any time that the circuit is active (on).

Figure 4-4 shows the addition of a selector switch, fuse, pilot light, and a second relay to the circuit shown in Figure 4-3. When the selector switch is rotated to the ON position, electrical energy is available at the two vertical sides of the circuit. The red light is energized, indicating that the control circuit is active. One additional relay contact is added in the circuit from relay 1CR. This contact closes when the relay 1CR is energized and it, in turn, energizes a second relay coil, 2CR. All devices can be de-energized by pressing the STOP push-button.

A numbering system for the wires and the devices has been included in the circuits. This system is used in all the remaining circuits in the text. The two vertical sides of the diagram, representing the source of electrical energy (control voltage), are numbered 1 and 2. Although practice will vary with the manufacturer, this text uses 1 for the hot (protected) side and 2 for the common side (generally grounded). There are other variations of this numbering system in the industry. See Chapter 17 and Appendix E.

Note that starting at one side or the other on each line of the circuit, the wire number changes each time you move through a contact or coil. Thus,

each contact (push-button, selector switch, limit switch, pressure switch, temperature switch, and so on) will have two numbers, one for the incoming conductor and one for the outgoing conductor. Two wire numbers are also used for all coils (relays, timers, contactors, motor starters, and so on).

Although this practice does not seem to have any importance in the operation of the circuit, it is invaluable when the components are wired on a machine in accordance with the specific circuit design.

This practice is not only a must for electro-mechanical control but also for solid-state and programmable controllers.

4.2 Timing Relays

A pneumatic (air controlled) timing head can be purchased and mounted directly to certain control relays. The timing head will operate in conjunction with the relay coil to which it is attached. A pneumatic timing relay is also available for purchase as a complete unit. The complete unit would be considered when building a circuit that requires a time delay. The timing head would be considered when an existing circuit needs to be modified to incorporate a time delay function.

The pneumatic timing relay consists of three units: a relay coil, instantaneous contacts, and time delay contacts. The relay coil and the instantaneous contacts are the standard relay as discussed in Section 4.1. The timed contacts will transition when a time delay has occurred. The timer may be arranged to begin the time delay when the relay coil is energized (ON delay timer) or when the relay coil is de-energized (OFF delay timer).

The adjustable timing range for a pneumatic timing range is on the order of 0.1 second to 60 seconds. Accuracy varies quite widely, depending on the size, type, and manufacturer. It is in the range of 10 percent. The pneumatic timing relay is a cost effective solution for providing a basic time delay in a circuit as long as the required delay is not too long and timing accuracy is not crucial to the functionality of the circuit.

The symbols for the timing relay coil and the instantaneous contacts are shown in Figure 4-5. Note that the timing relay coil is identified as TR where the standard relay coil is identified as CR. Therefore, when a timing head is added to a standard relay coil in the field, the control drawings must be modified to incorporate the TR designation.

The symbols for the timed contacts are shown in Figure 4-6. The drawing symbol used and the functionality of the timed contacts will vary between an ON delay and an OFF delay timer.

The ON delay timer will begin timing when the relay coil is energized. The symbols used for the timing contacts show the normal (relay coil off) condition for each contact. When the time delay has expired, the contacts will transition states (NO contacts will close and NC contacts will

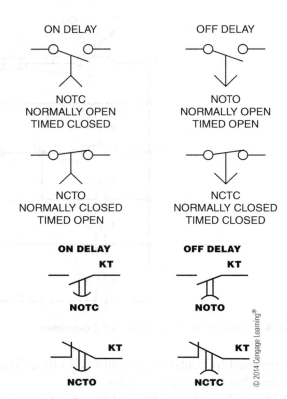

Figure 4-6 Electrical symbols for the timing contacts.

open). When the coil is de-energized, the contacts will return to their normal state. Therefore, the normally open timed closed (NOTC) contact will be open when the coil is off. The contact will remain in this state until the relay energizes and the time delay expires. At this time, the NOTC contact will close. Similarly, the normally closed timed open (NCTO) contact will be closed when the coil is off. The contact will remain in this state until the relay energizes and the time delay expires. At this time, the NCTO contact will open. Figure 4-7 details the operation of an ON delay timer by providing a timing diagram for the condition of the pilot lights for various stages of operation.

The OFF delay timer is different from the ON delay timer in the following ways:

- The symbols used to represent the time delayed contacts.
- The timing begins when the relay coil is de-energized (instead of energized).
- The transitioning of the time delayed contacts.

The OFF delay timer will begin timing when the relay coil is de-energized. The symbols used

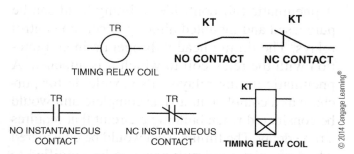

Figure 4-5 Electrical symbols for the timing relay coil and instantaneous contacts.

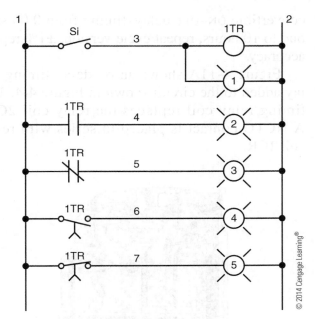

Figure 4-7A Control circuit

timing will begin. When the time delay expires, the time delayed contacts will transition again and return to their normal state. Note that the operation of the time delayed contacts for the ON delay and the OFF delay timer are quite different. Figure 4-8 details the operation of an OFF delay timer.

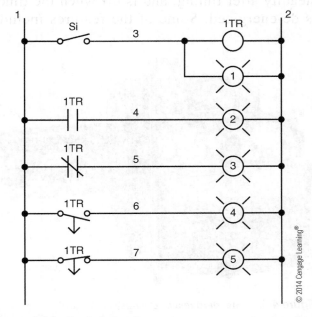

Figure 4-8A Control circuit

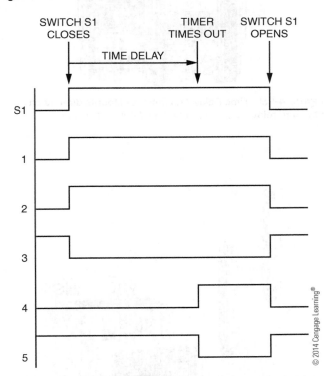

Figure 4-7B Timing diagram showing state of switch S1 and pilot lights

for the time delayed contacts show the normal (relay coil off) condition for each contact. When the relay coil energizes, the contacts will transition states immediately. The contacts will remain in this state as long as the relay coil stays energized. When the relay coil is *de-energized*, the

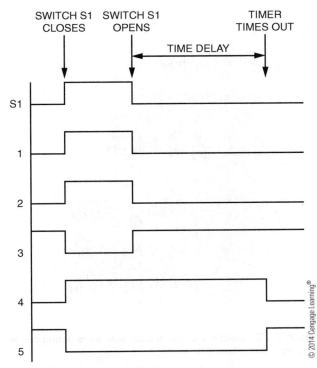

Figure 4-8B Timing diagram showing state of switch S1 and pilot lights

Figure 4-9A shows a standard relay, and Figure 4-9B shows a time delay attachment that mounts directly on the standard relay.

Figures 4-9C, 4-9D, and 4-10 display solid-state timing relays. Timing indication is provided by an LED that flashes during timing, glows steadily after timing, and is off when the timer is de-energized. Some of the features include convertible ON–OFF delay, timing from 0.05 second to 10 hours, repeat cycle version, ±1% repeat accuracy.

Figure 4-11A shows an on delay timing relay added to the circuit shown in Figure 4-4. The timing relay coil replaces the relay coil 2CR. A NCTO contact is placed in series with relay coil 1CR.

Figure 4-9A **Standard relay.** *(Courtesy of Rockwell Automation, Inc.)*

Figure 4-9B **Time Delay Attachment - Mounts directly on standard relay.** *(Courtesy of Rockwell Automation, Inc.)*

Figure 4-9C **Solid-state timing relay with fixed timing range.** *(Courtesy of Rockwell Automation, Inc.)*

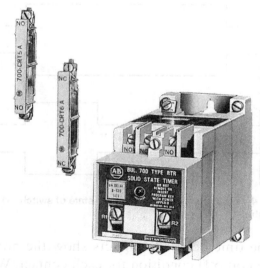

Figure 4-9D **Solid-state timing relay with replacement contacts.** *(Courtesy of Rockwell Automation, Inc.)*

Figure 4-10 Solid-state timing relay with replacement timing units. *(Courtesy of Rockwell Automation, Inc.)*

The sequence of operations proceeds as follows:

1. Press the START push-button.
2. Relay coil 1CR is energized.
   a. Relay 1CR contact 1CR-1 closes, sealing around the START pushbutton.

b. Relay contact 1CR-2 closes, energizing the green pilot light.
   c. Relay contact 1CR-3 closes, energizing ON delay timer relay coil 1TR.
3. After a preset time on 1TR, the NCTO time-delay contact 1TR opens, de-energizing relay coil 1CR.
   a. Relay 1CR contact 1CR-1 opens, opening the seal circuit around the START push-button.
   b. Relay 1CR contact 1CR-2 opens, de-energizing the green pilot light.
   c. Relay 1CR contact 1CR-3 opens, de-energizing the ON delay timer relay coil 1TR.

Figure 4-11B shows another timing relay and control relay added to the circuit shown in Figure 4-11A. The new OFF delay timing relay 2TR will have one NOTO contact arranged for time delay after de-energizing coil 2TR. The second relay coil is 2CR.

The sequence of operations proceeds as follows:

1. Press the START push-button.
2. Relay coil 1CR is energized.
   a. Relay 1CR contact 1CR-1 closes, sealing around the START pushbutton.
   b. Relay 1CR contact 1CR-2 closes, energizing ON delay timing relay coil 1TR.

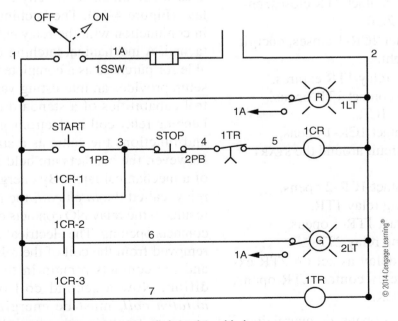

Figure 4-11A Control circuit from Figure 4-4 with a timing relay added.

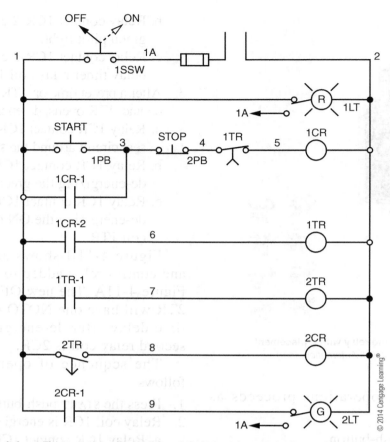

Figure 4-11B Control circuit from Figure 4-11A with additional timing and control relays.

c. Instantaneous contact 1TR-1 on timing relay 1TR closes, energizing OFF delay timing relay coil 2TR.

d. The NOTO timing contact 2TR closes, energizing relay coil 2CR.

3. The 2CR relay contact 2CR-1 closes, energizing the green pilot light.

4. Preset time on timing relay 1TR expires.

a. The NCTO timing contact 1TR opens, de-energizing relay 1CR.

b. The 1CR relay contact 1CR-1 opens, opening the seal circuit around the START push-button.

c. The 1CR relay contact 1CR-2 opens, de-energizing timing relay 1TR.

d. Instantaneous contact 1TR-1 opens, de-energizing timing relay 2TR.

5. After a preset time delay as set on 2TR expires, NOTO time-delay contact 2TR opens, de-energizing relay 2CR.

a. Relay contact 2CR-1 opens, de-energizing the green pilot light.

4.3 Latching Relays

A mechanical latching attachment can be purchased and mounted directly to certain control relays (Figure 4-12). The latching unit will operate in conjunction with the relay coil to which it is attached. A mechanical latching relay is also available for purchase as a complete unit. This latching setup provides an interesting variation of the control capabilities of a standard control relay. The latching relay coil is electromagnetically operated and functions the same as a standard relay coil. However, the contacts are held in place by means of a mechanical latch. By energizing a coil of the relay, called the *latch coil,* the relay operates and results in the relay NO contacts closing and the NC contacts opening. The electrical energy can now be removed from the coil of the relay (de-energized), and the contacts remain in their operated condition. Now a second coil on the relay, the *unlatch coil,* must be energized to return the contacts to their unoperated condition. This arrangement is often referred to as a *memory relay.*

Figure 4-12 Latching attachment for a relay. *(Courtesy of Rockwell Automation, Inc.)*

It is important to remember that the contacts on a latch coil will NOT return to their un-operated condition unless the unlatch coil is energized. In the event of a power failure to the control circuit, the contacts will remain in their current state during the power-off period and upon power-up of the circuit. Therefore, the use of latch coils to control motion must be carefully considered to avoid unintentional motion when powering-up the equipment. The latch coil is ideally suited for applications where the memory of a circuit must be maintained even if loss of power occurs. Part status information is an excellent example where the memory retention of the latch coil may be utilized.

The symbol for the latching relay is shown in Figure 4-13. To further explain the symbols:

- Contacts should be shown in the unlatched condition; that is, as if the unlatch coil were the last one energized.

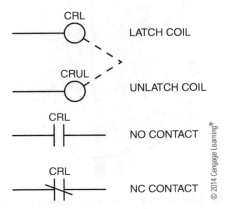

Figure 4-13 Symbols for the latching relay.

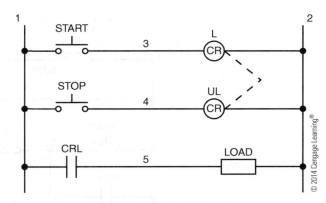

Figure 4-14 Control circuit for a latching relay.

- The latch and unlatch coils are on the same relay and always have the same reference number.
- The contacts are always associated with the latch coil as far as reference designations are concerned.

Figure 4-14 illustrates the symbols for a typical latching relay. Figure 4-15 expands on the circuit of 4-11B to incorporate a latch coil in the design.

The latch coil of a latching relay replaces the coil of control relay 2CR.

In Figure 4-15, the closing of timing contact 2TR energizes the latch coil 2CRL. NO contact 2CRL-1 closes, energizing the green pilot light. When the timing relay 2TR times out and the NOTO timing contact opens, the latch coil is de-energized. However, the light remains energized until the RESET push-button switch is pressed, energizing the unlatch coil 2CRUL.

4.4 Plug-In Relays

Some industrial machine relays are available as plug-in types. They are designed for multiple switching applications at or below 240 V. Coil voltages span standard levels from 6 V to 120 V. They are available for AC or DC. Mounting can be obtained with a tube-type socket, square-base socket mounting, or flange mounting using slip-on connectors.

A typical plug-in relay is shown in Figure 4-16.

The plug-in relay has a distinct advantage in allowing relays to be changed without disturbing the circuit wiring. In critical operations in which the relay service is very hard and downtime is a premium,

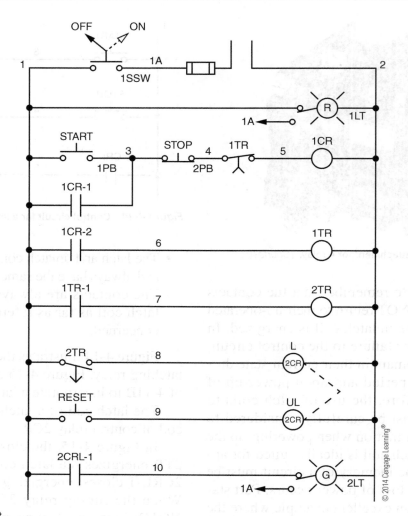

Figure 4-15 Control circuit.

the plug-in relay may have some advantages. Assessing the actual operating conditions in specific cases is the best way to determine their need.

4.5 Contactors

The *contactor,* in general, is constructed similarly to the relay. Like the relay, it is an electromechanical device. The same coil conditions exist in that a high inrush current is available when the coil of the contactor is energized. The current level drops to the holding or sealed level when the contacts are operated. Generally the contactor is supplied in two-, three-, or four-pole arrangements. The coil of the contactor may be purchased to operate at most standard voltages from 24 V to 600 V (AC or DC). The major difference between a standard relay and

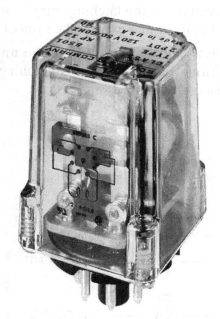

Figure 4-16 Plug-in relay. *(Courtesy of Square D/Schneider Electric.)*

Figure 4-17A Size 1 contactor. *(Courtesy of Rockwell Automation, Inc.)*

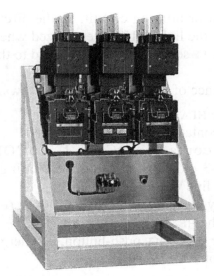

Figure 4-17B Size 9 contactor. *(Courtesy of Rockwell Automation, Inc.)*

a contactor is the current carrying capability of the contacts. Contactors capable of carrying current in the range of 9 A through approximately 2250 A are available. For example, a size 00 contactor is rated at 9 A (200–575 V); a size 9 contactor is rated at 2250 A (200–575 V).

One normally open auxiliary contact is generally supplied as standard on most contactors. This contact is used as a holding contact in the circuit, for example, around a normally open push-button. Additional normally open and normally closed

auxiliary contacts can be obtained as an option. They can be supplied with the contactor from the manufacturer or ordered as a separate unit and mounted in the field.

Figures 4-17A and 4-17B show two typical electromechanical contactors. Figure 4-17A is a size 1 with a 27-A rating; Figure 4-17B is a size 9 with a 2250-A rating.

The use of the auxiliary contact, which is generally rated at 10 A, is shown in Figure 4-18. It can be seen that the auxiliary contact is used like the

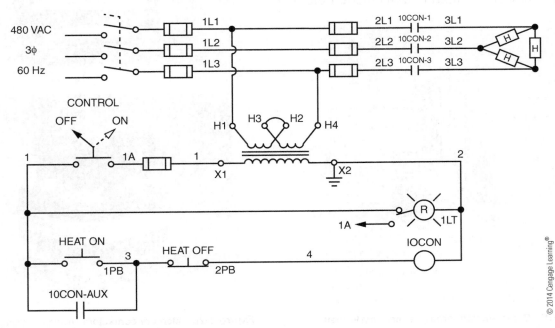

Figure 4-18 Control circuit for a heater contactor.

© 2014 Cengage Learning®

relay seal-in or holding circuit. In the circuit in Figure 4-18, the light (1LT) is energized when the control ON–OFF selector switch is rotated to the ON position.

The balance of the sequence is as follows:

1. Press the HEAT ON push-button.
2. Coil of contactor 10CON is energized.
3. Power contacts 10CON-1, 10CON-2, 10CON-3 close, energizing the heating elements at line voltage.
4. Auxiliary contact 10CON-AUX closes, sealing around the HEAT ON push-button.
5. Press the HEAT OFF push-button, de-energizing the coil of 10CON contactor.

   a. Contacts 10CON-1, 10CON-2, 10CON-3 open, de-energizing the heating elements.
   b. Auxiliary contact 10CON-AUX opens, opening the seal circuit around the HEAT ON push-button.

In addition to the conventional electromechanical contactor, a mercury-to-metal contactor is available. Figure 4-19 shows a cross section through such a contactor. The load terminals are isolated from each other by the glass in the hermetic seal. The plunger assembly, which includes the ceramic insulator, the magnetic sleeves, and related parts, floats on the mercury pool. When the coil is powered, creating a magnetic field, the plunger assembly is pulled down into the mercury pool, which in turn is displaced and moved up to make contact with the electrode. This action closes the circuit between the top and bottom load terminal, which is connected to the stainless steel can.

Figure 4-20 shows a group of one-, two-, and three-pole contactors. They are available up through 100-A capacity.

Figure 4-20 Mercury contactors. *(Courtesy of Eagle Signal Controls.)*

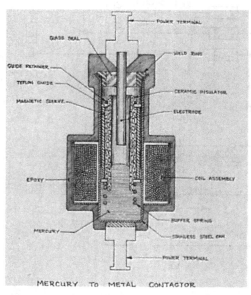

Figure 4-19 Cross section through a mercury-to-metal contactor. *(Courtesy of Eagle Signal Controls.)*

The basic use for the contactor is for switching of power in resistance heating elements, lighting, magnetic brakes, or heavy industrial solenoids. Contactors can also be used to switch motors if separate overload protection is supplied.

Recommended Web Links

Students are encouraged to view the following Web sites as a supplement to the concepts presented in this textbook. Review and analyze the array of products that are available for electrical control applications. Many of these sites offer technical information that can help in converting the principles to practical applications. To view catalogs, your PC may require Adobe Acrobat Reader software.

Relays

1. Allen-Bradley
   www.rockwellautomation.com
   Review: Products/Product Catalogs/Industrial Controls/Relays and Timers

2. General Electric
   www.geindustrial.com
   Review: Products/Relays & Timers-Control

3. Moeller Electric Ltd.
   www.eaton.com
   Review: Products & Services/Electrical Distribution & Control/Products & Services/ Automation and Control/Relays and Timers

Contactors

1. Allen-Bradley
   www.rockwellautomation.com
   Review: Products/Product Catalogs/Industrial Controls/Motor Control

2. Mercury Displacement Industry

   http://www.mdius.com
   Review: Relays

Achievement Review

1. What is the difference between inrush current and holding current in the coil of a control relay?

2. What are the three ratings generally published concerning control relay contacts?

3. Explain contact wipe, and explain how contact bounce can affect the operation of an electrical control circuit.

4. Draw the symbols for a control relay coil, normally open relay contact, and normally closed relay contact.

5. Show the use of a control relay contact in a seal-in circuit. Use a control relay, two push-button switches, and a red push-to-test pilot light.

6. What advantages can be obtained from the use of a time-delay relay as compared to a standard control relay?

7. Identify the following time-delay relay contacts. Specify whether they operate based on a time delay after the relay coil is energized or de-energized.

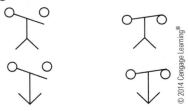

8. Follow the instructions in question 5, but substitute a time-delay relay for the standard control relay. In this case, the red pilot light is to energize at some predetermined time after the relay coil is energized.

9. Describe how a latching relay operates.

10. How many poles are generally available on contactors? What is the range in current-carrying capacity?

11. What is one advantage in using the plug-in relay?
    a. No circuit numbers are required.
    b. The best application of the plug-in relay is in power circuits.
    c. Relays can be changed without disturbing the circuit wiring.

12. Coils on relays are designed not to drop out (de-energize) until the voltage drops to what percentage of the rated voltage?

    a. 60%
    b. 85%
    c. 100%

13. Complete the chart for the following time delay relay contacts. Indicate whether the contact is open or closed.

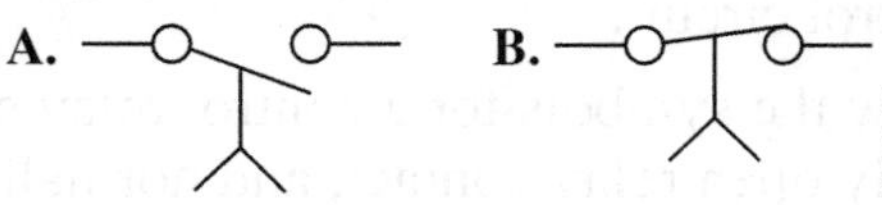

    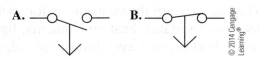

Timer coil	Contact A	Contact B
off		
on (initial)		
on (time delay expired)		

14. Complete the chart for the following time delay relay contacts. Indicate whether the contact is open or closed.

Timer coil	Contact A	Contact B
on		
off (initial)		
off (time delay expired)		

15. Redraw the circuit diagram shown in Figure 4-3 using IEC 617 symbols.

16. Redraw the circuit diagram shown in Figure 4-4 using IEC 617 symbols.

17. Redraw the circuit diagram shown in Figure 4-11A using IEC 617 symbols.

18. Redraw the circuit diagram shown in Figure 4-11B using IEC 617 symbols.

5
CHAPTER

Solenoids

OBJECTIVES

After studying this chapter, you should be able to:

- Draw the symbol for the solenoid.

- Explain why it is necessary for the plunger in a solenoid to complete its stroke.

- Know the difference between sealed current and inrush current in a solenoid.

- Describe the application of solenoids to operating valves.

- Explain the operation of a valve by describing the relationship between the spool positions and the porting of the pressure and tank ports.

- Draw a control circuit showing the energizing of a solenoid through the closing of a relay

- contact, using a control relay, two push-button switches, and a solenoid.

- List different configuration options for center positions on a three position valve.

- Explain the similarities and differences between a spring return and a detented valve.

- Draw a simple hydraulic or pneumatic circuit showing a valve connected to a cylinder.

- Know the function of the proportional solenoid.

- Know the function of the force motor in a servo valve.

5.1 Solenoid Action

A relay is used to control electrical circuits. The relay coil is energized causing the plunger to pull in and transition the contacts of the relay. The contacts are then used to control (open or close) other devices in the electrical circuit.

The general principle of solenoid action is very important in machine control. A solenoid-controlled valve is used to control air or hydraulic circuits. When the solenoid coil is energized, the spool in the valve will shift. The location of the spool will determine the porting of the pressure and tank lines to the cylinder, thus controlling motion on the machine. The solenoid valve and cylinder will be shown in the appropriate drawing set (air, hydraulic, gas, coolant, etc.).

Like the relay and contactor, the solenoid is an electromechanical device. In this device, electrical energy is used to magnetically cause mechanical movement. Since the solenoid is operated from electrical energy, the symbol for the solenoid must also be shown in the electrical control drawings. The symbols for the solenoid are shown in Figure 5-1.

As shown in Figure 5-2, the solenoid is made up of three basic parts:

1. Frame
2. Plunger
3. Coil

The frame and plunger are made up of laminations of high-grade silicon steel. The coil is wound with an insulated copper conductor.

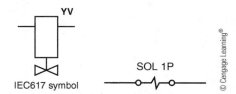

Figure 5-1A Solenoid symbol (Electrical).

Figure 5-1B Solenoid symbol (Air/Hydraulic).

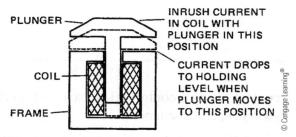

Figure 5-2 Cross section through air-gap solenoid.

When the solenoid coil is initially energized, the plunger is in an out position. Because of the open gap in the magnetic path (circuit), the initial current in the coil is high. As the plunger moves into the coil, closing the gap, the current level drops to a lower value.

Review the explanation given in Section 4.1 for the relationship between plunger travel and coil current for a relay. A similar condition exists for the solenoid. It is important that the plunger completes its stroke when the solenoid is energized. Otherwise, the current in the coil will be high, resulting in damage to the coil. Figure 5-3 shows a typical curve of the relationship between current and stroke and shows what happens to the current when the stroke is not completed to zero.

The current in amperes at the open position is called *inrush current*. The current in amperes at the closed position is called *sealed*, or *holding, current*. The ratio of inrush current to sealed current generally varies from approximately 5:1 in small solenoids to as much as 15:1 in large solenoids.

Typical industrial solenoids are shown in Figure 5-4.

When the coil of a solenoid is energized, a magnetic field is created around the coil. This magnetic field produces a force that acts on the

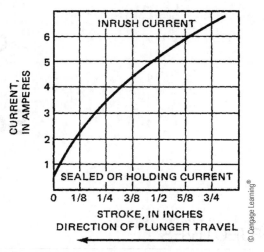

Figure 5-3 Graph of current versus stroke.

solenoid plunger. Because of this force, the plunger moves into the coil. This force on the plunger is called *pull*. The pull in solenoids varies widely. It may be as low as a fraction of an ounce or as high as nearly 100 pounds.

Electrical connections to the coil may be supplied in the following ways:

- Pigtail leads
- Terminals on the coil
- Terminal blocks
- Plug-in connections

There are two important problems to be considered in the application of a solenoid:

1. The pull of the solenoid must at all times exceed the load. If the pull is a little less, the solenoid action will be sluggish and may not complete the stroke. There are also conditions that the user may not always be able to control, such as low voltage or increased loading through friction or pressure. Therefore, it is generally advisable to overrate the solenoid by 20–25%. **Caution:** Too much pull will result in the plunger slamming, resulting in damage to the plunger and frame.

2. The duty cycle of the work load should be known. Some applications require a duty cycle with only an occasional operation. In other cases it may require up to several hundred operations per minute. **Caution:** The operation of the solenoid above its maximum cycling rate will result in excessive heating and mechanical damage.

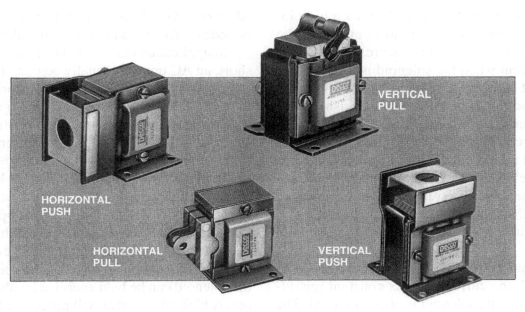

Figure 5-4 **Industrial solenoids.** *(Courtesy of Detroit Coil Company, Inc.)*

5.2 Solenoid Force and Voltage*

The pull-in force of a solenoid decreases rapidly as the voltage decreases below the coil nominal rating. As the voltage increases over the nominal value, the pull-in force will also increase. However, this situation may result in a rapid temperature rise in the solenoid.

5.3 Low Voltage

From a low-voltage standpoint, solenoid size selection should allow for adequate force at some arbitrary low voltage level. This allowance will ensure adequate solenoid force even during periods of low voltage and will prevent failure to pull-in and consequent coil burn out.

Design practices vary but low voltage levels are usually set at 85% or 90% of rated or nominal levels.

Forces at other reduced voltages can be closely approximated by means of the following formula:

$$F_1 = F \times \left(\frac{V_1}{V}\right)^2$$

where F_1 = solenoid force at a reduced
voltage V_1
F = solenoid force at rated voltage V

5.4 Overvoltage

The extra force resulting from overvoltage (or voltage above the coil nominal rating) will normally last for short periods of time. Under these conditions the mechanical life of a solenoid will not be seriously affected.

With the coil temperature rise, the consequent ultimate temperature of the solenoid will increase. Unless the ultimate temperature exceeds the class "A," 105°C rating of the insulation, the application will be satisfactory. The solenoid temperature can be lowered by mounting on a surface, such as an aluminum plate that will conduct the heat away.

5.5 AC Solenoids on DC*

A solenoid designed to operate on alternating current can also be operated on direct current. There are, however, some limitations.

*Courtesy of Detroit Coil Company, Inc.

A solenoid operating on AC draws a high inrush current when the solenoid is open. As the solenoid closes, this current decreases to a low holding current when the solenoid is fully closed. This current characteristic of an AC current is extremely important because the high inrush current provides a high initial force that is usually desirable to overcome the load on the solenoid. When the solenoid is held closed, the current is at the low holding level. This low holding current generates very little heat, so the solenoid remains cool. In short, an AC solenoid has a built-in current valve, which provides for a high force pull-in and cool holding.

Now suppose an AC solenoid is operating on DC. On DC, the current flow is constant regardless of whether the solenoid is open or closed. The inrush and holding currents are the same. Because of this constant current feature of DC, a compromise between the pull-in force and holding temperature must be made. Specifically, if enough current is provided to give the same pull-in force as AC would provide, the solenoid may overheat if held energized. If the current is reduced to a level that will prevent overheating when held closed, the pull-in force will be greatly reduced. Therefore, the question of whether a given AC unit performs satisfactorily on DC depends on how the pull-in force and overheating can be balanced.

Some applications may not require that the solenoid be held energized. In this case, an AC designed unit might be operated on either AC or DC power. A DC current could be supplied to give the high pull-in force, equal to the force obtained from AC power. Overheating during the brief energized periods on DC would not be a consideration.

In applications in which the stroke is extremely short, an AC unit can usually be operated successfully on DC because at short strokes on AC power, the inrush or open current is only slightly greater than the holding current. The constant current characteristic of DC will not be different enough to cause problems of pull-in force or overheating.

In many cases, AC solenoids can be operated on DC power with the addition of a switch and resistor. The switch is arranged to be opened when the solenoid closes. When the switch opens, the resistor is in series with the solenoid coil. The addition of this resistance reduces the coil current so the solenoid can be held energized without burning out. A high current that will produce a high pull-in force is then possible.

5.6 DC Solenoids on AC*

Within limits, DC solenoids can be operated on AC. See Figure 5-5 for typical DC solenoids. For reasons of economy and flexibility, DC solenoids are usually made with solid iron parts. When operated on AC, eddy current and eddy current losses are introduced. These losses in solid iron parts are high, and high temperatures can be developed.

Therefore, to prevent overheating, the use of AC power on DC solenoids should be limited to applications where a low current is adequate. The use of AC power is also practical where the solenoid is on and off in so short a time that the eddy current losses cannot generate excessive heat.

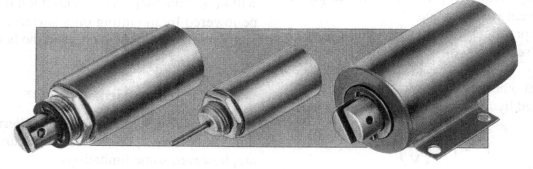

Figure 5-5 **Tubular DC solenoids.** *(Courtesy of Detroit Coil Company, Inc.)*

*Courtesy of Detroit Coil Company, Inc.

Also, if a DC unit is to be used on both AC and DC power, it should be equipped with shading coils. These coils, common on AC designs, keep the solenoid from buzzing when the AC sine wave goes through zero. A DC-design solenoid with shading coils is sometimes termed an AC-DC unit.

5.7 50- and 60-Cycle Solenoids*

With the exceptions of North America and most South American countries, nearly all countries operate on a 50-cycle power supply.

Coils for solenoids can be wound so that they may be used with either 50- or 60-cycle current. However, for best performance it is recommended that coils be wound for the specific frequency on which they will be used.

The so-called *dual frequency* (50–60) coils are actually wound for 50 cycles, and their use on 60 cycles is limited to applications in which their reduced force is adequate to operate the mechanism.

Sometimes one coil can serve as a dual frequency coil if the 50-cycle operating voltage is lower than the 60-cycle nominal voltage, for example, 120 V at 60 cycles, 100 V at 50 cycles. In this case, the winding for both specifications is the same so the same coil can be used successfully in both applications. In another way, the twenty extra volts available at 60 cycles will provide enough extra power to offset the power loss at 50 cycles mentioned in the previous paragraph.

Tapped or three-lead dual frequency (50–60 cycle) coils are a more expensive but practical solution. These coils have one common lead, with a 60-cycle lead tapped in near the end of the coil and a 50-cycle lead at the coils end. Such coils are usually wound for 60-cycle voltage at a multiple of 115 or 120 V, with a 50-cycle tap at a multiple of 110 V. For example, a coil could be wound for 50 cycles at 110 V with a tap or third lead for 60 cycles, 115-V operation.

Ideally, a 50-cycle solenoid should be manufactured with more laminations in both plunger and field, but because of the limited demand for 50-cycle solenoids, manufacturers use a 60-cycle plunger and field, and alter the coil only. This compromise results in a 50-cycle solenoid with slightly less than normal force. Force reduction is roughly 10% at $\frac{1}{2}$-inch stroke, and 5% at $\frac{1}{4}$-inch stroke. Holding force is affected very little.

A 50-cycle power supply is not always available for preshipment testing of equipment before export. If this happens, 50-cycle power can be simulated on 60 cycles by adjusting your 60-cycle voltage to a level of 6/5 rated voltage.

5.8 Solenoid Temperature Rise*

When current flows in a solenoid coil, resistance in the copper wire causes the coil temperature to rise in both AC and DC solenoids. In an AC solenoid, eddy currents and hysteresis losses in the iron caused by the alternating magnetic field also generate heat. The resulting temperature increase of the solenoid is called *heat rise*. The ultimate temperature of a solenoid is the ambient temperature surrounding the solenoid plus the solenoid temperature rise. It is important to know this ultimate solenoid temperature for two reasons: First, it must be kept below the temperature rating of the solenoid's electrical insulation. Otherwise the solenoid will burn out. Second, the solenoid's force decreases as its temperature increases.

According to Ohm's law, a change in applied voltage changes the current flow. So heat is directly related to the voltage input. It is known that changes in frequency affect current flow, so heat rise also varies with frequency. Specifically, a 60-Hz coil will have a higher heat rise when run on 50 Hz, for a given voltage.

Heat rise also depends on ambient temperature. The same solenoid in a high ambient environment will have less heat rise than in a lower ambient because in a higher ambient the coil resistance is higher, restricting the current flow. Lower current flow means lower temperature rise.

Solenoid temperature rise will also be affected by the solenoid mounting. A heavy metal mounting

*Courtesy of Detroit Coil Company, Inc.

will serve as a heat dispenser or heat sink and will conduct heat away from the solenoid, thus reducing the temperature rise.

Solenoid temperature rise can be measured in two ways: (1) by a thermocouple or (2) by the change-in-resistance method.

In method (1) we know that the highest temperatures are found at the heart of the coil. Therefore, to get an accurate reading the thermocouple or other temperature-sensing element must be placed near the center of the coil. You can see that

this method is impractical as it would require a special coil, with a thermocouple wound in to get an accurate reading.

Method (2) is based on the fact that a coil's resistance is directly related to its temperature. Therefore, the temperature rise from one condition to another can be computed if the resistance at these two conditions is known. These two conditions are usually termed *hot and cold resistances*. This is the easiest and most practical method by which you can measure temperature rise. See Figure 5-6.

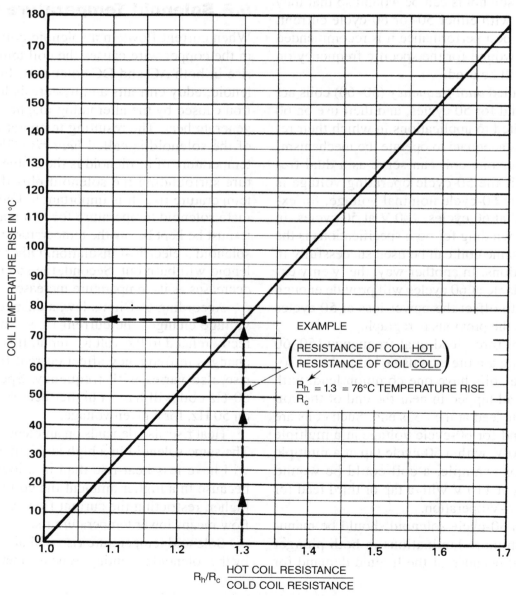

Figure 5-6 Measuring temperature rise by change in resistance. *(Courtesy of Detroit Coil Company, Inc.)*

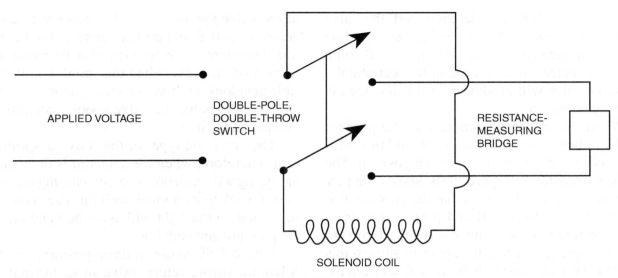

Figure 5-7 **Circuit for checking resistance.** *(Courtesy of Detroit Coil Company, Inc.)*

A convenient circuit that permits fast de-energizing of the coil and resistance measurement is shown in Figure 5-7. The solenoid is connected to the center of a double-throw, double-pole switch. The applied voltage is connected to one side and the bridge to the other. The switch can be quickly thrown to disconnect the power supply and connect the resistance measurement bridge.

When measuring temperature rise, be sure that the solenoid temperature is stabilized in the ambient so the cold resistance (R_c) is accurate. Similarly, energize the solenoid for about two hours so its temperature will stabilize before measuring (R_h).

By far the most satisfactory resistance measuring device is an accurate bridge. Volt-ohmeters are usually not accurate enough.

5.9 Valve Operation

The solenoid, as applied to valves, is covered several times in this book. Therefore, a brief explanation of basic valve operation and cylinder motion is in order. The pressure referred to can be either hydraulic (oil) or pneumatic (air). The pressure is ported (directed) to the end of a cylinder to cause motion on the machine.

The valve is identified by the number of spool positions, the number of solenoids, and the type of operation. Figure 5-8A shows a two-position, single-solenoid, spring return valve in its normal (de-energized) position. The symbols for the solenoid and valve are also shown. From the cutaway drawing, the operation of the valve may be illustrated.

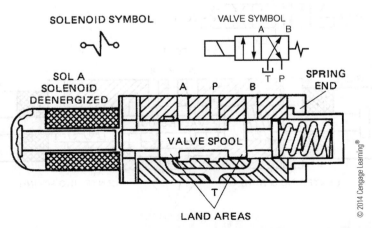

Figure 5-8A **Cross section through solenoid in normal, de-energized condition. A and B are ports, P stands for pressure, and T is the tank.**

When the solenoid is de-energized, the valve spool is in a position that allows the pressure line to be connected to port A, and the port B to be open (connected) to tank. Similar to electrical devices, the valve will be shown on the drawings in its normal position.

When the solenoid is energized, the plunger will be pulled into the solenoid coil and the valve spool will be shifted to the operated position. The cutaway drawing, in Figure 5-8B, shows the new position of the spool. Notice that the pressure line is now directed to port B and port A is open to tank. The force of the plunger is required to compress the spring to allow the spool to shift to this position. This situation may be visualized from the valve symbol by thinking of the solenoid pushing against the spring to shift the spool so that the left spool configuration is placed in line with the pressure and tank lines. In this position, the arrow lines

in the valve symbol show the pressure line connected to port B and port A connected to the tank line. Therefore, the valve symbol will represent the porting of the valve when the spool is in its possible positions. In this case, there are only two possible positions for the valve spool since this is a two-position valve.

The operation type for this valve is spring return. Therefore, when the solenoid is de-actuated the spring will push on the spool, causing the spool to shift back to its normal position. The spool configuration on the right will again be lined up with the pressure and tank lines.

Figure 5-8C shows a three-position, double-solenoid, spring return valve in its normal (de-energized) position. The symbol for the valve is also shown. With both solenoids de-energized, the centering spring will keep the spool in this position (center position lined up with the pressure

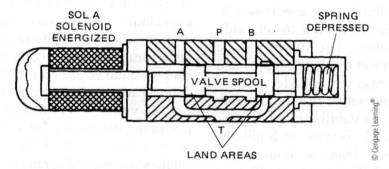

Figure 5-8B **Cross section through solenoid in energized condition. A and B are ports, P stands for pressure, and T is the tank.**

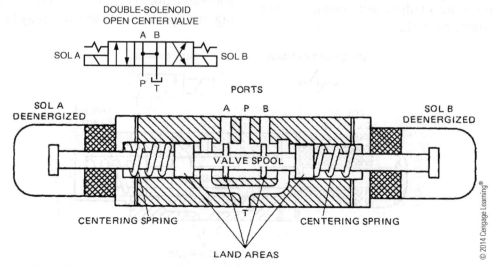

Figure 5-8C **Cross section through double-solenoid valve with both solenoids de-energized. A and B are ports, P stands for pressure, and T is the tank.**

and tank lines). This valve symbol and cutaway drawing shows all ports connected when the spool is in the center position (solenoids de-energized). This type of port configuration is referred to as an open-center spool since all ports are connected. With this configuration, pressure will not build in either port A or port B since the pressure will escape around the land areas of the valve spool to the tank port. Several options for the center position of the spool are shown in Figure 5-9.

Figure 5-9E shows the double-solenoid operating valve with solenoid A energized. The force or pull exerted on the plunger in solenoid A moves the valve spool to the right. The spring in solenoid B is compressed (loaded). The land areas on the valve spool are now located so that pressure available at port P is free to flow into port A. Port B is open to the tank port T. As there is generally little or no pressure connected at the tank port, it follows that there will be little or no pressure at port B.

Figure 5-9F shows the double-solenoid operating valve with solenoid B energized. The force or pull

Figure 5-9A Closed center.

Figure 5-9B Open center.

Figure 5-9C Float center.

Figure 5-9D Tandem center.

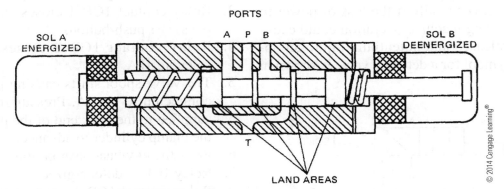

Figure 5-9E Cross section through double-solenoid valve with solenoid A energized and solenoid B de-energized. A and B are ports, P stands for pressure, and T is the tank.

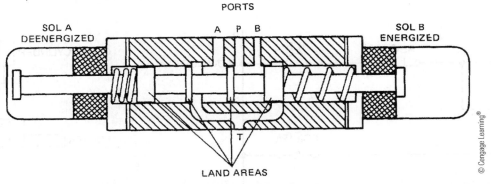

Figure 5-9F Cross section through double-solenoid valve with solenoid A de-energized and solenoid B energized. A and B are ports, P stands for pressure, and T is the tank.

exerted on the plunger in solenoid B moves the valve spool to the left. The spring in solenoid A is compressed (loaded). The land areas on the valve spool are now located so that pressure available at port P is free to flow into port B. Port A is open to the tank port T. As there is generally little or no pressure connected at the tank port, it follows that there will be little or no pressure at port A.

Spring return valves are very common in industry. Another type of operation that is used in specific cases is the detented valve. A detented valve requires two solenoids to operate. The valve spool will shift when the appropriate solenoid is energized, similar to the spring return valve. However, when the solenoid is de-energized the spool will be mechanically held in position until the opposing solenoid is energized. Once a solenoid is energized, the spool will continue to shift and remain shifted even if electrical power is removed from the machine. Therefore, safety considerations must be taken into account when using detented valves. Clamping circuits with a short stroke cylinder are potential applications for the use of a detented valve; especially if the loss of power to the clamps during machine operation could cause parts to be released and potentially injure an operator. The symbol for a detented valve is shown in Figure 5-10.

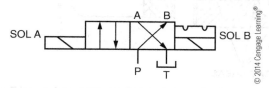

Figure 5-10 Symbol for detented valve.

5.10 Circuit Applications

Refer to the basic relay circuit shown in Figure 4-3. The same circuit is shown in Figure 5-11A, except that a solenoid is substituted for the green pilot light.

The pneumatic circuit for the two-position, single-solenoid, spring return valve is shown in Figure 5-11B.

The solenoid is now identified with a label SOL 1PA. The P in this designation represents a pneumatic valve. An H is used to indicate a hydraulic valve. The A represents the fact that pressure is connected to the A port when the solenoid is energized.

With the solenoid de-energized, pressure will be connected to port B. Port B is piped to the front end of the cylinder. Therefore, pressurized air will be supplied to the front end of the piston, causing the cylinder to be returned.

The sequence of operations is as follows:

1. Press the START push-button.
2. The coil of relay 1CR is energized.
3. Relay contact 1CR-1 closes, sealing around the START push-button.
4. Relay contact 1CR-2 closes, energizing solenoid 1PA.
5. The valve spool shifts causing pressure to be connected to port A. Pressurized air will be supplied to the back end of the piston, causing the clamp cylinder to advance.
6. Press the REVERSE push-button.
7. Relay 1CR is de-energized.
8. Relay contact 1CR-1 opens, opening the seal circuit around the START push-button.

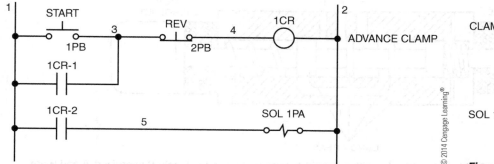

Figure 5-11A Control circuit for single-solenoid, spring-return valve.

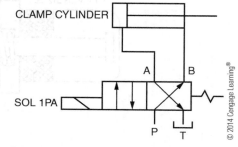

Figure 5-11B Pneumatic circuit for single-solenoid, spring-return valve.

9. Relay contact 1CR-2 opens, de-energizing solenoid 1PA.
10. The spring returns the spool to its normal position. The pressure line will be connected to port B, pressurizing the front end of the piston, and returning the cylinder.

In Figure 5-12A, a time-delay relay and an additional solenoid are added to the electrical circuit. Figure 5-12B shows the addition of another two-position, single-solenoid, spring return valve to the existing pneumatic circuit. Notice the additional cylinder is shown vertically instead of horizontally. It is customary to show a cylinder in the normal (home) position and to show the proper orientation of the cylinder (horizontal or vertical).

The sequence of operation is as follows:

1. Press the START push-button.
2. The coil of relay 1CR is energized.

3. Relay contact 1CR-1 closes, sealing around the START push-button.
4. Relay contact 1CR-2 closes, energizing solenoid 1PA.
5. The valve spool shifts causing pressure to be connected to port A. Pressurized air will be supplied to the back end of the piston, causing the clamp cylinder to advance.
6. Relay contact 1CR-3 closes, energizing timing-relay coil 1TR.
7. After a preset time on 1TR expires, the timing contact 1TR closes, energizing solenoid 2PA.
8. The valve spool shifts causing pressure to be connected to port A. Pressurized air will be supplied to the back end of the piston, causing the press cylinder to advance.
9. Press the REVERSE push-button, de-energizing relay coil 1CR.

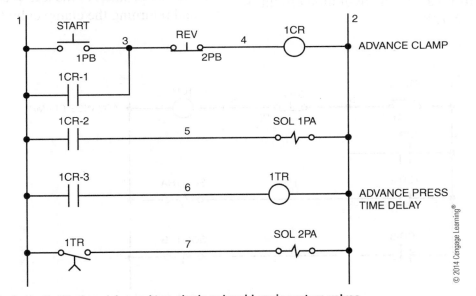

© 2014 Cengage Learning®

Figure 5-12A Control circuit with time delay and two single-solenoid, spring return valves.

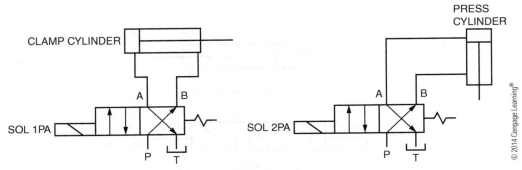

© 2014 Cengage Learning®

Figure 5-12B Pneumatic circuit for two single-solenoid, spring-return valves.

10. Relay contact 1CR-1 opens, opening the seal circuit around the START push-button.
11. Relay contact 1CR-2 opens, de-energizing solenoid 1PA.
12. Relay contact 1CR-3 opens, de-energizing the coil of timing relay 1TR.
13. Timing-relay contact 1TR opens, de-energizing solenoid 2PA.
14. The springs in both valves return the spools to their normal positions. The pressure line will be connected to port B on both valves, pressurizing the front end of the pistons, and returning both cylinders.

Note that if the REVERSE push-button is pressed before the preset time expires on 1TR, solenoid 2PA will not be energized.

In Figure 5-13A, a simplified electrical circuit is shown to control a three-position, double-solenoid, spring return valve. The hydraulic circuit is shown in Figure 5-13B.

The sequence of operation is as follows:
1. Press the START push-button.
2. The coil of relay 1CR is energized.
3. Relay contact 1CR-1 closes, sealing around the START push-button.
4. Relay contact 1CR-2 closes, energizing solenoid 1HA.
5. The spool shifts to the right, pressurizing port A and advancing the clamp cylinder.
6. Press the STOP push-button.
7. Relay 1CR is de-energized.
8. Relay contact 1CR-1 opens, opening the seal circuit around the START push-button.
9. Relay contact 1CR-2 opens, de-energizing solenoid 1HA.
10. The spool shifts to the center position.
11. Relay contact 1CR-3 closes, energizing solenoid 1HB.
12. The spool shifts to the left, pressurizing port B, and returning the clamp cylinder.

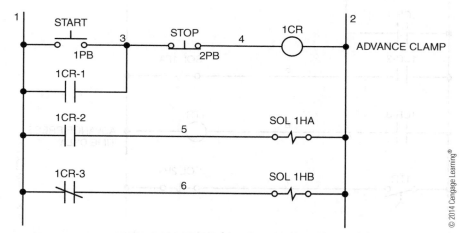

Figure 5-13A Control circuit for double-solenoid, spring-return valve.

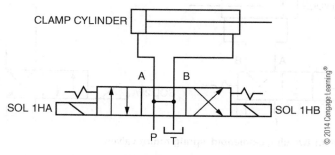

Figure 5-13B Hydraulic circuit for double-solenoid, spring-return valve.

Note from this design that solenoids 1HA and 1HB will not be energized simultaneously. If both coils on a double-solenoid valve are energized together, the valve can be damaged. Therefore, it is important in the electrical control design to ensure that only one solenoid is energized at a time.

When both solenoids are de-energized, the spool will return to the center position. Note that for the circuit in Figure 5-13, the valve will not remain in the center position since one of the solenoids will always be energized.

5.11 Variable Solenoids*

To this point the solenoids that have been discussed are of the ON/OFF type. These are used in the traditional type of solenoid valve to position the spool into one of the two or three discrete positions.

With proportional and servo valves, it is necessary to position the spool at an infinite number of intermediate positions. In the case of directional valves, spool position will provide both directional and flow control since the amount of spool movement away from center (valve opening) determines the flow rate through the valve (assuming constant pressure drop).

5.12 Proportional Valves*

Proportional valves use a proportional solenoid to directly or indirectly position the main spool and are normally used for open-loop speed control of actuators. Proportional solenoids operate in much the same way as ON/OFF DC solenoids, the main difference being that the solenoid current is varied to determine the solenoid force. A cross-sectional view of a proportional solenoid is shown in Figure 5-14.

A force is produced by the solenoid in relation to the current passing through the solenoid coil, which pushes the spool across in the valve body. The spool will continue to move until the spring compression force balances out the solenoid force. By varying the solenoid current, therefore, the spool can be moved to a greater or lesser amount in the valve body.

Proportional solenoids can be used where relatively large amounts of spool movement are required (up to 5 mm), but they will operate in only one direction of movement; that is, if the spool has to be positioned either side of center, then two are required, one at each end.

A cross-sectional view of a proportional directional control valve, single-stage design, without integral feedback transducers is shown in Figure 5-15.

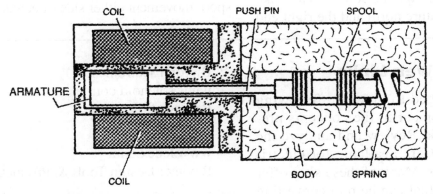

Figure 5-14 **Proportional solenoid.** *(Courtesy of Eaton Corporation.)*

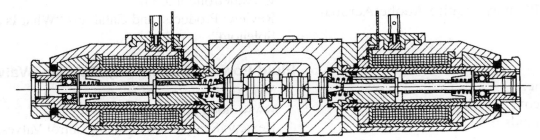

Figure 5-15 **Proportional directional control valve, single-stage design without integral feedback transducers.** *(Courtesy of Eaton Corporation.)*

*Information courtesy of Eaton Corporation.

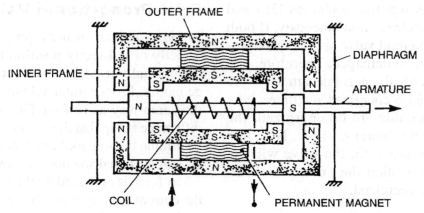

Figure 5-16 Force motor. *(Courtesy of Eaton Corporation.)*

5.13 Servo Valves*

Servo valves use a force motor to indirectly position the main spool and are used mainly for high-performance, closed-loop control of speed or position.

A cross-sectional view of a force motor is shown in Figure 5-16. The coil of the force motor is wound onto an armature, which is surrounded by an inner and outer nickel-iron frame. A permanent magnet magnetizes the two frames such that the inner one is, say, south and the outer one north. If a current is passed through the coil so as to make the left end of the armature north and the right end

south, a force is created (by attraction and repulsion of the poles) toward the right. The armature is supported on diaphragms that provide the spring force, so the armature will again move until the magnetic force and spring force balance out.

Force motors give a more linear force/current relationship than proportional solenoids, but they are normally confined to short-stroke applications (typically ±0.5 mm). Reversing the direction of current through the coil will reverse the armature polarity and, hence, the force direction. Unlike proportional solenoids, therefore, force motors can be used for spool movement either side of center (push/pull).

Recommended Web Links

Students are encouraged to view the following Web sites as a supplement to the concepts presented in this textbook. Review and analyze the array of products that are available for electrical control applications. Many of these sites offer technical information that can help in converting the principles to practical applications. To view catalogs, your PC may require Adobe Acrobat Reader software.

Solenoids

1. Cliftronics, Inc.
   www.gwlisk.com
   Review: Solenoids

*Information courtesy of Eaton Corporation.

2. Bicron Electronics
   www.solenoid.com
   Review: Solenoids

3. Ledex
   www.ledex.com
   Review: Design Tools & Products

4. Detroit Coil
   www.detroitcoil.com
   Review: Products, and Catalogs/ "What Is a Solenoid"

Fluid Power—Electrically Operated Valves

1. Moog
   www.moog.com
   Review: Search/Directional Control Valves

Solenoid Valves

1. IQS
   www.solenoid-valves.net
   Review: The links to more than 10 producers
   of solenoid valves

Proportional Valves

1. Moog
   www.moog.com
   Review: Products/Servo Valves & Servo-
   Proportional Valves

Electrohydraulic Proportional Valve

1. AxioMatic
   www.axiomatic.com
   Review: Products/Hydraulic Valve Drivers,
   Electronic Controllers

Achievement Review

1. Draw the symbol for the solenoid coil.

2. What happens in a solenoid if the plunger is prevented from completing its stroke?

3. At what position of the solenoid plunger does inrush current appear?

4. Why must the pull of a solenoid always exceed the load?

5. What will happen in a solenoid if the operation exceeds the solenoid design?

6. In the application of a solenoid to a spring-return operating valve, assume that the solenoid has been energized. What happens to the valve spool when the solenoid is de-energized?

7. Draw a control circuit using two pushbutton switches, a control relay, and a solenoid. The solenoid is to be energized when a normally open push-button switch is operated. It is to be de-energized when a normally closed push-button switch is operated.

8. What is the solenoid current in amperes with the plunger in the closed position?
   a. Maximum current
   b. Zero (no current will be indicated)
   c. Sealed or holding current

9. Which of the following statements is true of the spring when a solenoid is energized in a single solenoid, spring-return valve?
   a. It remains in the same condition as it was before the solenoid was energized.
   b. It is depressed (loaded).
   c. It aids the solenoid in moving the valve piston.

10. What causes the force acting on the plunger of a solenoid when the solenoid is energized?
    a. Hydraulic pressure
    b. Gravity
    c. A magnetic field that is produced about the coil

11. What is the difference between the operation of a proportional solenoid and the ON/OFF DC solenoid?

12. In using the proportional solenoid, how can the spool be moved to a greater or lesser amount?

13. In a force motor, how can the force direction be reversed?

14. What can you say about the linear force/current relationship in the proportional solenoid as compared to the force motor?

15. What will happen to the solenoid coil temperature if it is subjected to extended periods of overvoltage?

16. What is the relationship between inrush and holding current when an AC solenoid is operated on DC?

17. When a DC solenoid is used on both AC and DC, why should it be equipped with shading coils?

18. What two methods are used to measure solenoid temperature rise? Which is the more practical?

19. For the circuit described in problem 7, draw the pneumatic circuit to advance a cylinder when the solenoid is energized.

20. Explain the similarities and the differences between a spring return valve and a detented valve.

21. Draw a control circuit to control two slides. A normally open push-button will be used to advance each slide. That is, when the push-button is pressed and held, the slide will advance. When the push-button is released, the slide will return. Note: You will need two NO push-buttons to accomplish this task. The solenoids may not be allowed to be energized simultaneously.

22. Draw the pneumatic circuit for the control circuit described in problem 21.

23. What changes in the circuit described in problem 22, if hydraulic valves were to be used instead of pneumatic valves?

24. Draw a control circuit to control two slides. A normally open push-button will be used to advance both slides. That is, when the push-button is pressed, both slides will advance. The slides will stay advanced until a normally closed push-button is pressed. Note: Use one NO push-button, one NC push-button, and one relay to accomplish this task.

25. Draw a control circuit to control two slides. A normally open push-button (1PB) will be used to advance both slides. That is, when the push-button 1PB is pressed, both slides will advance. The slides will stay advanced until a different normally open push-button (2PB) is pressed. Note: Use two NO push-buttons and two detented valves to accomplish this task.

26. Redraw the circuit diagram shown in Figure 5-12A using IEC 617 symbols.

6

CHAPTER

Types of Control

OBJECTIVES

After studying this chapter, you should be able to:

- Understand the basic control system.

- Explain the advantages of closed-loop control.

- Explain the use of a sensor.

- Describe the difference among proportional, derivative, and integral types of control.

- Explain steady-state error and what can be done to correct it.

- Explain the use of a position transducer.

6.1 Open-Loop Control

Open-loop control is the simplest form of machine control. In an open-loop system, an output device is either turned *on* or *off*. A command signal may be given to the output device. However, there is no feedback provided to the control system to monitor the device and to compensate for any error in the system. An example of an open-loop control system is shown in Figure 6-1.

When the START push-button is pressed, the coil of 1M will energize, NO contact 1M-1 will close, and a seal-in circuit will be created around the START push-button. The motor (controlled by 1M) will drive a pump that fills a reservoir. The pump will continue to run until the STOP push-button is pressed.

In Chapters 7 through 9, we will be discussing control involving motion, pressure, and temperature. These sensors provide information to the control system. In Figure 6-2, a limit switch

is used to automatically control a hydraulically powered piston. The piston is to be at position A to start the cycle. When the CYCLE START push-button is pressed, the piston is to move to position B, stop, and return to position A. The limit switch is used to return the piston. In many cases this type of open-loop control is accurate enough to satisfy a given production requirement. Figure 6-3 shows a block diagram of a basic control system. However, there are many potential problems, involving things such as friction, oil temperatures in hydraulic systems, valve-shifting time, material being processed, and the like. All these conditions may lead to errors that cannot be tolerated in a given production machine. It then becomes necessary to account for the errors and adopt some method to minimize (hopefully eliminate) the errors. This may be accomplished by using closed-loop control.

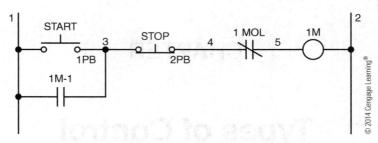

Figure 6-1 **Control of motor-pump system.**

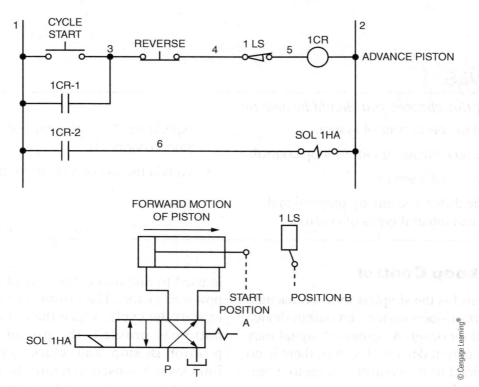

Figure 6-2 **A basic control circuit.**

Figure 6-3 **Block diagram of basic control system.**

6.2 Closed-Loop Control

In closed-loop control, a sensor is utilized to create a feedback loop for the control system. Any deviation from the desired output will be detected and compensated for through the use of the feedback loop. Figure 6-4 shows a block diagram of a basic closed-loop system.

The sensor will monitor the actual output of the system. This information will be compared

to the desired output and the control system will respond accordingly.

A *sensor* is a piece of hardware that measures system variables. Normally the sensor changes or transduces the measured quantity into another form of energy. It is the instrument or device that senses process condition in any process control system. Sensors provide the feedback signals by detecting the present condition, state, or value of

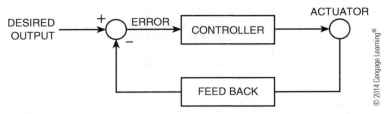

Figure 6-4 Block diagram of closed-loop system.

a process variable. They receive and respond to a stimulus or signal. An antenna, load cell, or photoelectric cell designed to detect and respond to some force, change, or radiation for purpose of information or control are good examples of sensors.

In any automatic process the corresponding components of sensing, information processing, and action can be distinguished. A simple industrial process might be the filling of a water tank. Figure 6-5 expands on the basic system of Figure 6-1 to automatically control the pump through the use of two sensors. The sensors detect the minimum and maximum permissible levels and signal the information processor. In this system, the information processor is a relay that determines from the signals when the actuator (pump) should be switched on and off.

This automated process demonstrates the simplest and most inexpensive form of closed-loop control—ON/OFF control. In this type of system, the output of the system is either ON or OFF

depending on the information received from the sensor(s). ON/OFF control almost always results in overshoot (going past the desired value). In addition, continuous cycling of the output is inherent in ON/OFF control systems.

For these reasons, the capabilities of ON/OFF control systems are limited. For the filling control system shown in Figure 6-5, the sensors provide information about the level of water in the tower. When the tank is at a high level the control system shuts off the pump and when the tank is at a low level the pump is started. The sensors do not have the intelligence to provide any indication as to the actual level of water in the tank. Therefore, more accurate control of the pump to maintain a consistent level of water in the reservoir is not possible. However, the ON/OFF control will maintain the water level between the two limits as defined by the placement of the two sensors. This simplified system also lacks other features that may be required in a real-life application.

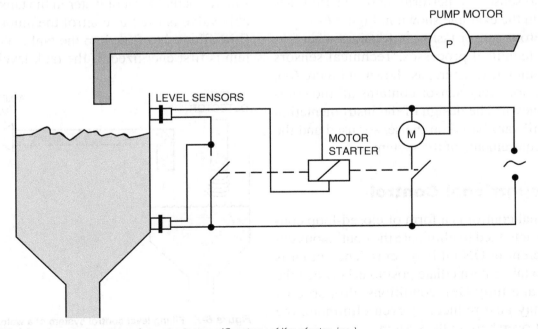

Figure 6-5 Filling level control system of a water tower. *(Courtesy of ifm efector, inc.)*

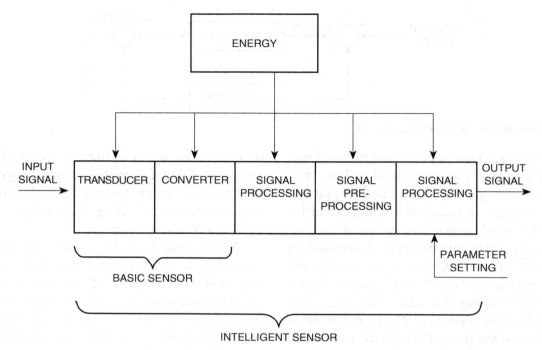

Figure 6-6 **Structure of a sensor.** *(Courtesy of ifm efector, inc.)*

For example, if no water is available to the pump, the "intelligence" is not sufficient to switch off the pump. If only dirty water were available, the control system would be equally incapable of interrupting the filling process. The information processor does not recognize conditions such as no water being available or dirty water in the line because no sensors to perform these functions are provided in the example shown in Figure 6-5.

Numerous sensor tasks are required when automating technical processes. Technical sensors share a common structure, as shown in Figure 6-6. However, not every sensor contains all the components shown. The design of the instrumentation system will vary based on the sensor used and the process requirements of the system.

6.3 Proportional Control

Proportional control is a form of closed-loop control that is intended to eliminate the continuous cycling inherent in ON/OFF systems. The output is allowed to take intermediate positions between the fully ON and fully OFF conditions. Proportional control may also reduce, or even eliminate, the amount of overshoot in the system.

The control signal can be thought of as the error (difference) between the desired output and the actual output (refer to Figure 6-4). The filling control system for the water tower may be modified to incorporate proportional control as shown in Figure 6-7.

In this system, a level detector is used to measure the actual level of water in the tank. An adjustable valve is used to control the amount of water flow allowed to flow into the tank. When the system is first energized, if the tank level is low, the

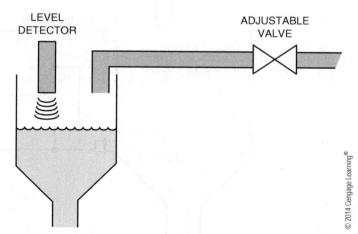

Figure 6-7 **Filling level control system of a water tower using proportional control.**

valve will be fully open to allow maximum flow of water to the tower. As the water level rises, the difference between the desired level and the actual level (error) will decrease. Therefore, the valve will close off proportionally to limit the amount of water flow. When the water level reaches the desired level, the error will become zero and the valve will be fully closed. The actual level, error, and desired level are shown in Figure 6-8. This type of control is known as proportional control because the control signal (valve opening) is proportional to the difference between the desired level and the actual level (error). As the error decreases, the valve opening decreases, thus slowing down the flow of water into the tank.

Now consider a servo valve being used to control a load. Although an error exists, the servo valve will operate to move the load. When the load reaches the desired position, the error signal will be zero. The servo valve centers and the load stops moving.

In examining the response of such a system, a step change in input signal may produce a characteristic as shown in Figure 6-9. Obviously the load current cannot move instantaneously to follow the input signal, so there will be a time delay between applying the input signal and the load reaching the desired position. During the first part of the load movement (a), the system is overcoming the inertia of the load and accelerating it to the required speed. There then may be a period of constant speed movement (b) in which the servo valve is wide open and the system is moving the load as fast as it is able. As the load approaches the required position (c), the speed reduces (curve flattens out), and the load gradually moves to the demand position at point (d). The flattening out at the end is because of the fact that the control signal to the servo valve is basically the error between input and output; as the error decreases, the control signal reduces, thus slowing down the movement of the load (Figure 6-10).

It may be possible to speed up the response of the system by increasing the gain of the amplifier so that for a given error, the amplifier produces a larger control signal. Increasing the gain, however, may produce a result like that shown in Figure 6-11.

Although the output arrives at the desired position faster, it overshoots because the load is traveling so fast that it cannot be stopped quickly enough when it reaches the required position. There may in fact be several overshoots and undershoots before it settles down to the final position. In many

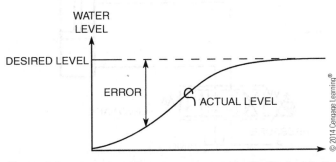

Figure 6-8 Error signal for water tower system in Figure 6-7.

© 2014 Cengage Learning®

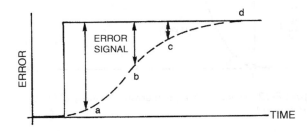

Figure 6-10 Change in error signal. *(Courtesy of Eaton Corporation.)*

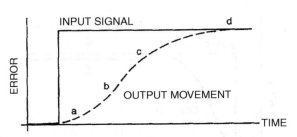

Figure 6-9 Graph of input signal versus output movement. *(Courtesy of Eaton Corporation.)*

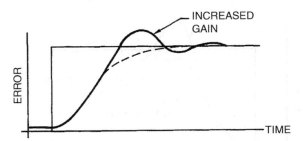

Figure 6-11 Response to increased gain. *(Courtesy of Eaton Corporation.)*

applications, an overshoot may be undesirable, so an alternative method of improving the system response is to modify the control signal such that it is proportional not only to the error, but also to the rate at which the error is changing (error-rate):

Control signal = [(Error signal) × Gain] + [Rate of change of error × Gain]

When the error is reducing, for example, the load is moving toward the required position; the error-rate term will then be negative and tend to reduce the control signal (Figure 6-12).

Between points (a) and (b) on the output curve, the error is changing relatively slowly so the large error term outweighs the small error-rate term and provides a large control signal to accelerate the load. From (b) to (c), however, the error is reducing quickly, so the error-rate term predominates and starts to reduce the control signal, which means the load starts to decelerate earlier, hence the overshoot can be eliminated (Figure 6-13).

Mathematically, the rate of change of a quantity is known as the derivative, so this type of control is termed *proportional plus derivative* and is expressed as:

Control signal
= [Proportion signal (error) × Proportional gain]
+ [Derivative signal (error rate) × Derivative gain]

Consider now a situation in which the load has to be positioned in a vertical plane or in which the load is subjected to some external force (Figure 6-14). Imagine that the load has moved to its required position. When the actual position and demand position are the same, the error will be zero, and the servo valve will center. However, the load on the piston and leakage across the servo valve spool will cause the piston to creep down. To counteract this, therefore, the servo valve spool must be displaced slightly away from center to prevent movement of the load. To offset the valve spool, a small control signal is required. For there to be a control signal, an error must exist between input and output; that is, the load can never be exactly in the required position. This condition is known as the *steady-state error* (Figure 6-15).

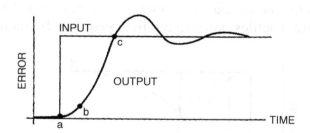

Figure 6-12 **Relative change in error.** *(Courtesy of Eaton Corporation.)*

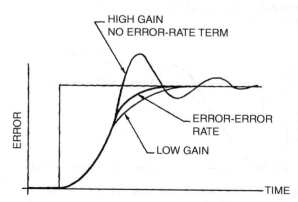

Figure 6-13 **Proportional plus derivative.** *(Courtesy of Eaton Corporation.)*

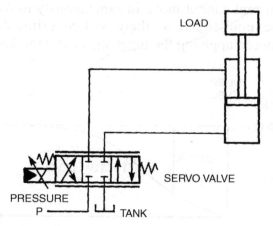

Figure 6-14 **Vertical load situation.** *(Courtesy of Eaton Corporation.)*

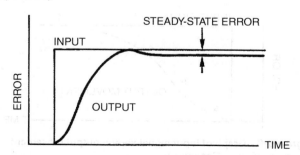

Figure 6-15 **Steady-state error.** *(Courtesy of Eaton Corporation.)*

6.4 Proportional-Integral

The steady-state error can be reduced by again modifying the control signal such that it is now proportional to the error plus an *error × time* term. The error × time component is known as the *integral term* and is effectively equal to the error multiplied by the length of time the error has existed. Thus, the longer the time, the larger the term becomes. This component can, therefore, be added to the proportional signal to virtually eliminate the steady-state error:

Control signal =
[Proportional signal (error) × Proportional gain]
+[Integral signal (error time) × Integral gain]

If a steady-state error exists, the integral term will increase with time until it is large enough to generate the control signal required to correct the error (Figure 6-16). Adjustment of the proportional and integral gains allows the system to be tuned for the best results. In some applications it may be necessary to reset the integral term to zero each time a new demand signal is applied to the system.

6.5 Proportional-Integral-Derivative

The performance of a basic (proportional) control system can be significantly improved by adding in:

- A derivative (error-rate) term to improve the dynamic response and reduce overshoot.
- An integral (error-time) term to reduce the steady-state error.

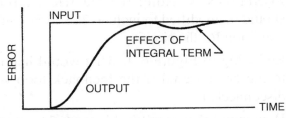

Figure 6-16 **Effect of integral term.** *(Courtesy of Eaton Corporation.)*

In fast-acting systems, the integral term tends to make the system less stable, so it is often used in conjunction with the derivative term, which tends to improve the stability. The system would then be described as having proportional, integral, and derivative (PID) control.

6.6 Proportional-Integral-Derivative Examples

The following examples provide basic guidelines to follow when tuning a system for maximum performance.

Example 1

A heater is used to bring the oil temperature in a hydraulic test system up to 135°F prior to starting the test. Measurements of the oil temperature are taken and recorded as shown in Figure 6-17. Production requirements reveal that the temperature must be brought up to the set-point in 30 seconds. What can be done to the PID controller to accomplish this task?

Solution

The proportional gain for the system should be increased because an increase in the proportional gain will decrease the rise time of the system. However, the amount of overshoot will increase with only a small change to the time required for the system to settle to the set-point value. The system response may look similar to the dashed line in Figure 6-18.

The derivative value may then be increased to reduce the amount of overshoot. An increase in the derivative value will also decrease the time required for the system to settle to the set-point value with only a minor change to the rise time of the system.

Example 2

A servo valve with a PID control circuit is used to control the pressure in a system by adjusting the opening of an orifice. A small leak has occurred in the system. Maintenance personnel are containing the leaking fluid until a replacement part arrives. Can anything be done temporarily to the PID control to eliminate the error?

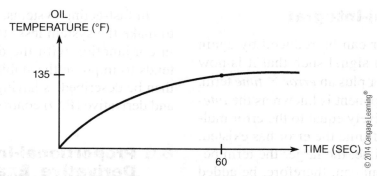

Figure 6-17 Oil temperature measurements for Example 1.

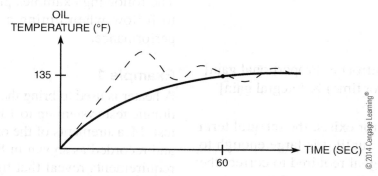

Figure 6-18 Increase in proportional gain for system in Figure 6-17.

Solution

The integral value may be increased to compensate for the loss (error) in pressure from the leak. An increase in the integral value will typically decrease the rise time, increase the overshoot, and increase the settling time. However, the steady state error will be eliminated.

Recommended Web Links

Students are encouraged to view the following Web site as a supplement to the concepts presented in this textbook. Review and analyze the array of products that are available for electrical control applications. Many of these sites offer technical information that can help in converting the principles to practical applications. To view catalogs, your PC may require Adobe Acrobat Reader software.

1. Control Solutions
   www.csimn.com
   Review: Applications/PID for Dummies

Achievement Review

1. Explain how the transducer is used in a closed-loop control system.

2. Explain proportional control.

3. In a closed-loop control system, what is the effect of increasing the gain of an amplifier?

4. In a closed-loop control system, how do you compensate for leakage across the valve spool?

5. Explain the term *error × time*.

6. What is meant by PID control?

7. Modify the electrical and hydraulic circuits in Figure 6-2 to include a limit switch (2LS) that will be actuated when the cylinder is at START POSITION A. Note: the electrical circuit should be modified to include 2LS as an initial condition for the circuit.

8. Referring to Figure 6-4, what would happen to the error signal if the feedback loop was disconnected?

9. How could the circuit in Figure 6-5 be modified to incorporate a sensor to indicate there is water available to the pump?

7

Motion Control Devices

OBJECTIVES

After studying this chapter, you should be able to:

- List the three basic classes of limit switches.
- Explain where rotating-cam limit switches are used.
- Explain the following terms relative to limit switches:
  - Operating force
  - Release force
  - Pretravel or trip travel
  - Overtravel
  - Differential travel
- Discuss the proximity limit switch and explain how it is used.
- Describe how the vane limit switch operates.
- Draw the limit switch symbols for four different conditions.

- Design an electrical operating circuit showing how the operation of a normally closed limit switch contact can be used to de-energize a solenoid, using two push-button switches, a normally closed limit switch contact, a relay, and a solenoid.
- List several methods for achieving proximity switching.
- List some uses for the linear transducer.
- List the information required for AC synchronous motor application in angular position control.
- List some of the advantages in using a DC stepper motor.
- Discuss the four types of photoelectric transducers.
- Explain the principle used in flow monitors.

7.1 Importance of Position Indication and Control

In the subject of electrical control for machines, position plays an important part. The problem is to accurately and reliably supply position information by providing an adequate electrical signal. This information may be used for both indication and control. Indication becomes a powerful tool in troubleshooting a machine or process.

Position information may be used to perform various tasks, such as:

- Sensing the presence of a part
- Sensing the position of a cylinder on a machine
- Measuring a characteristic of a part while being processed
- Sensing the rotation of a cam
- Sensing the relative position of a machine part

In many cases this position information is not critical to the machine or the process. However, in some cases position information must be reliable to .001 inch or less. The machine designer must know the operation of the machine and understand the process specifications to determine the required degree of accuracy. These factors must also be taken into account when replacing an existing sensor with another device.

There are numerous components that are used to obtain position information. Starting with the mechanical limit switch, many of the components are covered in this chapter.

7.2 Limit Switches—Mechanical

To describe the various limit switches available in the market today and their uses would require an entire book. Thus, we cover here only a few widely used units.

Mechanical-type switches can be subdivided into those operated from linear motion and those operated from rotary motion. Figure 7-1A is activated by a linear motion whereas Figure 7-1B is activated by a rotary motion.

There are large and small switches that operate from linear motion. The *precision limit switch* is a small switch. This switch varies from the larger size mainly in a lower operating force and shorter stroke. The operating force may be as low as

Figure 7-1B **Rotary activated limit switches.** *(Courtesy of Rockwell Automation, Inc.)*

1 pound. The stroke may be only a few thousandths of an inch.

Limit switches operated by rotary motion are generally called *rotating-cam limit switches.* These are control-circuit devices used with machinery having a repetitive cycle of operation in which motion can be correlated to shaft rotation. They are used to limit and control the movement of a rotating machine and to initiate functions at various points in the repetitive cycle of the machine. A rotating cam limit switch arrangement is shown in Figure 7-2.

The switch assembly consists of one or more snap-action switches. The cams are assembled

Figure 7-1A **Linear activated limit switches.** *(Courtesy of Rockwell Automation, Inc.)*

Figure 7-2 **Rotating-cam limit switch.** *(Courtesy of Rockwell Automation, Inc.)*

on a shaft. The shaft, in turn, is driven by a rotary motion on the machine. This connection may be made using either a direct coupling or a sequence of gears to adjust the speed ratios between the components.

The cams are independently adjustable for operating at different locations within a complete 360-degree rotation. In some cases the number of total rotations available is limited. In most cases the rotation can continue indefinitely. The rotation speed is controlled by the driving force and may be at a speed of up to 600 revolutions per minute (rpm).

In selecting a limit switch, it is important to determine its application in the electrical circuit. The following factors must be considered:

- Contact arrangement
- Current rating of the contacts
- Slow or snap action
- Isolated or common connection
- Spring return or maintained
- Number of NO and NC contacts required

In most cases the switch consists of double-break, snap-action, silver-tipped to solid silver contacts. The contact current rating will vary from 5 A to 10 A at 120 V AC continuous. The make-contact rating will be much higher, and the break-contact rating will be lower. Isolated NO and NC contacts are available. In some cases, multiple switches in the same enclosure operated by the same mechanical action are used.

A second important factor is the type of mechanical action available to operate the switch. Here the *operator* is the major decision. Length of travel, speed, force available, accuracy, and type of mounting possible are some of the considerations.

In discussing the action of limit switches, specific terms are used. Knowledge of these terms is helpful:

- *Operating force*—the amount of force applied to the switch to cause the "snap over" of the contacts
- *Release force*—the amount of force still applied to the switch plunger at the instant of "snap back" of the contacts to the unoperated condition
- *Pretravel or trip travel*—the distance traveled in moving the plunger from its free or unoperated position to the operated position
- *Overtravel*—the distance beyond operating position to the safe limit of travel; usually expressed as a minimum value
- *Differential travel*—the actuator travel from the point where the contacts snap over to the point where they snap back
- *Total travel*—the sum of the trip travel and the overtravel

Figure 7-3 shows pretravel, overtravel, differential travel, and total travel in diagram form.

Most manufacturers of limit switches list force and travel information for certain switches in their specifications. Accuracy of switch operators at

A. Pretravel B. Overtravel
C. Total travel D. Differential travel

1. Actuator — free position
2. Actuator — operating position
3. Overtravel — limit position
4. Actuator — release position

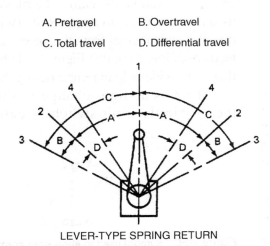

LEVER-TYPE SPRING RETURN

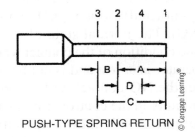

PUSH-TYPE SPRING RETURN

Figure 7-3 Limit switch operating movement.

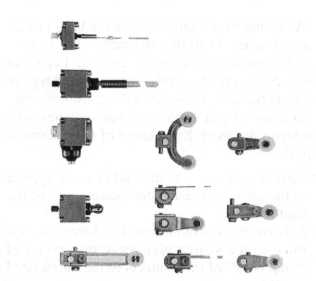

Figure 7-4 Operators for limit switches. *(Courtesy of Rockwell Automation, Inc.)*

the point of snapover varies with different types and manufacturers. In general, it is in the range of 0.001 inch to 0.005 inch.

The operator that probably has the greatest use is the roller lever. It is available in a variety of lever lengths and roller diameters. The next most frequently used operator is the push rod. It can consist of only a rod, or it can be supplied with a roller in the end. In most cases, particularly with the oil-tight machine-tool limit switch, the head carrying the operator can be rotated to four positions, 90 degrees apart. It can also be either top- or side-mounted. Two other operators used in machine control are the fork lever and the wobble stick. Some of the various operators available are shown in Figure 7-4.

7.3 Limit Switch Symbols

The symbols for the mechanical limit switch are shown in Figure 7-5A. There are two types of contacts—normally open (NO) and normally closed (NC). The normal state of a limit switch refers to the condition of the contact when the limit switch operator is at rest and is not being actuated. Therefore, the NO contact will be open and not allow continuity in the control circuit when the operator is at rest. It will allow continuity when the switch is actuated. Likewise, the NC contact will

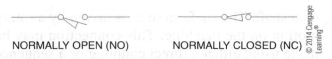

NORMALLY OPEN (NO)   NORMALLY CLOSED (NC)

Figure 7-5A Electrical symbols for limit switch.

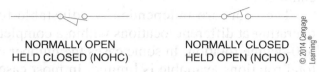

NORMALLY OPEN   NORMALLY CLOSED
HELD CLOSED (NOHC)   HELD OPEN (NCHO)

Figure 7-5B Electrical symbols for limit switch.

be closed (allowing continuity) when the operator is at rest. It will open when the switch is actuated.

Machine control drawings will generally show the devices on the machine as they exist in the home (returned) position of the machine, with the power off to the machine. Limit switches may be actuated when the machine is in its returned position. Therefore, the symbols shown on the drawings may include the actuated condition of the NO and NC contacts. The remaining set of symbols for limit switches is shown in Figure 7-5B.

Limit switches may be purchased with different combinations of NO and NC contacts. In some cases, additional contact blocks may be purchased and added to an existing switch.

There are times when more than one contact on a given limit switch is used in a circuit. Under this condition, the two contact symbols will be joined with a broken line. This line indicates that they are contacts on the same limit switch (Figure 7-5C).

To further illustrate the operation of the switch in relation to the symbol, consider the case in Figure 7-5D. Assume the cam to be moving to the right as shown. As the cam contacts the switch, the switch operates and changes from the unoperated to the operated condition. Note that Figure 7-5D has no use other than to provide help in remembering the symbols.

There is a slight change in the symbol for the proximity limit switch as compared to the

Figure 7-5C Symbol used to show both normally open and normally closed contacts on the same limit switch.

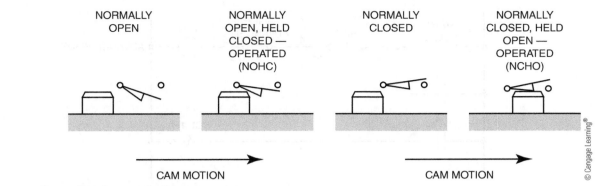

NORMALLY OPEN

NORMALLY OPEN, HELD CLOSED — OPERATED (NOHC)

NORMALLY CLOSED

NORMALLY CLOSED, HELD OPEN — OPERATED (NCHO)

CAM MOTION

CAM MOTION

Figure 7-5D **Operation of the switch in relation to the symbol.**

PROXIMITY LIMIT SWITCH

CLOSED

OPEN

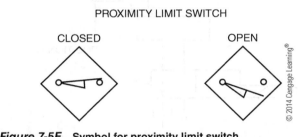

Figure 7-5E **Symbol for proximity limit switch.**

SQ

SQ

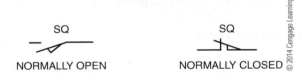

NORMALLY OPEN

NORMALLY CLOSED

Figure 7-5F **IEC 617 limit switch symbols.**

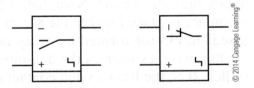

Figure 7-5G **IEC 617 proximity switch symbols (NO and NC).**

mechanical limit switch (Figure 7-5E). Note that only the two cases—NO and NC—exist for the proximity switch. This is because with power off to the machine, the proximity switch contacts may not be actuated. Therefore, the NOHC and the NCHO symbols will not be shown in the machine control drawings for the proximity switch.

7.4 Circuit Applications

Limit switches in machine-tool electrical control are used to gather information relative to the position of a machine part. To illustrate this, it is helpful to show a means of providing motion. In Chapter 5, the operating valve was introduced. The valve, along with a cylinder-piston assembly, is used in the limit switch application circuits.

The limit switch circuit shown in Figure 7-6 is built on the basic solenoid circuit shown in Figure 5-11. A cylinder-piston assembly is shown as a means of moving a cam on the cylinder rod from position A to position B. The cam (also referred to as a dog) moves with the cylinder rod and is used to actuate a limit switch when the cylinder is in the proper position. If solenoid 1PA is de-energized, the cylinder is in position A. If solenoid 1PA is energized, the cylinder moves to position B. Limit switch 1LS contact is normally closed.

The sequence of operations proceeds as follows:

1. Press the START push-button.
2. Relay coil 1CR is energized.
   a. Relay contact 1CR-1 closes, sealing around the START push-button.
   b. Relay contact 1CR-2 closes, energizing solenoid 1PA.

The valve spool shifts, permitting pressure to enter the main cylinder area behind the piston (left-hand end). The air pressure in the right end of the cylinder (rod end) is free to return to the tank. The piston moves from its start position at A to a new position at B.

3. At position B, limit switch 1LS is actuated, opening its normally closed contact.

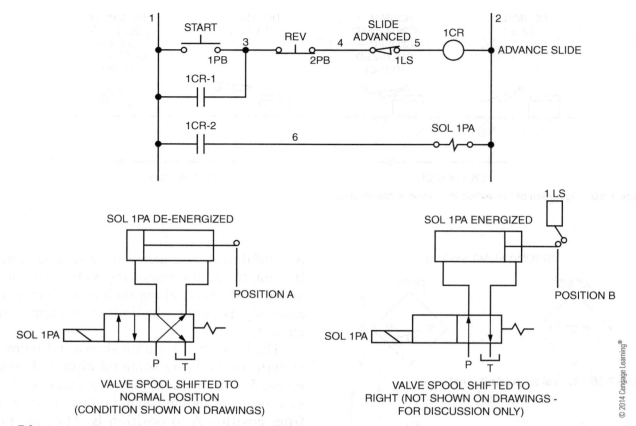

Figure 7-6 Control circuit with one cylinder and advanced limit switch (P-pressure, T-tank).

4. Relay coil 1CR is de-energized.

   a. Relay contact 1CR-1 opens, opening the seal circuit around the START push-button.
   b. Relay contact 1CR-2 opens, de-energizing solenoid 1PA.

With solenoid 1PA de-energized, the valve spring returns the valve spool to its initial position. This permits pressure to enter the rod end (right end), returning the piston to position A.

In the circuit shown in Figure 7-6, it is assumed that the piston was at position A to start the cycle. The only requirement to initiate the circuit is that the piston is not at the advanced position, actuating limit switch 1LS. To ensure that the piston is at position A to start a cycle, a second limit switch, 2LS, is placed at position A (Figure 7-7). A normally open limit switch contact is used. It is held operated (contact closed)

by a dog on the cylinder rod. Note that 2LS limit switch contact is placed inside the START push-button seal-in circuit formed by relay contact 1CR-1. Otherwise, as soon as the dog moves off limit switch 2LS, the limit switch contact opens, de-energizing the circuit.

The operation of this circuit is identical to that shown in Figure 7-6, except the piston must be in position A, actuating limit switch 2LS for start conditions.

A third circuit is shown in Figure 7-8. This circuit uses a three-position, double-solenoid, spring-return-to-center valve. A limit switch (2LS) performs a double duty. The NO contact is held closed in position A and ensures that the piston is at position A for start conditions. The NC contact is held open at the start position. This contact opens the circuit to solenoid 1PB again once the cylinder is fully returned.

© 2014 Cengage Learning®

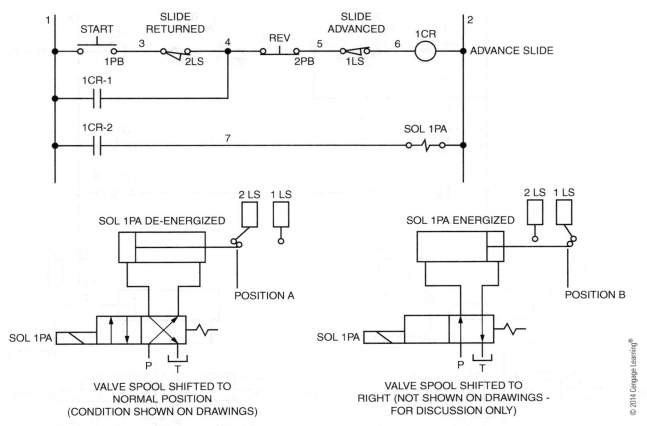

Figure 7-7 **Control circuit with one cylinder and 2 limit switches (advanced and returned) P-pressure, T-tank.**

The sequence of operations for the circuit shown in Figure 7-8 is as follows:

1. Press the START push-button.
2. If the piston is returned, 2LS will be actuated, 1LS will be deactuated, and relay 1CR is energized.
   a. Relay contact 1CR-1 closes, sealing around the START push-button.
   b. Relay contact 1CR-2 closes, energizing solenoid 1PA.
   c. Relay contact 1CR-3 opens.

As the piston moves off limit switch 2LS, the normally open contacts open and the normally closed contacts close. However, since relay contact 1CR-3 is now open, solenoid 1PB remains de-energized.

3. Piston moves to position B, actuating limit switch 1LS, opening its normally closed contact.

4. Relay 1CR is de-energized.
   a. Relay contact 1CR-1 opens, opening the seal circuit around the START push-button.
   b. Relay contact 1CR-2 opens, de-energizing solenoid 1PA.
   c. Relay contact 1CR-3 closes, energizing solenoid 1PB.

The piston now returns to position A, actuating limit switch 2LS.

5. The normally closed limit switch contact 2LS opens, de-energizing solenoid 1PB. Terminology is added to the control drawings for limit switch identification. The words chosen should indicate the condition when the switching device is actuated. If this format is followed, the drawings will be easier to decipher as to how a circuit functions.

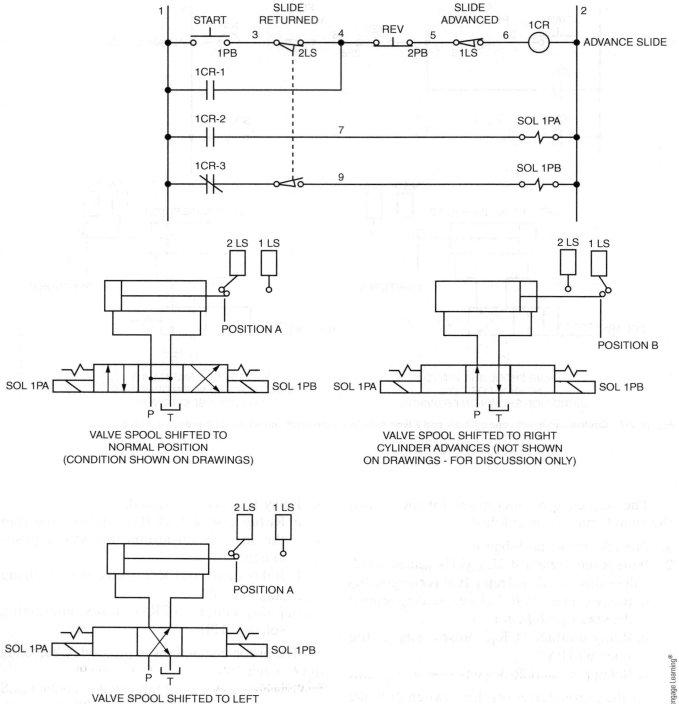

Figure 7-8 **Control circuit with one cylinder and 2 limit switches P-pressure, T-tank.**

7.5 Proximity Limit Switches*

The proximity switch is a generic term; it is a device capable of acting as an electronic switch when in the presence or close proximity of an object. The important distinction that differentiates it from a mechanical switch is that it does

*Courtesy of ifm efector, inc.

not require physical contact with anything else to operate. The methods of achieving this operation have been wide-ranging. They have made use of the effect of objects on radio frequency (RF) fields, magnetic fields, capacitive fields, acoustic fields, and light rays. RF fields are usually altered by the presence of ferrous materials that absorb energy by any currents produced in them by the field. Magnetic fields have been utilized to simply close reed switches by bringing the magnet up close to the switch or by introducing a magnetic material between the magnet and the switch. Magnets have also been used to alter electrical fields in devices making use of the Hall effect. (This effect is produced when a magnetic field applied to a conductor carrying a current produces a voltage across the conductor. This voltage is known as the *Hall voltage*. The Hall voltage is proportional to the product of the current and the field. That is, a device that exhibits the Hall effect is a multiplier. With constant current, the Hall voltage will be proportional to the magnetic field. With the magnetic field constant, the Hall voltage will be proportional to the current flow.)

Capacitive devices make use of the change in capacity that occurs when the object to be sensed acts as a plate of a capacitor for which the sensor acts as the other plate when detecting metallic objects or as an alteration to the dielectric between plates when detecting nonmetallic objects. Sonic devices utilize sound fields that are either interrupted by the object to be detected or detect the reflection of sound from objects. Photoelectric devices work in a similar manner except that light rays are detected rather than sound waves.

All of these methods have inherent strengths and weaknesses in actual application. For example, 2-wire DC proximity switches (Figure 7-9) eliminate the potential for wiring mistakes during installation because they are not sensitive to voltage polarity. The proximity switches have two black wires, exactly like a mechanical switch. To install the switch, one wire is connected to the power supply and the other connected to the load.

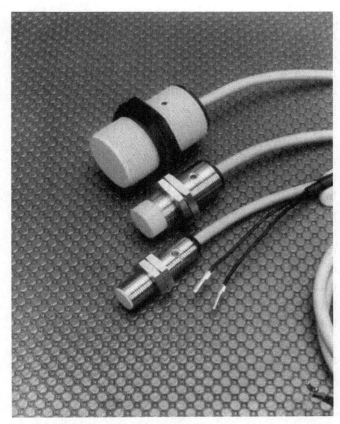

Figure 7-9 **Two-wire DC proximity switches** *(Courtesy of ifm efector, inc.)*

7.5.1 Inductive Sensors

The class of proximity switch using an RF field typically employs one half of a ferrite pot core whose coil is part of an oscillator circuit. When a metallic object enters this field, at some point the object will absorb enough energy from the field to cause the oscillator to stop oscillating. It is this difference between oscillating and not oscillating that is detected as the difference between an object present or not present. Obviously, there are several variables that determine the distance at which this detection will take place, including:

- The diameter of the pot core (distance varies directly with the core diameter)
- Size of object to be sensed (distance varies directly with size)
- Kind of metal (distance is greatest with iron, less for other metals)
- Circuit sensitivity (distance is established by circuit design)

Sensing distance will also be affected by voltage and temperature to some extent. However, in a well-designed switch, these effects can be minimized. In some units, the sensing distance can be adjusted; usually by means of a multi-turn trimmer potentiometer.

7.5.2 Capacitive Sensors

Capacitive sensors also contain oscillators. However, they usually begin oscillating in the presence of an object to be detected when that object creates enough capacitance in a critical part of the oscillator circuit to cause oscillation. In both detection methods (capacitive and inductive), the presence or absence of oscillation is detected and this information is used to do useful work by operating a load directly through a solid-state output circuit or indirectly through a relay.

Figure 7-10 shows two capacitive proximity limit switches. These two capacitive proximity limit switches are used with a separate control unit

to provide an intrinsically safe sensing system, approved for Factory Mutual Class I, II, III, Division 1, Group A-C hazardous locations. Typical applications include high/low level detection of liquids or granular solids in tank, hoppers, silos, and so on.

See Figures 7-11A through 7-11E for sketches showing typical industrial applications. Figure 7-12 shows proximity switches being used in a machine application.

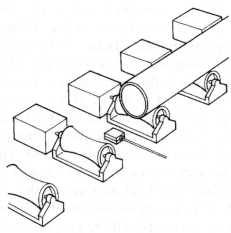

Figure 7-11A Pipes on a conveyor. *(Courtesy of ifm efector, inc.)*

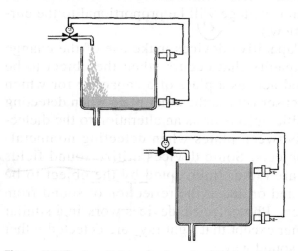

Figure 7-11B Levels in a plastic tank. *(Courtesy of ifm efector, inc.)*

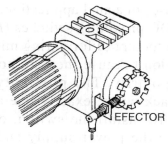

EFECTOR

Figure 7-11C Pulse pickup. *(Courtesy of ifm efector, inc.)*

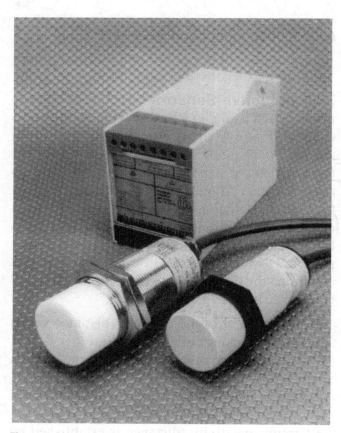

Figure 7-10 Capacitive proximity switches. *(Courtesy of ifm efector, inc.)*

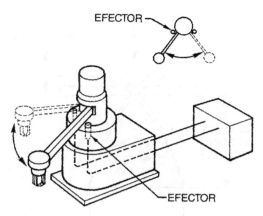

Figure 7-11D **Limit switch on robot arm.** *(Courtesy of ifm efector, inc.)*

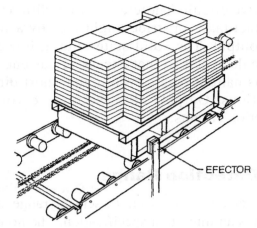

Figure 7-11E **Tiles on a conveyor system.** *(Courtesy of ifm efector, inc.)*

Figure 7-12 **Proximity switch in a machine application.** *(Courtesy of ifm efector, inc.)*

7.6 LED Indicators

LED indicators are usually available on the outside of the switch to show whether a switch is closed or open, whether an object to be sensed is present or not present, and whether there is line power available. It is possible to use three LEDs to indicate all three conditions, but it is most usual to use only one LED indicating that a switch is closed or open.

7.7 Solid-State Outputs

Solid-state output devices respond differently to excessive currents and wrong voltage polarities than electromechanical output devices do. Electromechanical contacts are not particularly affected by excessive currents if those currents are not sustained for too long a period of time, and the contacts do not have to break an inductive circuit. These current conditions that a contact takes in stride might destroy a solid-state output. The solid-state output, however, can be designed so that excessive current conditions cannot occur even under short circuit conditions. Outputs also use reverse voltage protection circuits.

7.8 Detection Range

When detecting an object, the sensing range of the switch is an important specification. The nominal sensing range given in a manufacturer's literature typically refers to the distance at which a reasonably large (compared to the sensing area of the switch) piece of mild steel is used as a target. All other metals will be sensed at distances less than this reference metal. The distance at which any target will be sensed depends on its positioning in front of the sensor area. The farther the object to be sensed is from the switch, the more closely it must be centered on the sensing area of the switch. For instance, if the material is very close to the switch, it need not cover the entire sensing area to be sensed. However, if the target is at the maximum sensing distance of the switch, the entire sensing area must be covered for the object to be detected.

7.9 Hysteresis

As a test target is brought from infinity toward the sensing area, the switch will transition at some distance from the sensing face within the attenuation range. As the target is removed to infinity, the switch will switch back at some distance from the sensing face (Figure 7-13). These two distances will be different by a small amount. This difference is called *hysteresis* and is usually expressed as a percentage of the switching distance. The percentage is the difference between the two sensing distances divided by the inbound sensing distance times 100. Hysteresis is introduced to prevent indecision and rapid on/off switching of the device when the target is right on the edge of the trip point.

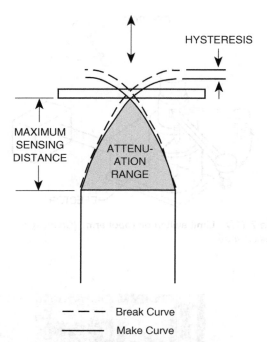

Figure 7-13 **Hysteresis.** *(Courtesy of ifm efector, inc.)*

7.10 Attenuation Range

The target can also approach the sensing area laterally in a plane parallel to the sensing area (Figure 7-14). A point will be reached in from the edge of the sensing area where the switch will trip. The closer the object is to the sensing surface, the closer to the edge of the surface the trip point will be. There will be an identical point on the other

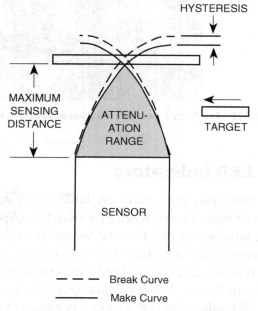

Figure 7-14 **Attenuation range.** *(Courtesy of ifm efector, inc.)*

side of the surface as the target leaves the sensing area. The distance between the pickup and drop-out point is called the *attenuation range*. Notice that the sensor picks up on the make curve on one side and drops out on the break curve on the other side. Obviously, the attenuation range decreases as the distance from the sensing area increases until a point is reached where the attenuation range is zero.

7.11 Speed

No electronic switch produces an output immediately on detecting an object. There is always some delay. The time that is required to detect the presence of an object and produce a full output to the load is called the *response time* of the switch.

The response time of a proximity switch is a function of the frequency of the oscillator in the switch. The higher the oscillator frequency, the shorter the response time and the higher the potential switching frequency. However, in AC switches, the limiting factor in the number of times that the output can switch depends on the frequency of the alternating voltage of the line. In general, DC units can switch their outputs at higher rates than AC units. Small switches with short distances usually have higher switching rate capability than larger units.

7.12 Magnet-Operated Limit Switch

The magnet-operated proximity limit switch (Figure 7-15) operates by passing an external magnet near the face of the sensing head to actuate a small, hermetically sealed reed switch. The 120-VAC pilot-duty model also includes an epoxy-encapsulated triac output.

The mercury proximity limit switch operates by passing a permanent magnet of sufficient strength past the switch. The sketch shown in Figure 7-16 displays the internal working parts and the relationship between the strength of the magnet and the operating distance of a mercury switch.

Figure 7-15 **Magnet-operated limit switch.** *(Courtesy of General Electric Company, General Purpose Control, Bloomington, IL.)*

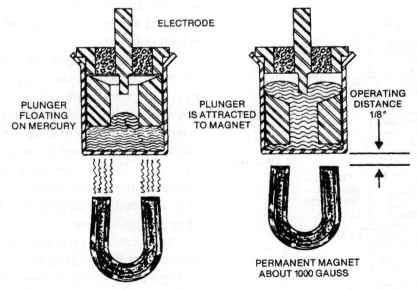

Figure 7-16 **Mercury proximity limit switch. This magnetic proximity switch requires a magnetic field of about 1000 gauss to attract the floating magnetic plunger down, thereby displacing mercury upward to the electrode. Rating: 1/3 hp @ 120 V ac.** *(Courtesy of Eagle Signal Controls.)*

7.13 Vane Switches

The vane-operated limit switch is actuated by the passage of a separate steel vane through a recessed slot in the switch (Figure 7-17). Either the vane or switch can be attached to the moving part of the machine. As the vane passes through the slot, it changes the balance of the magnetic field, causing the contacts to operate. The switch is available with either a normally open or normally closed contact.

The switch can detect very high speed of vane travel without detrimental effects such as arm or mechanism wear or breakage. There is no physical contact between vane and switch. Therefore, the upper limit on vane speed is governed by factors other than the switch.

The vane switch offers excellent accuracy and response time. Provided the path of the vane through the slot is constant, repeatability is constant within ±0.0025 inch or less. The time required for the switch to operate after the vane has reached the operating point (response time) is less than a millisecond.

Figure 7-18 shows an oil-tight and dust-tight vane-operated limit switch. It provides high reliability and long life, with an electrical rating capable of handling high inductive loads.

Figure 7-18 Vane-operated limit switch with indicating light. *(Courtesy of General Electric Company, General Purpose Control, Bloomington, IL.)*

7.14 Linear Position Displacement Transducers

In closed-loop control applications it is necessary to convert some physical property into an electrical signal to provide a feedback signal to the amplifier. The components that carry out this conversion are known as *transducers.*

There are many different devices available to measure linear movement. Factors such as accuracy required and environment will normally determine which one to use.

The simplest device would be a linear potentiometer in which the wiper is connected to the moving component and provides a voltage proportional to its position along the potentiometer winding. Problems of linearity, limited resolution, mechanical wear, and so on may occur in practice. The use of linear potentiometers as feedback transducers is, therefore, rather limited, though they are very inexpensive devices and may be suitable for simpler applications.

To overcome problems of mechanical wear, a noncontact device is required such as that shown in Figure 7-19, which illustrates a linear variable differential transformer (LVDT). It consists of one primary and two secondary coils surrounded by a

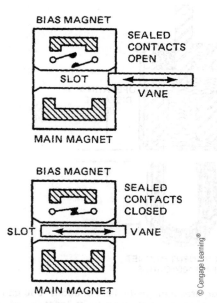

Figure 7-17 Cross section of vane-operated limit switch.

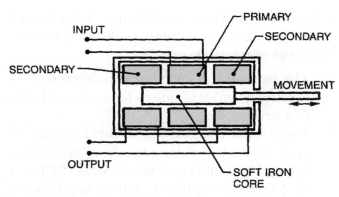

Figure 7-19 Linear variable differential transformer. *(Courtesy of Eaton Corporation.)*

soft iron core connected to the moving component. The primary coil is connected to a high-frequency AC supply, and voltages are induced in the two secondaries by transformer action. If the two secondary coils are connected in opposition, with the core centralized, the induced voltages in each coil will cancel out and produce zero output. As the core is moved away from the center, the voltage induced in one secondary will increase, whereas that in the other will reduce. This condition now produces a net output voltage, the magnitude of which is proportional to the amount of movement and the phase determined by the direction. The output can then be fed to a phase-sensitive rectifier (known as a *demodulator*), which will produce a DC signal proportional to movement and polarity, depending on direction.

The solid-state transducer shown in Figure 7-20, by precisely sensing the position of an external magnet, is able to measure linear displacements with infinite resolution. Since there is no contact between the magnet and the sensor rod, there is no wear, friction, or degradation of accuracy. The outputs of a solid-state transducer represent an absolute position, rather than an incremental indication of position change. Either digital or analog output is available.

The measuring principle by which this transducer operates can be explained as follows: when a current pulse is sent through a wire (which has been threaded through a tube and returned outside), the resulting magnetic field is concentrated in the tube, which acts as a wave guide. If this tube is then passed through a doughnut-shaped magnet, the two magnetic fields interact. A tube made of magnetostrictive material will experience a local rotary strain where these fields interact. This strain will continue for the duration of the electrical pulse. The rotary strain pulse travels along the wave guide element at ultrasonic speed and can be detected at the end of the tube.

By measuring the time from the generation of the initial electrical pulse until the ultrasonic pulse is detected, the distance of the external magnet from reference point can be determined.

This transducer is used for either position readout or closed-loop control. It may be mounted inside hydraulic cylinders or externally on machines.

The applications are very broad, covering small winding machines, machine tools, plastic forming machines, etc.

7.15 Angular Position Displacement Transducers

In many industrial applications, angular displacement is an important consideration or requirement. It may range from a few degrees to many revolutions of a shaft. The motion required for a specific application must be accurate in very small increments of change.

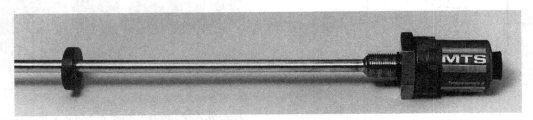

Figure 7-20 Linear displacement transducer. *(Courtesy of MTS Sensors Division.)*

The applications for motion control are many. They include machining, positioning of tools, testing, inspection, welding, and assembly. In each application a rotary variable differential transformer or drive motor and control combination is designed to satisfy the requirements.

As with linear motion, rotary potentiometers can be used for measuring angular position, but the same problems of wear, and so on apply, so again they are confined to relatively simple applications. The rotary equivalent of an LVDT is known as rotary variable differential transformer (RVDT) (Figure 7-21).

A specially shaped cam of magnetic material is rotated inside the primary and secondary coils by the input shaft. The profile of the cam determines the amount of magnetic coupling between primary and secondary; hence, as with the LVDT, an output signal proportional to shaft rotation is provided. The limitation in this case is the maximum angular movement that can be achieved and still produce a linear output, which in practice may be of the order of ±60 degrees.

7.15.1 Rotary Encoder Another approach to position control that involves the transfer of angular position to a usable signal is the *rotary encoder,* which can be either electromechanical, electronic,

optical, or a combination of all three. It can be used to monitor the rotary motion of a device. There are two types of rotary encoder:

1. The *incremental* encoder, which transmits a specific quantity of pulses for each revolution of a device.
2. The *absolute* encoder, which provides a specific code for each angular position of the device. It may be in binary coded decimal (BCD) or Gray code (see number system in Appendix H). The *Gray code* is defined as sequential numbers by binary values in which only one value changes at a time.

The rotary encoder shown in Figure 7-22 is an optical incremental rotary shaft encoder. It uses an infrared LED and precision optical and mechanical components to produce a series of pulses corresponding to shaft rotation. Output options include standard square wave, directional high/low, and quadrature. It also features reverse polarity and output short-circuit protection. The wiring circuit for this unit is shown in Figure 7-23A. The three optional outputs are shown in Figure 7-23B.

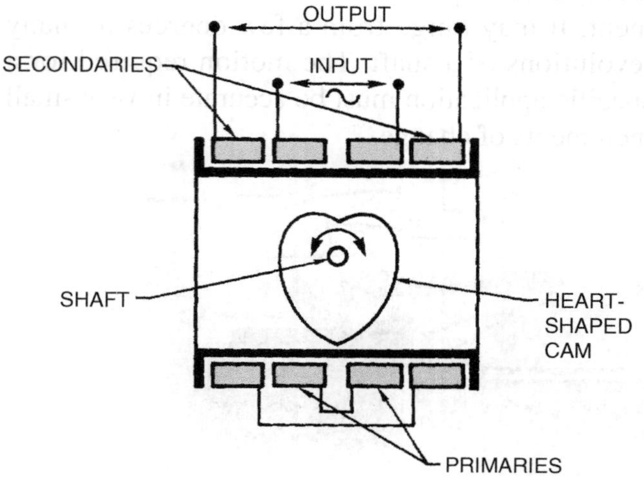

Figure 7-21 Rotary LVDT (RVDT). *(Courtesy of Eaton Corporation.)*

Figure 7-22 Rotary shaft encoder. *(Courtesy of Automatic Timing and Controls Company, Inc.)*

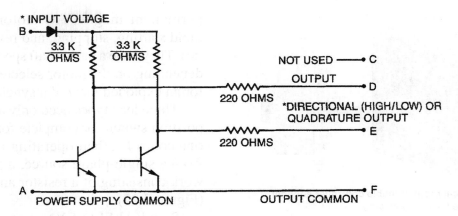

*Directional (High/Low) or quadrature are available output options

Concerning safety—ATC makes every effort to build a safe product.
We try to state specifications accurately. But every product made will eventually
fail, so we design our products into equipment so that they fail safely.

Figure 7-23A   **Wiring circuit for encoder.**   *(Courtesy of Automatic Timing and Controls Company, Inc.)*

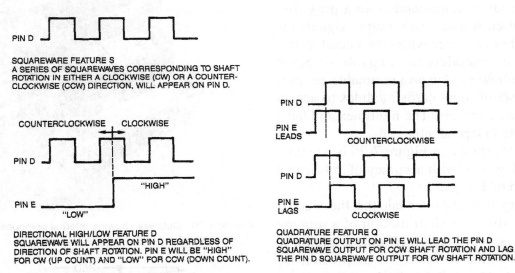

Figure 7-23B   **Three optional outputs for rotary encoder.**   *(Courtesy of Automatic Timing and Controls Company, Inc.)*

The unit shown in Figure 7-24 uses optical sensing to provide incremental conversion of rotary shaft motion into a digital output signal. An LED and phototransistor circuit with a slotted disk produces square wave output pulses. Some models with 10 pulses per revolution to 1200 pulses per revolution are available in unidirectional and bidirectional (quadrature) output types. Output pulse rate is 20,000 pulses per second maximum, and shaft speed is 4000 rpm continuous and 6000 rpm maximum.

The unit shown in Figure 7-25 offers a non-contacting optical design that allows high-speed,

Figure 7-24   **Rotary encoder.** *(Courtesy of Eagle Signal Controls.)*

Figure 7-25 Optical programmable controller encoder.
(Courtesy of Rockwell Automation, Inc.)

low-torque operation and includes an LED light source. This optical programmable controller encoder (absolute) is specifically designed to interface with programmable controllers and contains all the necessary electronics to provide a latched output on command from a programmable controller. A data-ready output signals the programmable controller when the encoder data is latched. The encoders are capable of being multiplexed, allowing one programmable controller with one set of input cards/modules to accept data from many encoders. The multiplex and latch circuits operate independently.

The encoder shown in Figure 7-25 is a single-turn (0–359 degrees), with accuracy of ±1 degree and resolution of 1 degree.

In summary, then, it can be said that the encoder electronically digitizes shaft motion of a rotating element; that is, it converts mechanical motion to an electronic digital format for a controller. By coupling the encoder shaft to a lead screw, rack and pinion, or rotating element, its electronic output can be used to measure distance from a known position and to provide the direction of rotation.

7.16 Use of AC Synchronous and DC Stepping Motors

One supplier of motion-control equipment offers an AC synchronous motor and a DC stepping motor made under the trademark of SLO-SYN®*. SLO-SYN AC synchronous motors are

permanent-magnet AC motors with extremely rapid starting, stopping, and reversing characteristics. They have a basic shaft speed of 72 or 200 rpm, depending on the motor selected. See Figure 7-26 for an exploded view of a synchronous motor.

Three lead types need only a single-pole, three-position switch for complete forward, reverse, and OFF control. When operating a SLO-SYN motor from a single-phase source, a phase-shifting network consisting of a resistor and capacitor is used (Figure 7-27).

Standard SLO-SYN motors are available in torque ratings of 22–1500 ounce-inches. The output torque varies with changes in the input voltage.

Figure 7-26 Cutaway showing internal components. *(Courtesy of Kollermorgen.)*

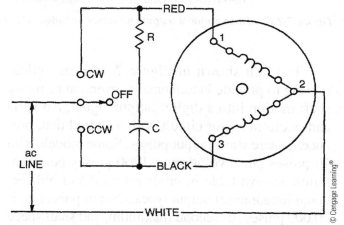

Figure 7-27 Input connections for ac operation.

*Information on motors is courtesy of Kollermorgen.

When a SLO-SYN motor is energized, AC current flows only through the windings. Current does not flow through the rotor or through brushes, since the motor is of a brushless construction. Therefore, it is not necessary to consider high in-rush currents when designing a control for a SLO-SYN motor, because starting and operating current are, for all practical purposes, identical.

The permanent-magnet construction of a SLO-SYN motor provides a small residual torque that holds the motor shaft in position when the motor is de-energized. When holding torque is required, a DC voltage can be applied to one winding when the AC voltage is removed. If necessary, DC voltage can be applied to both windings with a resulting increase in holding torque (Figure 7-28).

In determining the AC synchronous motor required for a specific application, the following information must be known:

1. *Motor shaft speed.* SLO-SYN motors are available in shaft speeds of 72 and 200 rpm at 60 Hz. Gearing can be used to obtain other speeds. It is also necessary to consider the effects of gears (if used) on torque and inertial load characteristics of the load.
2. *Load characteristics.* Examples that follow show how to determine the torque and moment of inertia characteristics of the load.

a. *Torque* in ounce-inches = Fr (Figure 7-29), where F is the force in ounces and r is the radius in inches. Example: Using a 4-inch-diameter pulley it is found that a 2-pound pull on the scale is required to rotate the pulley:

$$F = 2 \text{ pounds} = 32 \text{ ounces}$$

$$r = \frac{4}{2} = 2 \text{ inches}$$

The torque is, then, $32 \times 2 = 64$ ounce-inches

b. *Moment of inertia* in lb.-in.^{2} (Figure 7-30):

$$I(\text{lb.-in.}^2) = \frac{Wr^2}{2} \quad \text{for a disc}$$

$$I(\text{lb.-in.}^2) = \frac{W}{2}(r_1^2 + r_2^2) \quad \text{for a cylinder}$$

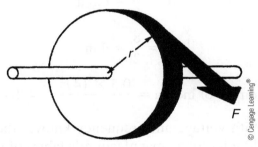

Figure 7-29 Torque.

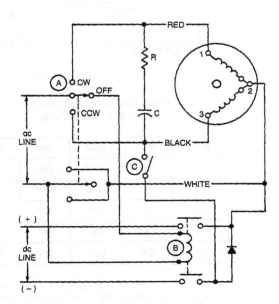

Figure 7-28 Typical circuit for utilizing dc voltage to provide holding torque.

SWITCHING SEQUENCE

MODE	Ⓐ	Ⓑ	Ⓒ
HOLDING	"OFF"	CLOSED	CLOSED
START	OPEN	OPEN	CLOSED
	OPEN	OPEN	OPEN
RUN	CW OR CCW	OPEN	OPEN
STOP	OPEN	OPEN	CLOSED
	"OFF"	CLOSED	CLOSED

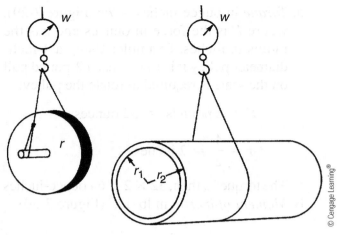

Figure 7-30 Moment of inertia.

where W is the weight in pounds and r is the radius in inches. Example: Using a load of an 8-inch-diameter gear weighing 8 ounces:

$$W = \frac{8}{6} = 0.5 \text{ lb}$$

$$r = \frac{8}{2} = 4 \text{ in.}$$

$$\text{Moment of inertia} = \frac{0.5 \times (4)^2}{2} = 4 \text{ lb.-in.}^2$$

With voltage and frequency known, the user should refer to the manufacturer's tables of torque and moment of inertia to select the motor that best suits the requirements.

Another motor is the SLO-SYN DC stepping motor. It is a permanent-magnet motor that converts electronic signals into mechanical motion. A picture of a DC stepping motor is shown in Figure 7-31.

SLO-SYN stepping motors operate on phase-switched DC power. Each time the direction of current in the motor windings is changed, the motor output shaft rotates a specific angular distance. The motor shaft can be driven in either direction and can be operated at very high stepping rates.

The motor shaft advances 200 steps per revolution (1.8 degrees per step) when a four-step input sequence (full-step mode) is used, and 400 steps per revolution (0.9 degrees per step) when an eight-step input sequence (half-step mode) is used. Power transistors connected to flip-flops or other logic devices are normally used for switching, as

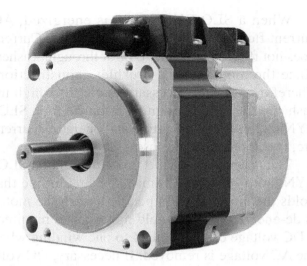

Figure 7-31 A DC stepping motor. *(Courtesy Baldor Electric Company.)*

shown in the wiring diagram in Figure 7-32. The four-step input sequence is also shown here. Since current is maintained on the motor windings when

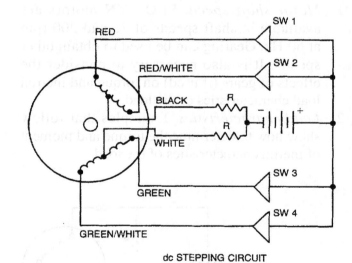

dc STEPPING CIRCUIT

FOUR-STEP INPUT SEQUENCE*
(FULL STEP SEQUENCE)

STEP	SW1	SW2	SW3	SW4
1	ON	OFF	ON	OFF
2	ON	OFF	OFF	ON
3	OFF	ON	OFF	ON
4	OFF	ON	ON	OFF
1	ON	OFF	ON	OFF

* Provides CW rotation as viewed from nameplate end of motor. To reverse direction of motor rotation perform switching steps in the following order; 1,4,3,2,1.

Figure 7-32 A dc stepping circuit.

the motor is not being stepped, a high holding torque results.

A SLO-SYN DC stepping motor offers many advantages as an actuator in a digitally controlled positioning system. It easily interfaces with a microcomputer or microprocessor to provide opening, closing, rotating, reversing, cycling, and highly accurate positioning in a variety of applications.

Mechanical components such as gears, clutches, brakes, and belts are not needed because stepping is accomplished electronically.

A SLO-SYN DC stepping motor applies holding or detent torque at standstill to help prevent an unwanted motion.

The motors are available in a range of frame sizes and with standard step angles of 0.72 degrees, 1.8 degrees, and 5 degrees. Step accuracies are 3% or 5%, noncumulative. They can be driven at rates to 20,000 steps per second with a minimum of power input.

In determining the motor required for a specific DC stepping motor application, the following information must be determined:

- Output speed in steps per second
- Torque in ounce-inches
- Load inertia in lb.-in.^{2}
- Required step angle
- Time to accelerate in milliseconds
- Time to decelerate in milliseconds
- Type of drive system to be used
- Size and weight considerations

1. *Torque* in ounce-inches = *Fr*, where *F* (in ounces) is force required to drive the load and *r* (in inches) is radius.
2. *Moment of inertia*:

$$I(\text{lb.-in.}^2) = \frac{Wr^2}{2} \quad \text{for a disc}$$

$$I(\text{lb.-in.}^2) = \frac{W}{2}(r_1^2 + r_2^2) \quad \text{for a cylinder,}$$

where *W* is the weight in pounds and *r* is the radius in inches.

3. *Equivalent inertia.* A motor must be able to:

a. Overcome any frictional load in the system.

b. Start and stop all inertia loads including that of its own rotor.

The basic rotary relationship is:

$$T = \frac{I\alpha}{24}$$

where T = torque in ounce-inches
  I = moment of inertia
  α = angular acceleration in radians per second^2
  24 = conversion factor from ounce-inch-sec$^2 \times 24$ lb.-in.^{2}

Measuring angles in radians. A radian is the angle at the center of a circle that embraces an arc equal in length to the length of the radius. The value of the radian in degrees equals 57.296 degrees. Then, 3.1416 radians denotes an angle of 180 degrees. Note that it is often convenient to measure angles in radians when dealing with angular velocity. If ω = angular velocity per second of the revolving body, in radians; V = velocity of a point on the periphery of the body, in feet per second; and r = radius in feet, then:

$$\omega = \frac{V}{r}$$

For example, if the velocity of a point on the periphery is 10 feet per second and the radius is 1 foot, then:

$$\omega = \frac{10}{1} = 10 \text{ radians per second}$$

Angular acceleration (α) is a function of the change in velocity (ω) and the time (t) required for the change:

$$\alpha = \frac{\omega_2 - \omega_1}{t}$$

Or, if starting from 0:

$$\alpha = \frac{\omega}{t}$$

where ω = angular velocity in rad/sec and t = time in seconds.
Since:

$$\omega = \frac{\text{Steps per second}}{\text{Steps per revolution}} \times 2\pi$$

angular velocity and angular acceleration can also be expressed in steps per second (ω') and steps per second^2 (α'), respectively.

SAMPLE CALCULATIONS

1. To calculate the torque required to rotationally accelerate an inertia load (Figure 7-33):

$$T = 2 \times I_o \left(\frac{\omega'}{t} \times \frac{\pi\theta}{180} \times \frac{1}{24} \right)$$

where T = torque required in ounce-inches
I_O = inertial load in lb.-in.^{2}
π = 3.1416
θ = step angle in degrees
ω' = step rate in steps per second

Example 1

Assume the following conditions:

- Inertia = 9.2 lb.-in.^{2}
- Step angle = 1.8 degrees
- Acceleration = 0−1000 steps per second in 0.5 seconds

Then:

$$T = 2 \times 9.2 \times \frac{1000}{0.5} \times \frac{\pi \times 1.8}{180} \times \frac{1}{24}$$

$$= 48.2 \text{ ounce-inches to accelerate inertia}$$

2. To calculate the torque required to accelerate a mass moving horizontally and driven by a rack and pinion or similar device (Figure 7-34):

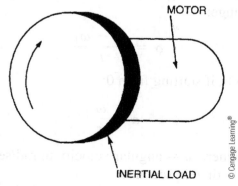

Figure 7-33 Torque problem 1.

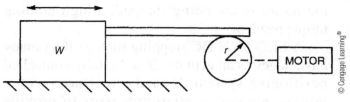

Figure 7-34 Torque problem 2.

The total torque that the motor must provide includes the torque required to:

a. Accelerate the weight, including that of the rack
b. Accelerate the gear
c. Accelerate the motor rotor
d. Overcome frictional forces

To calculate the rotational equivalent of the weight:

$$I_{eq} = Wr^2,$$

where W is the weight in pounds and r is the radius in inches.

Example 2

Assume the following:

- Weight = 5 lb.
- Gear pitch diameter = 3 inches
- Gear radius = 1.5 inches
- Velocity = 15 feet per second
- Time to reach velocity = 0.5 second
- Pinion inertia = 4.5 lb.-in.^{2}
- Motor rotor inertia = 2.5 lb.-in.^{2}

$$I_{eq} = Wr^2 = 5(1.5)^2 = 11.25 \text{ lb.-in.}^2$$
$$\text{Pinion} = \quad 4.5 \text{ lb.-in.}^2$$
$$\text{Rotor} = \quad 2.5 \text{ lb.-in.}^2$$
$$\text{Pinion} = 18.25 \text{ lb.-in.}^2$$

Velocity is 15 feet per second, with a 3-inch pitch diameter gear. Therefore:

a. Speed $= \dfrac{15 \times 12}{3 \times 3.1416} = 19.1$ revolutions per second

b. The motor step angle is 1.8° (200 steps per revolution)

c. Velocity in steps per second $\omega' = 19.1 \times 200$
= 3820 steps per second

d. To calculate torque to accelerate the system:

$$T = 2 \times I_\text{o} \times \frac{\omega}{t} \times \frac{\pi\theta}{180} \times \frac{1}{24}$$

$$= 2 \times 18.25 \times \frac{3820}{0.5} \times \frac{3.1416 \times 1.8}{180}$$

$$= \frac{1}{24} = 365 \text{ ounce-inches}$$

To calculate the torque needed to slide the weight, assume a frictional force of 6 ounces. Then, $T_\text{friction} = 6' \times 1.5 = 9$ ounce-inches. Therefore, the total torque required = 365 + 9 = 374 ounce-inches.

There are many ways in which these motors can be controlled. In general, the design of the control will depend on the particular application.

A typical system block diagram is shown in Figure 7-35. This approach allows the user to select the combination of options and features needed for a specific application.

7.17 Servo Positioning Control

The servo is controlled in a closed-loop configuration similar to Figure 7-35. Typical controllers are designed to control from one to three independent servo motors. For example, Figure 7-36 illustrates the ability of a three-axis controller to simultaneously monitor the position feedback of each axis's motor, and provide appropriate current and voltage output to the axis's motor power amplifier.

The role of the controller is to translate the profile of travel desired by the operator into the exact position of the servos. Typically, controllers are applied to:

- Machining centers with multiple cutting devices
- Robots with multiple coordinates of travel

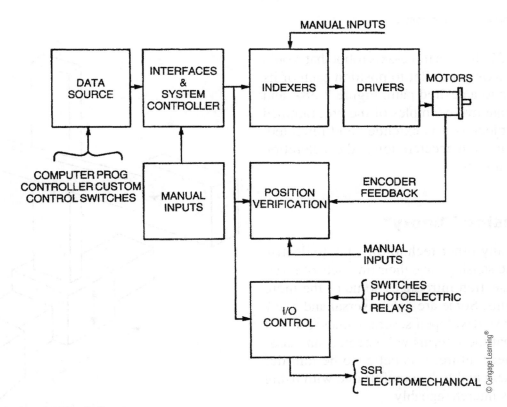

Figure 7-35 **Typical system block diagram.**

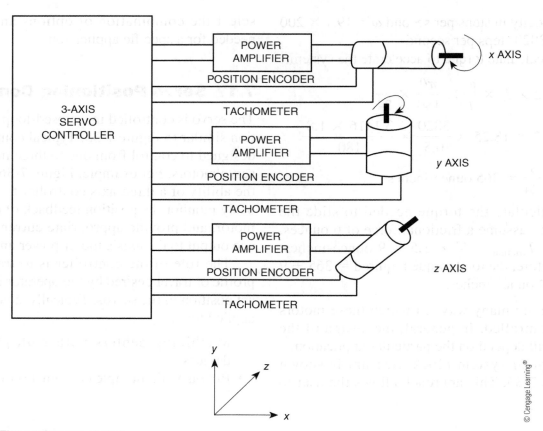

Figure 7-36 Three-axis servo control.

Figure 7-37 shows a three-axis robot that would need a three-axis controller to position each of its arms to move to a desired path. Figure 7-38A and 7-38B illustrate two examples of the programmed path of an object as it is extended and retracted. Note the changes in acceleration and deceleration as the object travels.

7.18 Sensing Theory*

Like nearly any other technology, photoelectric and ultrasonic sensing have their own sets of buzz-words. This section introduces some of the more important terms. Some are not universal and a few definitions have developed several names. Since it is unlikely that these terms will become standardized in the near future, it is better to be familiar with all of them and to be comfortable with using the synonyms interchangeably.

*Courtesy of Banner Engineering Corporation.

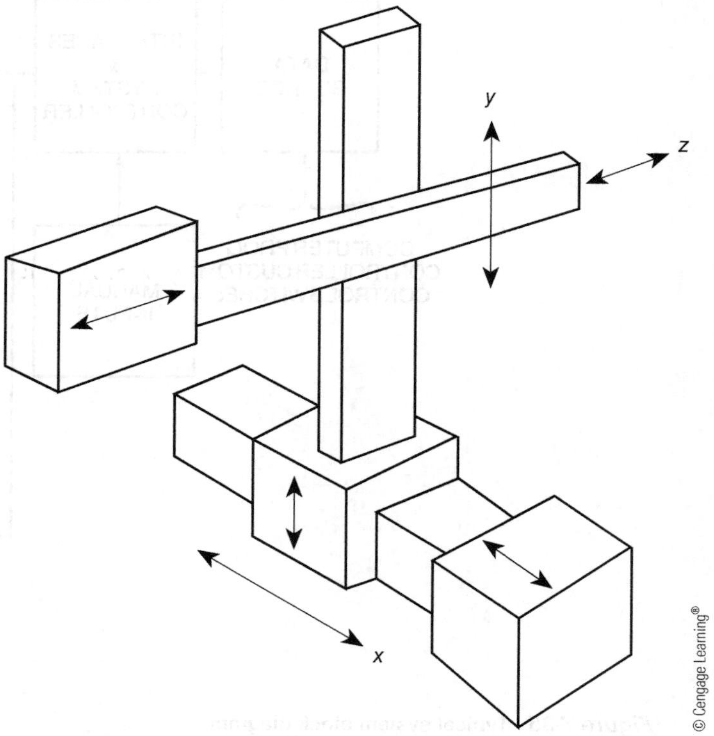

Figure 7-37 Three-axis robot.

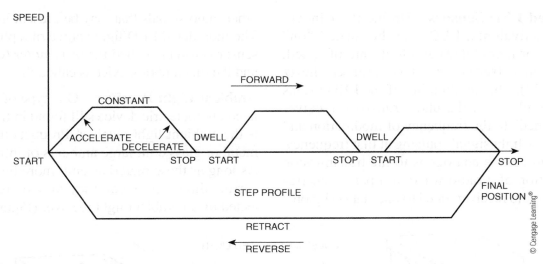

Figure 7-38A   **Movement profile involving a dwell time.**

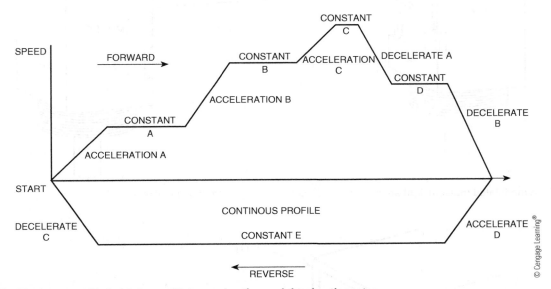

Figure 7-38B   **Movement profile involving multiple acceleration and deceleration rates.**

Photoelectric Sensor   A photoelectric sensor is an electrical device that responds to a change in the intensity of the light falling on it.

Light Emitting Diode (LED)   An LED is a solid-state semiconductor, similar electrically to a diode, except that it emits a small amount of light when current flows through it in a forward direction. LEDs can be built to emit green, yellow, red, or infrared light. (Infrared light is invisible to the human eye.) LEDs are usually built into sensors as a light source to change with process conditions.

Phototransistor   The phototransistor is the most widely used optoelement in industrial photoelectric sensor design. It offers the best trade-off between light sensitivity and response speed compared to photoresistive and other photojunction devices. Photocells are used whenever greater sensitivity to visible wavelengths is required, as in some color registration and ambient light detection applications. Photodiodes are generally reserved for applications requiring either extremely fast response time or linear response over several magnitudes of light level change.

Modulated LED Sensors   Unlike their incandescent equivalents, LEDs can be turned "on" and "off" (or modulated) at a high rate of speed, typically at a frequency of several kilohertz (Figure 7-39). This modulating of the LED means that the amplifier of the phototransistor receiver can be "tuned" to the frequency of modulation and amplify only light signals pulsing at that frequency.

This process is analogous to the transmission and reception of a radio wave of a particular frequency. A radio receiver tuned to one station ignores other radio signals that may be present in the room. The modulated LED light source of a photoelectric sensor is usually called the *transmitter* (or *emitter*), and the tuned photodevice is called the *receiver*.

Ambient Light Receiver   One type of nonmodulated photoelectric device still found in frequent use is the ambient light receiver. Products like red-hot metal or glass emit large amounts of infrared light. As long as these materials emit more light than the surrounding light level, they may be reliably detected by an ambient light receiver (Figure 7-40).

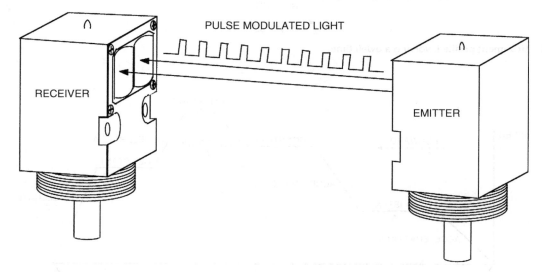

Figure 7-39   **A modulated (pulsed) light source.** *(Courtesy of Banner Engineering Corporation.)*

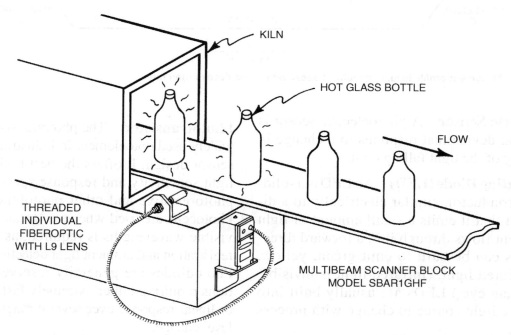

Figure 7-40   **Counting bottles during the manufacturing process by detecting infrared (heat) radiation with an ambient light receiver.** *(Courtesy of Banner Engineering Corporation.)*

Ultrasonic Sensors Ultrasonic sensors emit and receive sound energy at frequencies above the range of human hearing (above about 20 kHz). Ultrasonic sensors are categorized by transducer type: either electrostatic or piezoelectric. Electrostatic types can sense objects up to several feet away by reflection of ultrasound waves from the object's surface. Piezoelectric types are generally used for sensing at shorter ranges (Figure 7-41).

Remote Photoelectric Sensors Remote photoelectric sensors contain only the optical components of the sensing system. The circuitry for system power, amplification, and output are all at another location, typically in a control panel. Consequently, remote sensors are generally smaller and more tolerant of hostile sensing environments than are self-contained sensors.

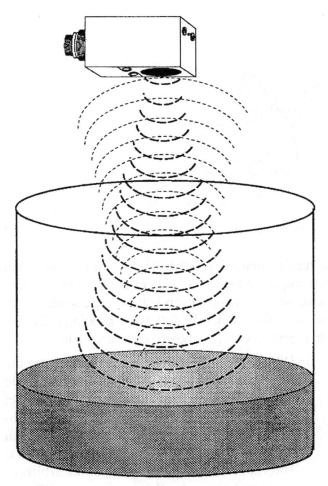

Figure 7-41 **Detection of ultrasonic sound energy.** *(Courtesy of Banner Engineering Corporation.)*

Self-Contained Photoelectric Sensors Self-contained photoelectric sensors contain the optics along with all the electronics. Their only requirement is a source of voltage for power. The sensor itself does all of the work, which includes modulation, demodulation, amplification, and output switching.

Fiber Optics There are many sensing situations where space is too restricted or the environment too hostile even for remote sensors. For such applications, photoelectric sensing technology offers fiber optics as a third alternative in sensor "packaging." Fiber optics are transparent strands of glass or plastic that are used to conduct light energy into and out of such areas.

Fiber-optic "light pipes," used along with either remote or self-contained sensors, are purely passive, mechanical components of the sensing system. Since fiber optics contain no electrical circuitry and have no moving parts, they can safely pipe light into and out of hazardous sensing locations and withstand hostile environmental conditions. Moreover, fiber optics are completely immune to all forms of electrical "noise" and may be used to isolate the electronics of a sensing system from known sources of electrical interference.

Sensing Modes The optical system of any photoelectric sensor is designed for one of three basic sensing modes: opposed, retroreflective, or proximity. The photoelectric proximity mode is further divided into four submodes: diffuse proximity, divergent-beam proximity, convergent-beam proximity, and fixed-field proximity. Ultrasonic sensors are designed for either opposed- or proximity-mode sensing.

Opposed Mode Opposed-mode sensing is often referred to as "direct scanning" and is sometimes called the "beam break" mode. In the opposed mode, the emitter and receiver are positioned opposite each other so that the sensing energy from the emitter is aimed directly at the receiver. An object is detected when it interrupts the sensing path established between two sensing components (Figure 7-42).

Retroreflective Mode The photoelectric retroreflective sensing mode is also called the "reflex" mode, or simply the "retro" mode (Figure 7-43).

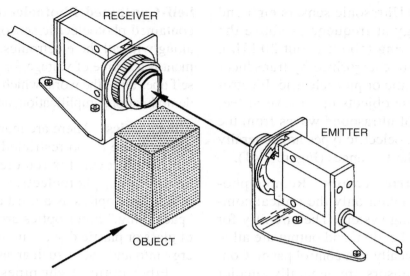

Figure 7-42 Opposed sensing mode. *(Courtesy of Banner Engineering Corporation.)*

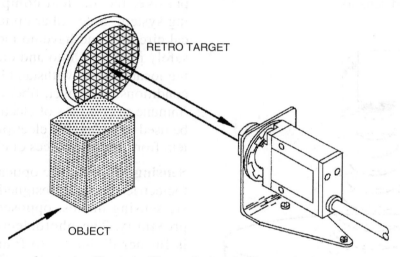

Figure 7-43 Retroreflective sensing mode. *(Courtesy of Banner Engineering Corporation.)*

A retroreflective sensor contains both emitter and receiver circuitry. A light beam is established between the emitter, the retroreflective target, and the receiver. Just as in the opposed-mode sensing, an object is sensed when it interrupts the beam.

Proximity Mode Proximity-mode sensing involves detecting an object that is directly in front of a sensor by detecting the sensor's own transmitted energy reflected back from the object's surface. For example, an object is sensed when its surface reflects a sound wave back to an ultrasonic proximity sensor. Both the emitter and receiver are on the same side of the object, usually together in the same housing. In proximity sensing modes, an object, when present,

actually "makes" (establishes) a beam, rather than interrupts the beam. Photoelectric proximity sensors have several different optical arrangements. They are described under the following headings: diffuse, divergent, convergent beam, and fixed-field.

Diffuse Mode Diffuse-mode sensors are the most commonly used type of photoelectric proximity sensor. In the diffuse sensing mode, the emitted light strikes the surface of an object at some arbitrary angle. The light is then diffused from that surface at many angles. The receiver can be at some other arbitrary angle and some small portion of the diffused light will reach it (Figure 7-44).

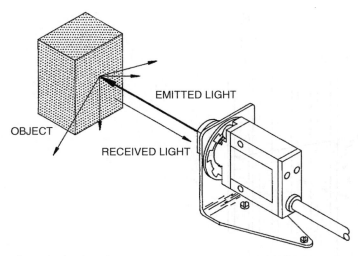

Figure 7-44 **Diffuse sensing mode.** *(Courtesy of Banner Engineering Corporation.)*

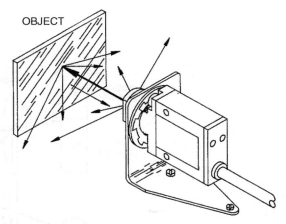

Figure 7-45 **Divergent proximity sensing mode.** *(Courtesy of Banner Engineering Corporation.)*

Divergent   To avoid the effects of signal loss from shiny objects, special short-range, un-lensed divergent-mode sensors should be considered. By eliminating collimating lenses, the sensing range is shortened, but the sensor is also made much less dependent on the angle of incidence of its light to a shiny surface that falls within its range (Figure 7-45).

Convergent Beam   Another proximity mode that is effective for sensing small objects is the convergent beam mode. Most convergent beam sensors use a lens system that focuses the emitted light to an exact point in front of the sensor and focuses the receiver element at the same

point. This design produces a small, intense, and well-defined sensing area at a fixed distance from the sensor lens (Figure 7-46).

Fixed Field   There is a photoelectric proximity mode that has a definite limit to its sensing range. Fixed-field sensors ignore objects that lie beyond their sensing range, regardless of the object surface reflectivity.

Fixed-field sensors compare the amount of reflected light that is seen by two differently aimed receiver optoelements. A target is recognized as long as the amount of light reaching receiver R2 is equal to or greater than the amount "seen" by R1. The sensor's output is cancelled as soon as the amount of

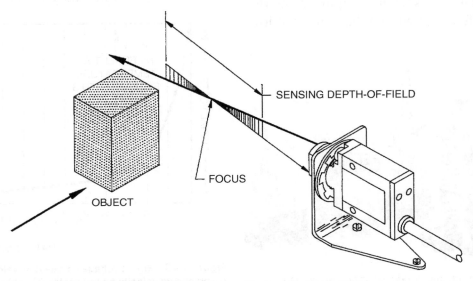

Figure 7-46 **Convergent beam sensing mode.** *(Courtesy of Banner Engineering Corporation.)*

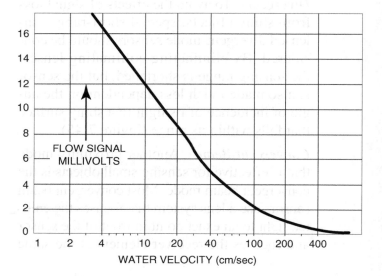

Figure 7-47 Fixed-field proximity sensing mode. Object is sensed if amount of light at R1 is greater than the amount of light at R2. *(Courtesy of Banner Engineering Corporation.)*

light at R1 becomes greater than the amount of light at R2 (Figure 7-47).

7.19 Flow Sensors*

The ifm efector flow sensors (Figure 7-48) are rugged devices constructed of corrosion resistant material. These sensors measure the flow rate of fluids or gases and provide an analog or switched output.

The operating principle of a flow sensor is based on the calorimetric principle. These sensors use the cooling effect of a flowing fluid or gas to monitor the flow rate. The amount of thermal energy that is removed from the tip determines the local flow rate. As shown in Figure 7-49, the temperature differential will be the highest at low

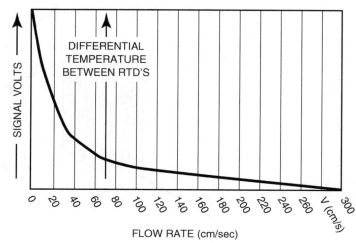

Figure 7-48 Flow products. *(Courtesy of ifm efector, inc.)*

Figure 7-49 Signal voltage versus flow velocity charts. Top graph shows velocity on log scale. *(Courtesy of ifm efector, inc.)*

*Courtesy of ifm efector, inc.

flow; therefore, a small increase in flow velocity will remove a large amount of heat from the tip. Tip temperature is lower at higher flow velocities, and so small changes in flow velocities will remove proportionally smaller amounts of heat from the tip. This temperature-based operating principle can reliably sense the flow of virtually any liquid or gas.

An example of how a flow sensor is used can be seen in Figure 7-50

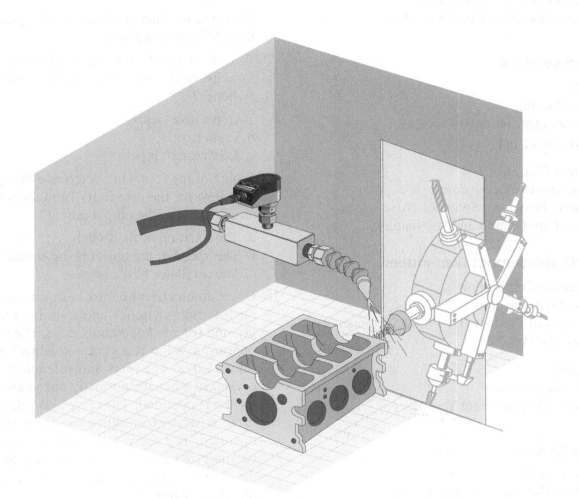

Figure 7-50 **Flow sensors are used in machine tool coolant flow applications.** *(Courtesy of ifm efector, inc.)*

Recommended Web Links

Students are encouraged to view the following Web sites as a supplement to the concepts presented in this textbook. Review and analyze the array of products that are available for electrical control applications. Many of these sites offer technical information that can help in converting the principles to practical applications. To view catalogs, your PC may require Adobe Acrobat Reader software.

Motion Control

1. Motion Control.com
   www.motioncontrol.com
   Review: Controllers/Motion Controller

Mechanical Limit Switches

1. Square D
   www.schneider-electric.com
   Review: Product/Limit Switches

2. Honeywell
   http://sensing.honeywell.com
   Review: Switches/Limit Switches

Position sensors

1. ATC
   www.flw.com
   Review: Distribution/ATC/Sensors

   Proximity switches

2. Baumer Electric
   www.baumerelectric.com
   Review: Proximity Switches (Also review areas of application and key features)

Linear Displacement Transducer

1. Daytronic Corporation
   www.daytronic.com
   Review: All Products/By Product/Transducers/LVDT Overview

2. Magtrol
   www.magtrol.com
   Review: Displacement Transducers

Stepper Motors

1. www.applied-motion.com
   Review: Products/Stepper Motors

Achievement Review

1. What are the three basic classes of industrial limit switches?

2. What is one use for the rotating-cam limit switch?

3. List some of the considerations when selecting a limit switch for operation in an electrical control circuit.

4. What is the difference between operating force and release force in a mechanical limit switch?

5. Draw a sketch to show the difference between pretravel and overtravel.

6. Discuss the accuracy and response time in a vane-type switch.

7. Draw the mechanical limit switch symbols for four different conditions.

8. Which of the following methods are used to achieve the operation of proximity limit switches?

   a. Capacitive fields
   b. Light rays
   c. Low current input

9. Which of the following determine the speed of response for the proximity limit switch using an oscillator in the output circuit?

   a. The location of the switch
   b. The speed of the object being sensed
   c. The oscillator frequency

10. Draw an electrical control circuit, using two push-button switches and a limit switch, relay, and solenoid operating valve. The solenoid operating valve is a single-solenoid, spring-return valve. The relay and solenoid are to be energized when the normally open push-button switch is operated. The relay and solenoid are to be de-energized when the limit switch is operated. The relay and solenoid can also be de-energized at any time during a cycle, before the limit switch is operated, by operating a second push-button switch.

11. Under what conditions would you use fiber optics in applying photoelectric control?

12. What thermal element is used to convert temperature differences into electrical signals in a flow monitor?

13. On what principle does the flow monitor operate?

    a. Velocity of flow
    b. Type of medium being measured
    c. Thermal conductivity

14. Explain the operation of a pulse modulated photoelectric sensor as shown in Figure 7-39. Also, explain the impact on its operation if the emitter is moved further away from the receiver.

15. Explain the fundamental differences between an ultrasonic sensor and a photoelectric sensor. Also, in what type of industrial environment are each best suited.

16. In a multiple-axis controller, describe the role of the position decoder. Would the controller need to be programmed? What is the purpose of the software program?

17. Figure 7-38A illustrates a motion profile (speed and direction) of some device traveling along some axis. Describe this profile in terms of a cylinder rod in motion.

18. Figure 7-38B illustrates a motion profile (speed and direction) of some device traveling along some axis. Describe this profile in terms of a cylinder rod in motion.

19. Modify the electrical and pneumatic circuits in Figure 7-6 to include a limit switch (2LS) that will be actuated when the cylinder is at START POSITION A. (Note: The electrical circuit should be modified to include 2LS as an initial condition for the circuit.)

20. Modify the electrical circuit in Figure 7-7 to include a "Machine Cycled" memory circuit. A pilot light should be included to notify the operator that the machine has been cycled and is ready for another cycle to be initiated. The "Machine Cycled" relay coil should energize when the slide is advanced (1LS actuated) and de-energize when the START push-button is pressed.

21. Modify the electrical circuit in Figure 7-8 to include an additional REVERSE push-button to the circuit.

22. Modify the electrical circuit in Figure 7-7 to replace the limit switches with proximity switches.

23. Modify the electrical circuit in Figure 7-8 to replace the limit switches with proximity switches.

24. Modify the electrical circuit in Figure 7-8 to replace the START push-button with a vane switch. A new cycle will be initiated any time the part slides through the part shuttle activating the vane switch.

8

CHAPTER

Pressure Control

OBJECTIVES

After studying this chapter, you should be able to:

- Define terms used with pressure switches.
- Discuss tolerance in a pressure switch.
- List four types of pressure switches.
- Explain how the piston-type pressure switch operates.
- Discuss why the rated operating pressure of a Bourdon-tube pressure switch should not be exceeded during use.
- Describe the operation of the diaphragm pressure switch.
- Explain how the pressure transducer operates.
- Draw the normally open and normally closed pressure switch symbols.
- Draw an electrical control circuit showing how a normally closed pressure switch contact can be used to de-energize the circuit.

8.1 Importance of Pressure Indication and Control

Pressure indication and control are important in many electrical control systems in which air, gas, or a liquid is involved. The control takes several forms. It may be used to start or stop a machine on either rising or falling pressure. It may be necessary to know that pressure is being maintained.

The *pressure switch* is used to transfer information concerning pressure to an electrical circuit. The electrical switch unit can be a single normally open or normally closed switch. It can also be both, with a common terminal or two independent circuits. Usually it is easier to use a switch with two independent circuits in the design of a control circuit. The *normal* condition for the pressure switch is the state of the switch when no pressure is applied to the circuit.

The pressure switch symbol is shown in Figure 8-1. The NC contact opens and the NO contact closes on reaching preset pressures.

Some terms associated with pressure switches are best explained in chart form. Figure 8-2 shows some of the terms we are concerned with and some of the most important points.

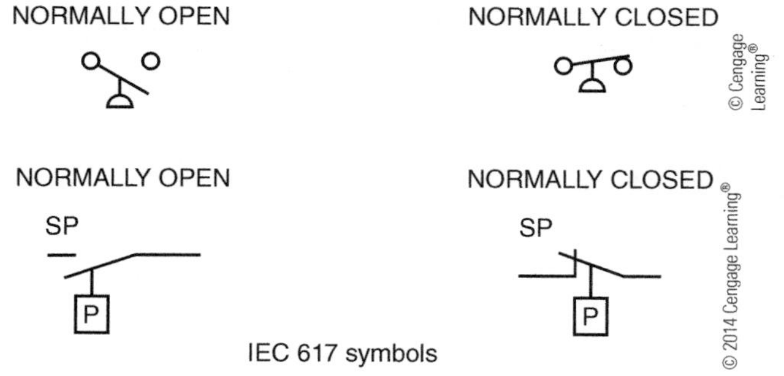
NORMALLY OPEN
NORMALLY CLOSED
NORMALLY OPEN
SP
P
NORMALLY CLOSED
SP
P
IEC 617 symbols
© Cengage Learning®
© 2014 Cengage Learning®

Figure 8-1 **Pressure switch symbols.**

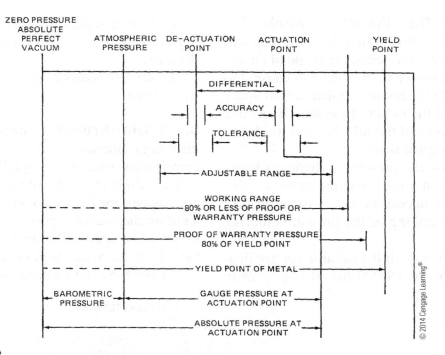

Figure 8-2 Pressure chart.

Tolerance in most switches is about 2%; *accuracy* is about 1%. These are expressed as percentages of the working range. In some cases, it is necessary to know and use a differential pressure. As used here, the *differential* is the range between the *actuation point* and the *de-actuation point*. For example, the electrical contacts transition at a preset rising pressure and remain actuated until the pressure drops to a lower level. This differential may be a fixed amount, generally proportional to the operating range, or it can be adjustable.

8.2 Types of Pressure Switches

There are several types of pressure switches available. Some of the most widely used types are:

- Sealed piston
- Bourdon tube
- Diaphragm
- Solid state

8.2.1 Sealed Piston Type   The sealed-piston type of pressure switch is actuated by means of a piston assembly that is in direct contact with the hydraulic fluid. The assembly consists of a piston sealed with an O-ring, direct acting on a

snap-action switch (Figure 8-3). An extremely long life can be expected from this switch where mechanical pulsations exceed 25 per minute and where high overpressures occur. The sealed piston saves the cost of installing return lines.

The piston-type pressure switch can withstand large pressure changes and high overpressures. These units are suitable for pressures in the range of 15 pounds per square inch (psi) to 12,000 psi.

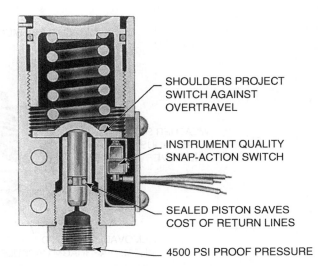

SHOULDERS PROJECT SWITCH AGAINST OVERTRAVEL

INSTRUMENT QUALITY SNAP-ACTION SWITCH

SEALED PISTON SAVES COST OF RETURN LINES

4500 PSI PROOF PRESSURE

Figure 8-3 Piston-type pressure switch. *(Courtesy of Barksdale Controls Division, Transamerica Delaval Inc.)*

8.2.2 Bourdon-Tube Pressure Switch

The pressure operating element in a Bourdon-tube unit is made up of a tube formed in the shape of an arc. Application of pressure to the tube will cause it to straighten out. Care should be taken not to apply pressures beyond the rating of the unit. Excessive pressures tend to bend the tube beyond its ability to return to its original shape.

The Bourdon-tube pressure switch unit is extremely sensitive. It senses peak pressures; for this reason, it may be necessary to use some form of dampening or snubbing of the pressure entering the tube.

This pressure switch is available for applications from 50 psi to 18,000 psi. It is generally used more in indicating service than in switching because of the inherent low work output of the unit.

Figure 8-4 shows a typical Bourdon-tube pressure switch.

8.2.3 Diaphragm Pressure Switch

A diaphragm pressure switch consists of a disk with convolutions around the edge. The edge of the disk is fixed within the case of the switch. Pressure is applied against the full area of the diaphragm. The center of the diaphragm opposite the pressure side is free to move and operate the snap-action electrical contacts. Units are available from vacuum to 150 psi.

Figure 8-5 shows a diaphragm pressure switch.

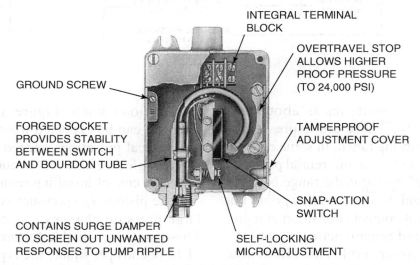

INTEGRAL TERMINAL BLOCK

OVERTRAVEL STOP ALLOWS HIGHER PROOF PRESSURE (TO 24,000 PSI)

GROUND SCREW

FORGED SOCKET PROVIDES STABILITY BETWEEN SWITCH AND BOURDON TUBE

TAMPERPROOF ADJUSTMENT COVER

SNAP-ACTION SWITCH

CONTAINS SURGE DAMPER TO SCREEN OUT UNWANTED RESPONSES TO PUMP RIPPLE

SELF-LOCKING MICROADJUSTMENT

Figure 8-4 **Bourdon-tube pressure switch.** *(Courtesy of Barksdale Controls Division, Transamerica Delaval Inc.)*

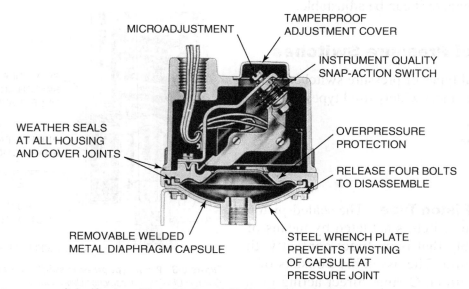

MICROADJUSTMENT

TAMPERPROOF ADJUSTMENT COVER

INSTRUMENT QUALITY SNAP-ACTION SWITCH

WEATHER SEALS AT ALL HOUSING AND COVER JOINTS

OVERPRESSURE PROTECTION

RELEASE FOUR BOLTS TO DISASSEMBLE

REMOVABLE WELDED METAL DIAPHRAGM CAPSULE

STEEL WRENCH PLATE PREVENTS TWISTING OF CAPSULE AT PRESSURE JOINT

Figure 8-5 **Diaphragm pressure switch.** *(Courtesy of Barksdale Controls Division, Transamerica Delaval Inc.)*

8.2.4 Solid-State Pressure Switch

A solid-state pressure switch is interchangeable with existing electromechanical pressure switches. Pressure sensing is performed by semiconductor strain gauges with proof pressures up to 15,000 psi acceptable. Switching is accomplished with solid-state triacs. An enclosed terminal block allows four different switching configurations:

1. Single-pole, single-throw (NC)
2. Single-pole, single-throw (NO)
3. Single-pole, double-throw
4. Double-make, double-break

See Figure 8-6 for a typical solid-state pressure switch.

8.2.5 Pressure Transducers

With the use of programmable controllers in industry (covered in Chapter 15), the measurement of pressure has become important.

A pressure transducer, such as shown in Figure 8-7, utilizes semiconductor strain gauges that are epoxy-bonded to a metal diaphragm. Four gauges (or two gauges with fixed resistors) form a Wheatstone bridge. Pressure applied to the diaphragm through the pressure port produces a minute deflection, which introduces strain to the gauges. The strain produces an electrical resistance change proportional to the pressure. An electrical signal may then be taken into an analog input card in the PLC or it

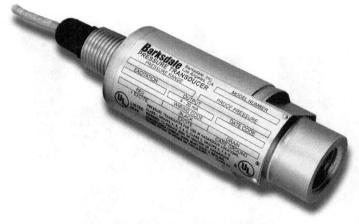

Figure 8-7 Transmitter and transducer. *(Courtesy of Barksdale Controls Division, Transamerica Delaval Inc.)*

may be taken to a strain gage conditioner card in an instrumentation system. The signal may be a voltage or a current level that is proportional to the pressure being read by the transducer. The pressure transducer, therefore, transforms a pressure signal into an electrical signal that may be interpreted by the system.

8.2.6 Pressure Sensors

Figure 8-8 shows sensor solutions for hydraulic and pneumatic applications. The pressure family includes pressure sensors, pressure transmitters, and vacuum sensors. These sensors incorporate the same ceramic pressure-sensing technology found in high-accuracy pressure transmitters. Ceramic pressure sensors are the preferred choice because they can withstand higher overpressure and are more resistant to pressure spikes. The base of a ceramic sensor has a measuring electrode plus a reference electrode. A diaphragm containing a second measuring electrode is attached to the base with a glass solder layer. The two measuring electrodes form the plates of a capacitor, and with no pressure applied to the sensor, the electrodes are approximately 10 μm apart. As pressure is applied to the diaphragm, this distance decreases and thus increases the capacitance value and produces a signal that is proportional to the applied pressure. When combined with its mechanical mounting, this sensing element provides overall long term mechanical stability. See Figure 8-9 for an application of a pressure transmitter. These pressure transmitters have all the features needed for

Figure 8-6 Solid-state pressure switch. *(Courtesy of Rockwell Automation, Inc.)*

Figure 8-8 **Pressure sensors.** *(Courtesy of ifm efector, inc.)*

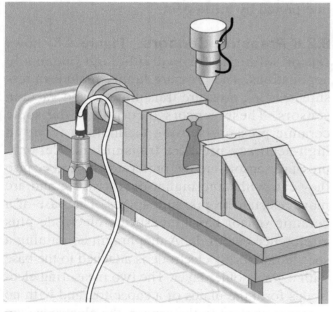

Figure 8-9 **Application of a pressure transmitter on an injection molding machine.** *(Courtesy of ifm efector, inc.)*

monitoring hydraulic clamping and ram pressures as, for example, those present in injection molding machines.

8.3 Circuit Applications

In application, the pressure switch contact may be substituted for the limit switch contact. The basic circuit showing the limit switch as a means of reversing the piston is shown in Figure 7-6. In Figures 8-10B and 8-10C, the pressure switch is connected into the fluid power circuit. This circuit uses a two-position, single-solenoid, spring return valve. A normally closed contact from the pressure switch is connected into the electrical circuit (Figure 8-10A).

The function of this circuit is the same as that shown in Figure 7-6, except now the piston reverses when a preset pressure is reached. Remember that a normally closed pressure switch contact opens on rising pressure when the pressure exceeds the trip setting on the switch.

In the circuit shown in Figures 8-11A, 8-11B, 8-11C, and 8-11D, a three-position, double-solenoid, spring return valve is used. Remember that in the de-energized condition of both solenoids, pressure is allowed to bypass through the center of the valve spool to the tank. The piston must be in position A for start conditions.

The sequence of operations proceeds as follows:

1. Press the START push-button.
2. If the returned limit switch (2LS) is actuated and the pressure switch (1PS) is deactuated, then relay coil 1CR is energized.
   a. Relay contact 1CR-1 closes, sealing around the START push-button.
   b. Relay contact 1CR-2 closes, energizing solenoid 1HA.
3. The valve spool now shifts to the right (Figure 8-11C), allowing pressure to enter the back end of the piston (left end).
   a. The piston moves forward.
   b. On reaching the workpiece, pressure builds against it.
   c. When the preset pressure is reached on pressure switch 1PS, its NC contact opens.

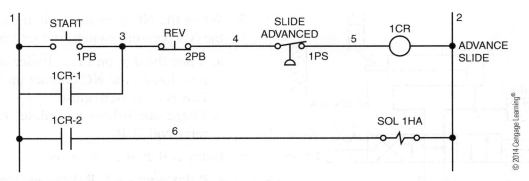

Figure 8-10A Control circuit with one cylinder and advanced pressure switch.

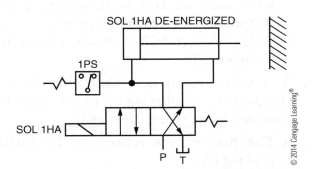

Figure 8-10B Hydraulic circuit-returned. Valve spool shifted to normal position (condition shown in drawings) P-pressure, T-tank.

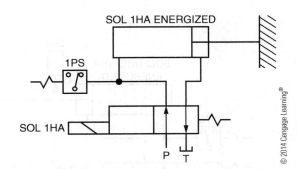

Figure 8-10C Hydraulic circuit-advanced. Valve spool shifted to right (not shown on drawings, for discussion only) P-pressure T-tank.

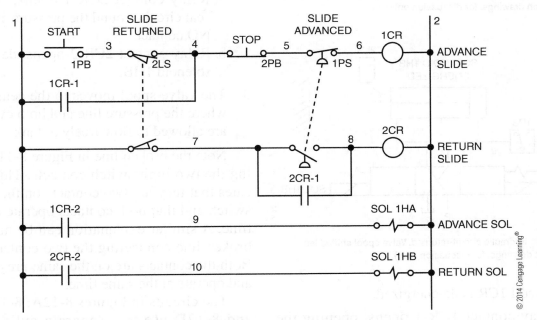

Figure 8-11A Control circuit with one cylinder, advance pressure switch, and returned limit switch.

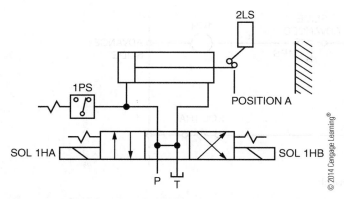

Figure 8-11B Hydraulic circuit showing returned (home) position (condition shown on drawings) P-pressure T-tank.

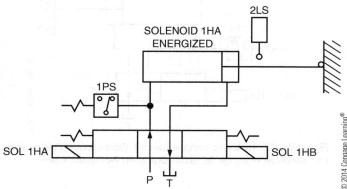

Figure 8-11C Hydraulic circuit-advanced. Valve spool shifted to right (not shown on drawings, for discussion only).

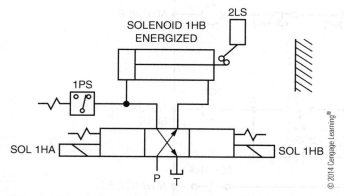

Figure 8-11D Hydraulic circuit-returned. Valve spool shifted left (not shown on drawings, for discussion only).

4. Relay coil 1CR is de-energized.

   a. Relay contact 1CR-1 opens, opening the seal circuit around the START push-button.

   b. Relay contact 1CR-2 opens, de-energizing solenoid 1HA.

5. When the NC pressure switch contact opens, the NO pressure switch contact closes.

   a. Since the dog on the cylinder is away from position A, the NC contact on limit switch 2LS is allowed to close.

   b. These conditions complete the circuit to relay coil 2CR.

6. Relay coil 2CR is energized.

   a. Relay contact 2CR-1 closes, sealing around the NO pressure switch contact. (Note: This is necessary because the pressure drops when solenoid 1HA is de-energized. The NO pressure switch contact opens.)

   b. Relay contact 2CR-2 closes, energizing solenoid 1HB.

7. The valve spool now shifts to the left (Figure 8-11D), allowing pressure to enter the front end of the piston (right end).

   a. The piston now returns to its initial position (A).

   b. On reaching this point, limit switch 2LS is actuated, opening its NC contact.

8. Relay coil 2CR is de-energized.

   a. Relay contact 2CR-1 opens, opening the seal circuit around the pressure switch (1PS) NO contact.

   b. Relay contact 2CR-2 opens, de-energizing solenoid 1HB.

9. The valve spool moves to the center position where the pressure line and both cylinder lines are allowed to flow freely to tank.

Note the broken line in Figure 8-11A connecting the two limit switch contacts. This line indicates that they are two contacts on the same limit switch and that both contacts operate at the same time. A similar explanation can be made for the broken line connecting the two contacts on 1PS. Both the contacts are on the same pressure switch and operate at the same time.

The circuits in Figures 8-12A, 8-12B, 8-12C, and 8-12D use two separate cylinder-piston assemblies. Each cylinder assembly is powered through a two-position, single-solenoid, spring-return valve.

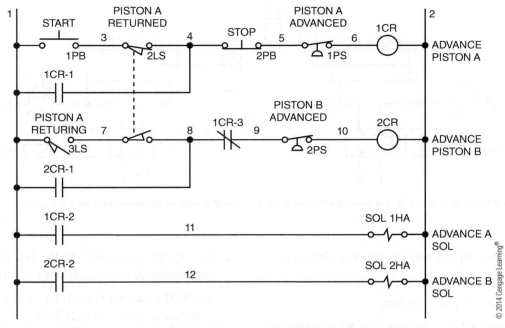

Figure 8-12A Control circuit with two cylinders (each with advanced pressure switch) and returned limit switch.

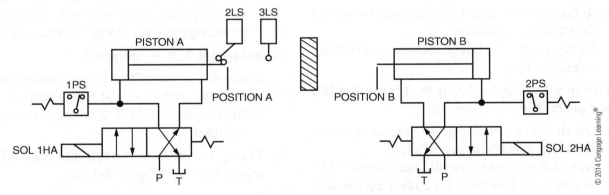

Figure 8-12B Hydraulic circuit showing returned (home) position. Both solenoids are de-energized (condition shown on drawings) P-pressure, T-tank.

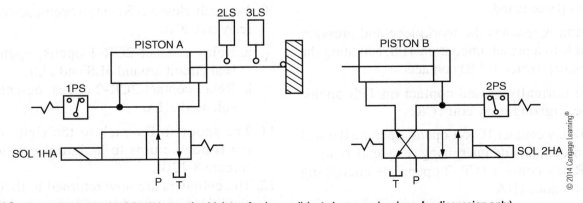

Figure 8-12C Hydraulic circuit with SOL1HA energized (piston A advanced) (not shown on drawings, for discussion only).

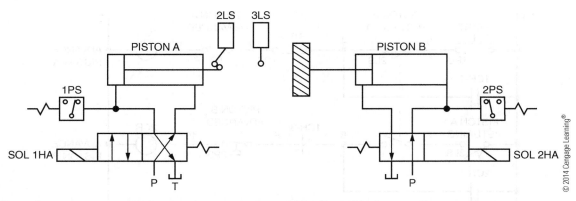

Figure 8-12D Hydraulic circuit with SOL2HA energized (Piston B advanced) (not shown on drawings, for discussion only).

The sequence of operations for this circuit is as follows:

1. Press the START push-button.

2. If the retuned limit switch (2LS) is actuated and the pressure switch (1PS) is de-actuated, then relay coil 1CR is energized.

   a. Relay contact 1CR-1 closes, sealing around the START push-button.

   b. Relay contact 1CR-2 closes, energizing solenoid 1HA.

3. The spool shifts to the right (Figure 8-12C) and piston A moves forward.

   a. On the forward stroke of piston A, limit switch 2LS is no longer actuated.

   b. As piston A moves forward, limit switch (3LS) is actuated; however, no circuit action will occur since NC relay contact 1CR-3 is open.

   c. When piston A has reached its fully advanced stroke, limit switch (3LS) will again be de-actuated.

4. Piston A reaches the workpiece and pressure builds to a preset amount on 1PS, actuating the pressure switch (1PS) contacts.

5. The normally closed contact on 1PS opens, de-energizing relay coil 1CR.

   a. Relay contact 1CR-1 opens, opening the seal circuit, and de-energizing relay coil 1CR.

   b. Relay contact 1CR-2 opens, de-energizing solenoid 1HA.

   c. Relay contact 1CR-3 closes.

6. The spool shifts back to the left (Figure 8-12B) and piston A now returns to position A.

   a. On the return travel, the cam on the piston actuates limit switch 3LS.

   b. Limit switch (3LS) is actuated, limit switch (2LS) is de-actuated, relay coil 1CR is off (contact 1CR-3 is closed), and pressure switch (2PS) is de-actuated. These conditions complete the circuit to relay coil 2CR.

7. Relay coil 2CR is energized.

   a. Relay contact 2CR-1 closes, sealing around 2LS NC contact and 3LS NO contact.

   b. Relay contact 2CR-2 closes, energizing solenoid 2HA.

8. The spool shifts to the left and piston B now moves forward (Figure 8-12D).

9. Piston B reaches the workpiece and pressure builds to the preset amount on pressure switch (2PS), actuating the pressure switch contact.

10. Normally closed 2PS contact opens, de-energizing relay coil 2CR.

    a. Relay contact 2CR-1 opens, opening the seal circuit around 3LS and 2LS.

    b. Relay contact 2CR-2 opens, de-energizing solenoid 2HA.

11. The spool shifts back to the right and piston B now returns to its initial position at B (Figure 8-12B).

12. The cylinders are now returned to their initial conditions and another cycle may be initiated.

Recommended Web Links

Students are encouraged to view the following Web sites as a supplement to the concepts presented in this textbook. Review and analyze the array of products that are available for electrical control applications. Many of these sites offer technical information that can help in converting the principles to practical applications. To view catalogs, your PC may require Adobe Acrobat Reader software.

1. Honeywell
   http://sensing.honeywell.com
   Review: Sensors/Pressure

2. Entran Sensors and Electronics
   www.entran.com
   Review: Pressure Sensors

3. Omega Engineering
   www.omega.com
   Review: Pressure, Strain, Force

Achievement Review

1. Draw the normally open and normally closed pressure switch symbols.

2. Explain differential in pressure switch operation.

3. List four major types of pressure switches.

4. In a piston-type pressure switch, what is used to prevent oil leakage past the piston?

5. Why is a form of dampening sometimes used with the Bourdon-tube pressure switch?

6. What is the usual pressure operating range for the diaphragm-type pressure switch?

7. Design and draw an electrical control circuit showing how the operation of a normally closed pressure switch contact can be used to de-energize an output solenoid. Use two push-button switches, a relay, pressure switch contact, and a solenoid.

8. Which of the following can a pressure transducer use to transfer pressure information?

   a. A low resistance coil
   b. A specially designed plug-in relay
   c. Semiconductor strain gauges

9. How is switching accomplished in the solid-state pressure switch?

10. List four different switching configurations available in the solid-state pressure switch.

11. How is the differential resistance in a pressure transducer measured?

   a. By applying a constant voltage to the bridge and deflection of the diaphragm
   b. Through calculation using data provided by the manufacturer
   c. By estimating the change on rising pressure

12. Modify the circuit shown in Figure 8-11 so the return solenoid is always energized until the conditions are present to advance the slide.

13. For the circuit shown in Figure 8-11, add a "cycled" memory relay coil that energizes when the slide is advanced and de-energizes when the START push-button is pressed.

14. For the circuit shown in Figure 8-12, explain how the operation of the circuit would change if 3LS were moved to the right so that it would not be actuated until piston A is fully advanced.

15. For the circuit shown in Figure 8-12, add a "cycled" memory relay coil that energizes when slide B is advanced and de-energizes when the START push-button is pressed.

16. Redraw the circuit diagram shown in Figure 8-10A using IEC 617 symbols.

17. Redraw the circuit diagram shown in Figure 8-11A using IEC 617 symbols.

18. Redraw the circuit diagram shown in Figure 8-12A using IEC 617 symbols.

9

CHAPTER

Temperature Control

OBJECTIVES

After studying this chapter, you should be able to:

- Identify five important factors to consider when selecting a temperature controller.

- Explain the difference between controller sensitivity and controller operating differential.

- Describe how the thermocouple, thermistor, and resistance unit obtain temperature information that can be used in a control circuit.

- Explain how time proportioning is used in a temperature controller.

- List the advantages and disadvantages of the potentiometric-type controller.

- Discuss bandwidth, automatic reset, and rate control in pyrometers.

- Name two types of temperature switches or thermostats and explain how each operates.

- Show how a temperature switch contact can be used in an electrical control circuit to prevent the operation of the circuit unless the temperature of the part being sensed is at or above a preset temperature.

9.1 Importance of Temperature Indication and Control

The use of temperature control and indication is important in at least three areas:

1. Safety
2. Troubleshooting
3. Means of processing material

As a safety issue, it should be noted that excessive heat can cause many problems that lead to unsafe operating conditions, including overheated conductors and excessive heat in operating components, such as circuit breakers and other operating electromechanical components. Overheating is often the result of loose connections in a circuit carrying power current. Motor bearings may overheat due to improper lubrication or defective bearings. Overheating in a control cabinet can create false tripping of protective devices. More on this subject is covered in Chapter 19 on troubleshooting. Indication of overheated components or conductors can lead to their correction before trouble starts.

In processing materials, it is important that the correct temperature be used and maintained. Both control and indication are important to result in a good product, particularly in the molding of plastics and the diecasting of metals.

This chapter is devoted entirely to the use of temperature control and indication in the processing of materials.

9.2 Selection of Temperature Controllers

Temperature controllers basically consist of two parts: a sensing element that responds to temperature and a switch consisting of NO and/or NC contacts. The contacts operate based on the reading from the temperature-sensing element. The operating point for the contacts is preset. The switch operation is usually accomplished when the temperature at the sensing point changes from any given level to the preset point. The main difference among temperature switches is the means by which the temperature information is transferred from the sensing element to the switching element.

There are several factors to consider when selecting a temperature controller:

- *Temperature range available.* Temperature range is the overall operating range of the controller. Not all controllers cover the entire temperature range used in industrial control. The controller is generally selected after the range is determined. The sensing element often becomes an important factor.
- *Type of sensing element.* There are three basic types: (1) electronic, (2) differential expansion of metal, and (3) expansion of fluid, gas, or vapor.
- *Response time.* Speed of response is a measure of the elapsed time from the instant the temperature change occurs at the sensing element until it is converted into controller action. The response to a temperature change may vary between one type of sensing element and another. Here the application may be the deciding factor. For example, if rapid response is not required, the slower-response and generally less-expensive types may be used.
- *Sensitivity.* Units vary with the amount of temperature change required to operate the switch. This amount is fixed in some switches; in others it is adjustable. It is usually desirable to have a controller that has a relatively high sensitivity. *Resolution sensitivity* is the amount of temperature change necessary to actuate the controller.
- *Operating differential.* Operating differential is the difference in temperature at the sensing

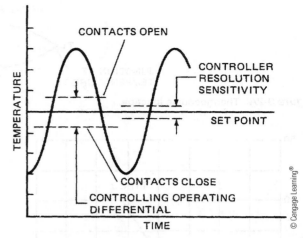

Figure 9-1 Temperature versus time curve.

element between make and break of the controller's contacts when the controller is cycled.

Figure 9-1 illustrates some of these operating characteristics in temperature controllers.

9.3 Electronic Temperature Controller (Pyrometer)

The electronic controller, or pyrometer, may have one of three different temperature sensing elements:

1. Thermocouple (approximate temperature range 200°F to 5000°F)
2. Thermistor (approximate temperature range 100°F to 600°F)
3. Resistance temperature detector (RTD) (approximate temperature range 300°F to 1200°F)

The *thermocouple* operates on the principle of joining two dissimilar metals, A and B, at their extremities (Figure 9-2A). When a temperature difference exists between the two extremity points, a potential is generated in proportion to the temperature difference. Combinations such as copper-constantan, iron-constantan, and chromel-alumel are used in thermocouple construction.

When the temperature at T1 is different than at T2, electromotive force (emf) exists. By the use of an indicating device that indicates emf or the flow of current, the temperature difference can be determined.

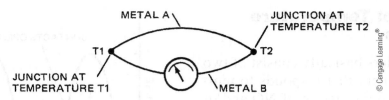

Figure 9-2A Thermocouple junction.

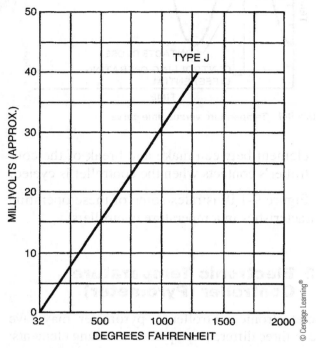

Figure 9-2B Temperature versus millivolt curve.

The relationship between temperature in degrees Fahrenheit and millivolts output for a type J thermocouple (iron-constantan) is shown in Figure 9-2B.

Three different physical arrangements of thermocouples are shown in Figures 9-3A, 9-3B, and 9-3C. Figure 9-3A shows the dual-element thermocouple. It can be used to sense temperature at two different physical levels. Figure 9-3B shows the single thermocouple. Both thermocouples are spring loaded. The single thermocouple uses an adaptor and small steel ferrule on a flexible armored cable to obtain the spring loading. The dual-element thermocouple uses a small steel plate that is secured to the deep couple with an adaptor and locking nut. The springs on both thermocouples are loaded against the steel plate. Figure 9-3C shows an adjustable-depth plastic melt-type thermocouple with a plug attachment.

The *thermistor* is a semiconductor whose resistance decreases with increasing temperature. This element is connected into a null-reading-type AC bridge circuit with the output fed to an amplifier. The output of the amplifier operates a relay that is used in the control circuit. Figure 9-4A shows a line drawing of a thermistor.

The *RTD unit* consists of a tube made of stainless steel or brass. It contains an element made in the form of a coil. The coil is generally wound of a fine nickel, platinum, or pure copper wire. As the temperature changes, the resistance of the coil changes proportionally. This change is converted to a voltage by its application in the electrical control circuit. The voltage level is compared to a reference value. Therefore, as the voltage changes with a temperature change, an output

Figure 9-3A Dual-element thermocouple. *(Courtesy of Noral, Inc.)*

Figure 9-3B Single thermocouple. *(Courtesy of Noral, Inc.)*

Figure 9-3C Adjustable-depth plastic melt-type thermocouple. *(Courtesy of Noral, Inc.)*

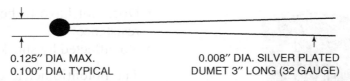

0.125″ DIA. MAX.   0.008″ DIA. SILVER PLATED
0.100″ DIA. TYPICAL   DUMET 3″ LONG (32 GAUGE)

Figure 9-4A **Line drawing of a thermistor.** *(Courtesy of Yellow Springs Instrument Co., Inc., Industrial Division.)*

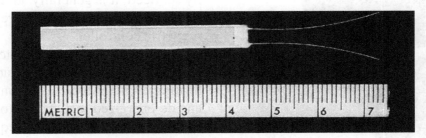

Figure 9-4B **Resistance bulb unit (RTD).** *(Courtesy of Noral, Inc.)*

proportional to the temperature change is obtained. Figure 9-4B shows an RTD unit.

The electronic temperature controller is usually the most expensive type. It is more sensitive and has a faster response. Generally, electronic controllers have a smaller sensing device. The sensing device can be remotely located with ease. Normal control accuracies of .25% to 5% of full-scale reading can be expected.

There are a wide range of applications that require temperature control. Temperature controllers may usually be purchased or configured to handle the diverse range of applications. Some of the terminology used in this field will be briefly explained to provide a general understanding for the setup and use of these products.

On/off control is the simplest and oldest method of temperature control. The output (heater or cooler) is simply switched ON when a temperature value is reached and is switched OFF when another temperature value is reached. The actual temperature in an on/off system will always fluctuate about some value. Depending on the time delay between the control signals and the actual thermal response of the system, the amplitude of the temperature swing using this type of control may exceed good operating control.

Time proportioning control turns the heating elements on or off based on the difference between the actual temperature and the set-point. There is a temperature below the set-point at which the power to

the heating elements is continuously *on*. Likewise, there is a temperature above the set-point at which the power is always *off*. The proportional band must be programmed into the controller. The output will be switched on at a rate that is proportional to the progression through the band. For example, at a temperature of 20% through the proportional band, the power will be 80% on; half way through the proportional band, the power will be 50% on. This method differs from on/off control in that it does not require an actual temperature change to cycle.

To accomplish on/off and proportioning control, pyrometers can be divided into two types. These are the millivoltmeter controller and the potentiometric controller.

The *millivoltmeter controller* is the oldest type in the market. It uses a millivoltmeter operated directly from a temperature sensor such as a thermocouple. The adjustment of the set-point establishes the position of either a photocell and light source or a set of oscillator coils. A small flag is connected to the indicating pointer for the actual temperature. As the temperature approaches the set-point, the flag moves between the photocell and light source or between the oscillator coils. Control action now starts, cutting back on heat.

One of the major sources of trouble with the millivoltmeter controller is the adverse effect of shock and vibration. Another problem involves the sensing device. For good indication accuracy, the external resistance of the thermocouple circuit should match that of the instrument. Some

Figure 9-5 Potentiometric controller. *(Courtesy of Barber-Colman Company.)*

- Does not have to be calibrated for external resistance
- Not affected by shock and vibration

The potentiometric controller has the disadvantage of more electronic circuitry that is not easily serviced; it is generally advisable to replace the entire plug-in controller. Figure 9-5 shows a potentiometric controller.

The millivoltmeter and potentiometric controllers both generally use a contactor to energize the heating element load. The contactor is energized and de-energized in response to a demand for heat. For example, when heat is required as sensed by a thermocouple, the contactor is energized by the controller output. Electrical power is then connected to the heating element load. When there are no heat requirements, the heating elements are completely isolated from the power source by the opening of the contacts on the contactor.

Mechanical contactors switch power in full ON and OFF cycles. Therefore, they should be used at cycle times of 15 seconds or longer for reasonable service life. Because of the full ON and OFF switching and the limited cycle time, contactor-controlled processes must have a higher tolerance for process temperature overshoot and undershoot (Figure 9-6).

Solid-state controllers do not use a contactor to energize and de-energize the heating element load. With this type of controller, the power remains connected to the load at all times. Just enough power is supplied to the heating elements to satisfy

manufacturers offer controllers with adjustable external resistance circuits. Such instruments can be calibrated after they are installed.

The *potentiometric controller* differs from the millivoltmeter type in that the signal from the temperature sensor is electronically controlled and compared to the set-point temperature. This instrument can be supplied with or without indication. The indicating method is completely separate from the temperature control circuit.

The potentiometric controller has the following advantages:

- No moving parts

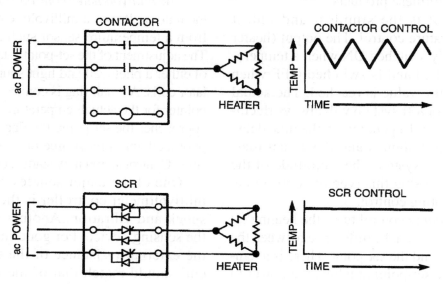

Figure 9-6 Contactor versus SCR control. *(Courtesy of Chromalox.)*

the temperature requirements. This type of controller is called a *stepless* or *proportional* controller and has a current output. It is generally used with silicon-controlled rectifiers (SCR) or triac solid-state power packs.

Two different outputs are available with this type of controller. Power is applied to the heating element load through:

- Phase control (phase-angle-fired)
- Zero crossing (zero-angle-fired)

With phase control, the output is continuous at the amount required for consistent temperature. Phase control is accomplished by "firing" the SCR at some point through the wave, depending on the amount of power required.

One problem with phase-fired control is the radio frequency interference (RFI) generated by the SCR switching at any place in the wave at an extremely fast rate. This switching is on the order of one-half microsecond (µs). Another problem is the resulting distorted wave shape which creates a problem for the power company.

These problems can be avoided by the use of zero-crossing SCR circuitry. Its characteristics are similar to the time-proportioning results. However, power is always switched off and on when its voltage is zero.

Zero-crossing or *zero-crossover-fired* power controllers are ideal for control of pure resistive loads that can accommodate rapid, full-power, ON/OFF cycling. Zero-crossover firing does not create RFI and will not adversely affect sensitive electronic equipment (computers, other SCR power controllers, logic controllers, etc.) located in the same area. Additionally, SCRs are protected from line voltage transients, making them more reliable in a wide variety of applications.

Zero-crossover-fired SCRs, when coupled with a time-proportioning control (firing package), will operate in a series of full ON and OFF cycles known as *time-proportional burst firing*. The time-proportioning control accepts the control output signal and converts it into a time-proportional signal, determining the amount of ON time and OFF time per duty cycle. The continuous, highly repetitious rate of full ON and full OFF produces a smooth power output to the load (heater) and a stable process temperature (Figure 9-7).

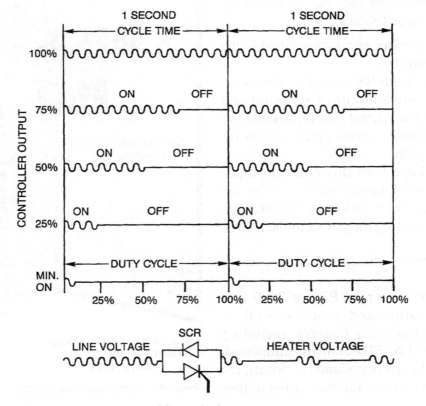

Figure 9-7 **Time-proportional burst firing.** *(Courtesy of Chromalox.)*

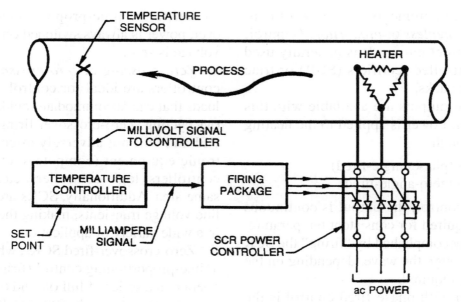

Figure 9-8 **Power control system.** *(Courtesy of Chromalox.)*

A typical power control system consists of an RTD or thermocouple, temperature controller, firing package, and SCR power controller. Often the firing package is part of the temperature controller and is not a separate component. Figure 9-8 shows a schematic drawing of a typical system.

These components work together to control the heating of the process as follows:

- The temperature sensor provides a signal to the temperature controller.
- The temperature controller compares the sensor signal with the predetermined set point and generates an output signal that represents the difference between the actual process temperature and the set point.
- The firing package uses this control output to generate a time-proportional signal for the SCR power controller, switching the SCR on and off, thus regulating the power to the heater.

Figure 9-9 shows a complete SCR control panel.

The instrument shown in Figure 9-10 is a ramp/soak temperature and process controller. This controller has many features, including ON/OFF, proportional or PID control, ramp/soak function, dual display (process and set point), relay, triac (4–20 mA), and solid-state drive output

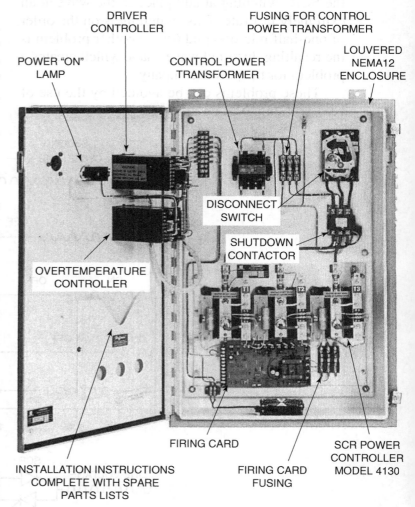

Figure 9-9 **SCR control panel.** *(Courtesy of Chromalox.)*

modules. It has a nonvolatile memory, thus requiring no battery backup and eight event outputs to initiate remote activity such as valve actuation, operator annunciation, output to programmable controller, and status indication. See Figure 9-11 for a more detailed description of the front panel.

One of the series of temperature controllers shown in Figure 9-12 has features of ON/OFF,

proportional or PID control, ramp/soak, dual display (process and set point), relay, triac, (4–20 mA), and solid-state relay drive output modules. There is a nonvolatile memory requiring no backup battery. The front panel displays and push buttons are shown in Figure 9-13.

A drawing showing the remote set point and process signal output feature of this controller

Figure 9-10 Ramp/soak temperature and process controller. *(Courtesy of Chromalox.)*

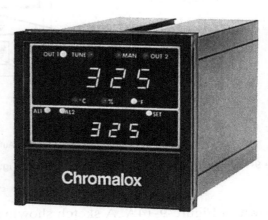

Figure 9-12 Temperature controller. *(Courtesy of Chromalox.)*

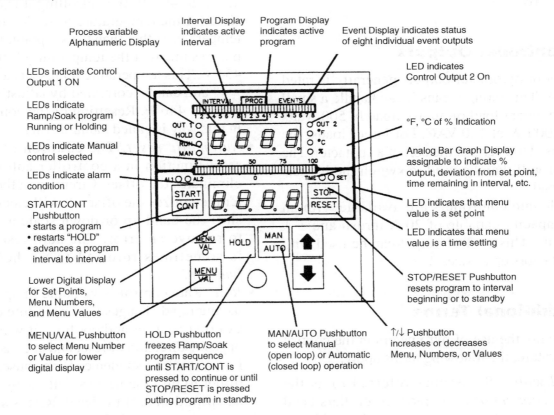

Figure 9-11 Front panel functions of Figure 9-10. *(Courtesy of Chromalox.)*

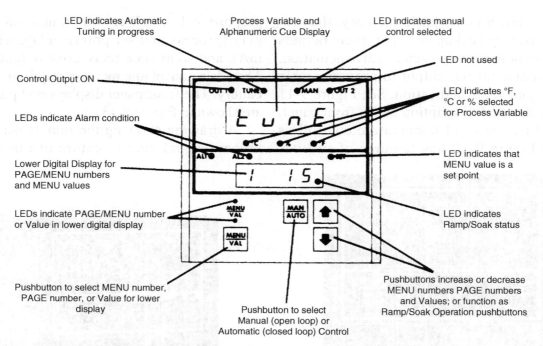

Figure 9-13 Front panel display and push buttons for Figure 9-12. *(Courtesy of Chromalox.)*

is shown in Figure 9-14A. A sketch showing the heat/cool feature of this controller is shown in Figure 9-14B.

9.4 Controller Outputs

The relay in pyrometer outputs generally is rated at 5–10 A. This rating means it can handle a small heating element load of approximately 500 watts (W) to 1000 W at 120 VAC. For larger loads, the relay is used to energize the coil of a contactor. The contactor will then be used to switch the higher-current load.

Solid-state controllers are available with an output capacity capable of 20 A inrush and 1 A continuous. This control output can be used with load contactors up to sizes 2 or 3.

9.5 Additional Terms

In discussing the use of pyrometers in the control of temperature, the following terms apply:

- *Band width.* Sometimes referred to as the *proportioning band*, in most controllers band width is adjustable (Figure 9-15). A band width

set for 3% is 24°F on an 800°F instrument. If the set point is at 450°F, the band extends from 437°F to 463°F. In controlling it is quite possible for the temperature to settle out at some temperature other than the set point. For example, assume that the temperature settles out at 454°F. This temperature represents an offset of +4° and can be corrected by adjusting the set point at 446°F. Resetting of the control point in this way is called *manual reset.*

- *Automatic reset.* The addition of the auto reset function in a controller automatically eliminates the offset. Circuitry in the controller recognizes the offset. It will electronically shift the band up or down scale as required to remove an error. The auto reset feature is sometimes referred to as the *integral function.*

- *Rate.* Rate control is valuable in applications having rapid changes in temperature caused by external heating and cooling. It works in the opposite direction of reset and at a faster speed. For example, a sudden cooling causes the controller to turn the heaters full ON by means of an upward shift of the band. Rate is sometimes referred to as the *derivative* function.

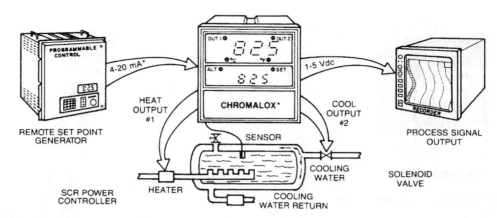

Figure 9-14A **Remote set point and process signal output.** *(Courtesy of Chromalox.)*

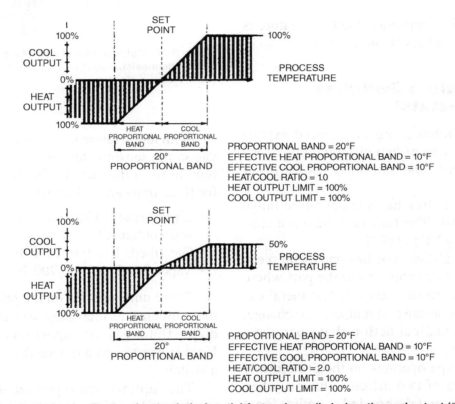

Figure 9-14B **Heat/cool ratio feature. The heat/cool ratio (gain ratio) feature is applied when the cool output (#2) control mode is PID. It balances the heating and cooling capacities of the system, yielding stable process control. Note that the heat/cool ratio creates an "effective cool output limit" of 50% in the second illustration.** *(Courtesy of Chromalox.)*

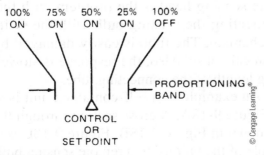

Figure 9-15 **Control or set point.**

- *Mode.* With reference to temperature controllers, mode refers to the operational functions that a controller has. A *two-mode controller* is one having proportioning and auto reset or proportional and auto reset. A *three-mode controller* is one having proportioning, auto reset, and rate or proportional, auto reset, and rate.
- *Analog and digital set point.* With an analog set point, the temperature is set on a scale. With the

Figure 9-16 **Digital solid-state controller.** *(Courtesy of Barber-Colman Company.)*

Figure 9-17 **Cutaway view of a bimetal-type temperature switch (thermostat).** *(Courtesy of Fenwal Controls, a Division of Kidde, Inc.)*

digital set point controller, the temperature is displayed in numbers (Figure 9-16).

9.6 Temperature Switches (Thermostats)

Temperature switches using differential expansion of metals may be of two different types. One uses a *mechanical link;* the other uses a *fused bimetal.*

The mechanical link has a temperature range of 100°F to 1500°F. The bimetal type has a temperature range of 40°F to 800°F.

The mechanical-link type has one metal piece directly connected or subjected to the part where the temperature is to be detected. The metal expands or contracts because of temperature change, producing a mechanical action that operates a switch.

The bimetal type operates on the principle of uneven expansion of two different metals when heated. With two metals bonded together, heat will deform one metal more than the other. The mechanical action resulting from a temperature change is used to operate a switch. In bimetal units the sensing element and switch are generally enclosed in the same container or are adjacent.

The bimetal type is usually the smallest and least expensive type of controller. It is not suited for high-precision work. However, its compact, rugged construction lends it to many uses. Accuracies of 1% to 5% of full scale can be expected. Figure 9-17 is a cutaway view of a bimetal unit.

With temperature switches using liquid, gas, or vapor, the sensing elements are generally located remote from the switch. The temperature ranges for these units are as follows:

- Liquid filled: 150°F to 2200°F (these may be self-contained)
- Gas filled: 100°F to 1000°F
- Vapor filled: 50°F to 700°F

These units operate on the principle that when the temperature increases, expansion of the medium (fluid, gas, or vapor) takes place. Thus, a force is exerted on a device that, in turn, operates a switch.

The liquid-filled type that is self-contained (switch and sensing element together) has a relatively fast response time.

A factor to consider in using units with a remote sensing head is the problem with the tube connecting the sensing bulb with the switching mechanism. The tube is easily damaged by mechanical abuse. Also, the response is slower with long length of the connecting tube.

An example of a self-contained unit is shown in Figure 9-18A. A cross section through the unit is shown in Figure 9-18B. Figure 9-18C is an example of the unit using a remote sensing bulb.

Figure 9-18A Dual-action, liquid-filled temperature controller with switch and sensing element together. *(Courtesy of Fenwal Controls, a Division of Kidde, Inc.)*

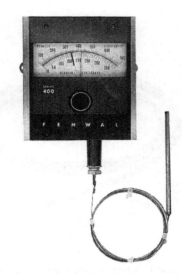

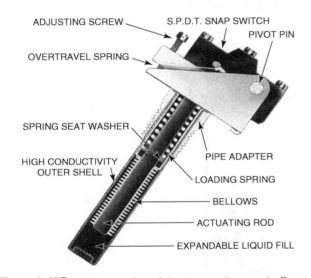

Figure 9-18B Cross section of the temperature controller shown in Figure 9-18A. *(Courtesy of Fenwal Controls, a Division of Kidde, Inc.)*

The switch contact ratings on temperature switches (thermostats) vary from 5 to 25 A at 120 VAC. They generally consist of one normally open and one normally closed contact. They may have a common connection point, or they may have isolated contacts.

9.7 Temperature Sensors

Temperature monitoring systems utilizing sensors provide reliable feedback in temperature control

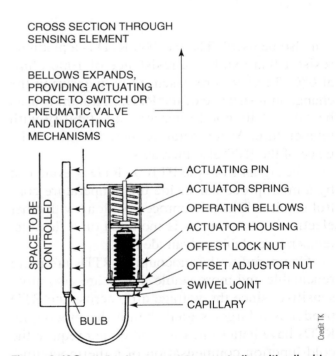

Figure 9-18C Indicating temperature controller with a liquid expansion, capillary-type remote sensing bulb. *(Courtesy of Fenwal Controls, a Division of Kidde, Inc.)*

applications. Such a system consists of an electronic control monitor and a variety of sensor probes (Figure 9-19). The control monitor is packaged in a compact stainless-steel casing that can be mounted either directly to a sensor probe or remotely using a cabled probe.

This temperature sensor utilizes a PT1000 RTD for temperature measurement (PT100 RTDs

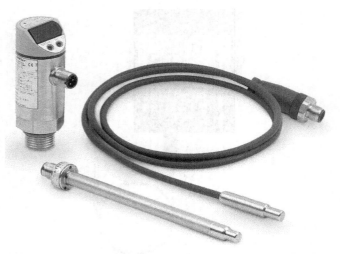

Figure 9-19 The imf efector monitoring system includes a control monitor and a variety of sensor probes. *(Courtesy of imf efector, inc.)*

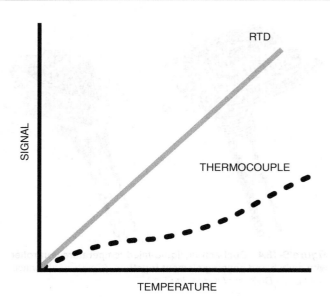

Figure 9-20 Temperature-signal chart. *(Courtesy of imf efector, inc.)*

can also be used). The PT1000 RTD is a platinum resistor that exhibits a resistance of 1000 ohms at 0°C. Temperature is sensed by measuring the change in resistance of the RTD. Platinum is used because of its good linearity and stability with temperature. As temperature increases, the resistance of the RTD also increases.

The signal from the PT1000 RTD is evaluated by a microprocessor inside the temperature control monitor. This microprocessor along with other electronic components is mounted on a flexible, temperature-stable polyamide film.

Compared to thermocouples, RTDs are more repeatable and more stable. RTDs are also more sensitive, since the voltage drop across the RTD produces a larger signal than a thermocouple. RTDs have better linearity and do not require the cold junction compensation of a thermocouple. See Figure 9-20.

An example of an application for a temperature sensor is shown in Figure 9-21, in which a temperature sensor is used to monitor tank temperature to ensure a safe operating range.

9.8 Circuit Applications

Refer to Figure 4-18 which shows a heating circuit with only heat OFF/ON push-button switches for control. These switches control the heat circuit through contactor 10CON.

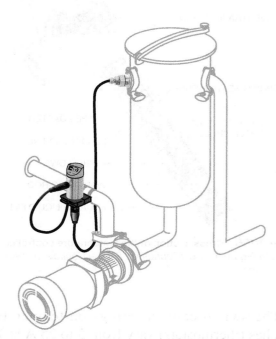

Figure 9-21 The imf efector sanitary temperature sensors can be used to monitor tank temperature ensuring a safe operating range. *(Courtesy of imf efector, inc.)*

In Figure 9-22A a two-position selector switch is substituted for the two push-button switches in Figure 4-18. A pyrometer is inserted in the circuit to control the contactor coil 10CON. Note that the symbol for the pyrometer shows only the relay contact in the pyrometer and the thermocouple connections. More details on internal circuit construction can be obtained from the manufacture of the pyrometer. Terminal connections, special wiring, and any jumpers or switch settings should be identified on the control drawings.

When the heat OFF/ON selector switch is rotated to the ON position, the circuit is complete to the NO relay contact. If the temperature of the part being sensed is below the set point on the pyrometer, the relay contact will be closed, completing the circuit to the coil of contactor 10CON. The coil of contactor 10CON is energized. Contactor contacts 10CON-1, 10CON-2 and 10CON-3 close, energizing the heating elements at power line voltage.

When the temperature of the part being sensed by the thermocouple reaches the set point, the pyrometer relay contact opens, de-energizing the coil of contactor 10CON. Contactor contacts 10CON-1, 10CON-2, and 10CON-3 open, de-energizing the heating elements. The pyrometer relay contact continues to close and open, depending on the thermocouple input. The rate of opening and closing depends on the type of pyrometer.

The temperature switch (thermostat) is generally used at lower temperatures than the pyrometer. It is also used where accuracy and sensitivity are not important factors in temperature control. For example, a tank containing a fluid of some type (oil, asphalt, or brine) is to be heated to a given temperature. Depending on the size of the tank, it may take considerable time to heat it safely to a given temperature. If the preset temperature is 200°F, it generally is not important whether the temperature is actually 204°F or 196°F.

Another use of the thermostat is to indicate and/or control the temperature of fluid used in a machine operation. For example, it may be required that the fluid in a hydraulic system should not exceed 125°F. Figure 9-22B illustrates this example.

The sequence of operations is as follows:

1. Temperature below set point of 125°F, circuit ready to operate.
2. Press the START push-button.
3. Energize relay coil 1CR.

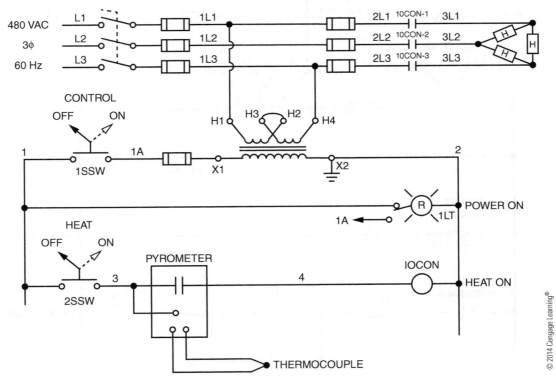

Figure 9-22A **Control circuit for controlling heater.**

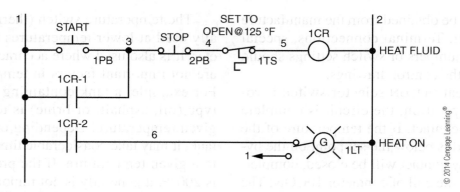

Figure 9-22B Heating circuit to control fluid temperature.

a. Relay contact 1CR-1 closes, sealing around the START push-button.
b. Relay contact 1CR-2 closes, energizing the green pilot light.
c. Heat is applied to the hydraulic fluid.
4. Temperature exceeds 125°F, NC temperature switch contact opens.
5. Relay coil 1CR de-energized.

a. Relay contact 1CR-1 opens, opening the seal circuit around the START push-button.
b. Relay contact 1CR-2 opens, de-energizing the green pilot light.
c. Heat is removed from the hydraulic fluid.

The temperature must drop below 125°F, allowing the NC 1TS contact to close before relay coil 1CR can again be energized.

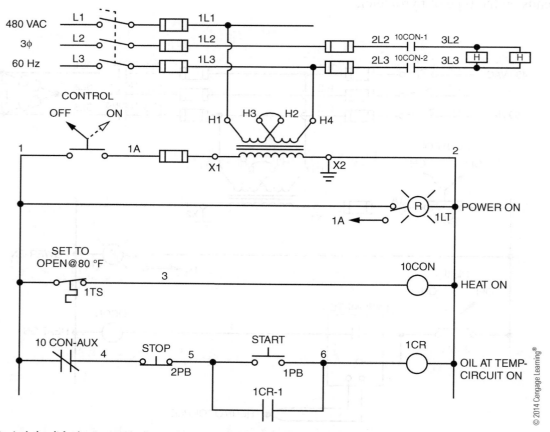

Figure 9-23 Control circuit for heat control, oil must be at proper temperature to enable circuit 1CR.

Figure 9-23 shows the control for two heating elements. With the temperature of the medium being heated (oil, asphalt, brine) below 80°F:

The sequence of operations is as follows:

1. Contactor coil 10CON is energized.
2. Contactor contacts 10CON-1 and 10CON-2 close, energizing the two heating elements.
3. Contactor auxiliary contact (10CON-AUX) opens, preventing the energizing of relay coil 1CR through the operation of the START push-button.
4. The heated medium now reaches 80°F, de-energizing 10CON contactor coil by the opening of 1TS.
5. Contactor contacts 10CON-1 and 10CON-2 open, de-energizing the two heating elements.
6. Contactor auxiliary contact (10CON-AUX) closes.
7. The START push-button can now be pressed, energizing relay coil 1CR.

    a. Relay contact 1CR-1 closes, sealing around the START push-button.

Relay coil 1CR now remains energized until such time as the temperature of the heated medium drops below 80°F. At that time contactor coil 10CON will again energize, energizing the heating elements and opening the 10CON auxiliary contact. Relay coil 1CR will be de-energized.

In the circuit shown in Figure 9-24, hot bearing control is added to the circuit as shown in Figure 9-22B.

Assume that a machine has four bearings. The machine should not operate with any of the bearings at a temperature in excess of 130°F. A thermostat can be placed on each of the four bearings. The thermostat contacts will be normally open, set to close (operate) at 130°F. The modified circuit functions as follows:

1. With all bearing temperatures below 130°F, relay coil 2CR is not energized.
2. When the temperature of one or more of the bearings exceeds 130°F, relay coil 2CR is energized.

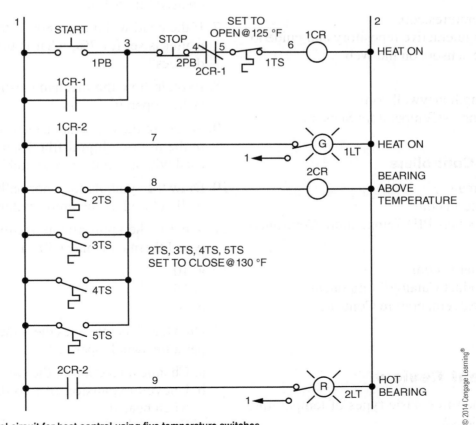

Figure 9-24 Control circuit for heat control using five temperature switches.

a. Normally closed relay contact 2CR-1 opens.
b. Relay coil 1CR is de-energized.
c. Relay contact 1CR-1 opens, opening the seal circuit around the START push-button.
d. Relay contact 1CR-2 opens, de-energizing the green pilot light.
e. Relay contact 2CR-2 closes, energizing the hot bearing light (red).

When the bearing problem has been corrected and all of the normally open thermostat contacts are open, the 1CR relay circuit can be energized through the START push-button if the temperature is below 125°F degrees F (1TS closed).

Recommended Web Links

Students are encouraged to view the following Web sites as a supplement to the concepts presented in this textbook. Review and analyze the array of products that are available for electrical control applications. Many of these sites offer technical information that can help in converting the principles to practical applications. To view catalogs, your PC may require Adobe Acrobat Reader software.

Temperature Sensors

1. www.temperatures.com
   Review this interactive repository and guide to temperature sensors on the Web

2. Honeywell
   http://sensing.honeywell.com
   Review: Sensors/Temperature Sensors

Temperature Controllers

1. Barber-Colman
   www.eurotherm.com
   Review: Products/PID Temperature Controllers

2. Chromalox
   www.chromalox.com
   Review: Product Catalog/Component Technologies/Temperature Controls/Controllers

Achievement Review

1. What are the three basic types of temperature sensing devices?

2. What is the speed of response in a temperature controller?

3. Explain the difference between controller resolution, sensitivity, and controller operating differential.

4. Explain the thermocouple operating principle.

5. Explain the function of time-proportioning control.

6. What size of loads can be handled with a relay output controller? Can this load capacity be increased? If so, how?

7. If the band width in an 800°F temperature controller is set for 2%, what is the band width in degrees?

8. Explain how the mechanical-link temperature switch operates.

9. What are the disadvantages or problems found in the use of a liquid-filled sensing element as used with a temperature switch?

10. Draw the symbols for a normally open and normally closed temperature switch contact.

11. What is the approximate millivolt output of a type J thermocouple at 500°F?
    a. 10
    b. 15
    c. 20

12. On what principle does the bimetal type of temperature switch operate?
    a. Chemical reaction of the two metals
    b. Uneven expansion of two different metals when heated
    c. Expansion of air, gas, or vapor

13. For the circuit shown in Figure 9-24, explain how the circuit would be modified to create a failsafe circuit design. Hint: Temperature switches (2TS–5TS) would need to be NC switches and 2CR would be energized to indicate "All Bearings below Hi Temperature."

14. For the circuit shown in Figure 9-24, add the controls necessary to sound a horn when the bearing temperature has been exceeded. The horn should remain on until an operator presses a push-button to acknowledge the fault condition.

<big>**10**</big>

CHAPTER

Time Control

After studying this chapter, you should be able to:

- Describe the differences between a timer and a time-delay relay.

- Explain where the timer has a definite use.

- List three major types of timers discussed in this chapter.

- Explain the arrangement of the contacts in a reset-type timer under the conditions of reset, timing, and timed out.

- Draw simple control circuits showing (1) clutch and motor energized at the same time, (2) clutch energized before the motor, and (3) motor energized before the clutch.

- Explain how the multiple-interval timer operates.

- Discuss the operation of the repeat-cycle timer.

- Describe some of the design features of the solid-state timers.

10.1 Selected Operations

Many types of timers are available for use on industrial machines. Their major function is to place information about elapsed time into an electrical control circuit. The method of accomplishing this function varies with the type of timer used. Time range, accuracy, and contact arrangement vary among types of timers. A few suggestions are made here for the proper use of timers.

It is true that time elapses during nearly every action taking place on a machine. This fact does not necessarily mean that a timer is always the best means of control.

For example, a machine part is to move from point X to point Y, a definite distance. It is observed that this motion requires approximately 5 seconds to complete. If the position of the machine part is important, a position control device (such as a limit switch) not dependent on elapsed time should be used to indicate that the part did arrive at point Y. Because of variables in the machine, one cycle could require 5.0 seconds, the next 5.1 seconds, and a third cycle 4.9 seconds. Thus, if a timer is used the machine part will not always stop at point Y.

In Figure 10-1 the timer is set for 5 seconds. In case A, time runs out before the part reaches point Y. In case B, time runs out after the part has passed point Y.

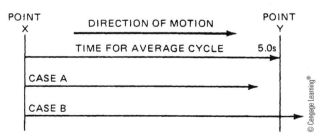

Figure 10-1 Timing for two different cases.

A similar situation may exist when a timer is substituted for a pressure indication. In one case pressure may build to a preset value before a preset time has elapsed. Also, time may elapse before the preset pressure is reached. Similar conditions may arise when a timer is substituted for a temperature controller to obtain temperature information.

Another important factor to consider when sensing position is verifying that the motion has occurred. Valve assemblies may be purchased that have position sensors integrated into the assembly. These sensors will provide feedback that the spool has properly shifted within the valve body. An issue with this method may occur if the tooling is broken. In this case, the valve spool may shift correctly, but the tooling may not actually move.

Another method of sensing position incorporates a dog attached to the end of a cylinder rod. This method will ensure the cylinder actually moved to its intended position. If it is crucial to the process that the machine tooling moves to the proper position, then the actual position of the tooling must be sensed. This is the only method that will ensure that the tooling physically moved to the proper location while processing the part.

The important point is to determine the critical condition to be met. Is it time, position, pressure, or temperature? Timers are very useful tools, but they must be applied properly.

Another point to clarify is reference to timers and time-delay relays. Generally, *time-delay relays* are devices having a timing function after the timer coil has been energized or de-energized. Time-delay relays have a normally open (NO) and a normally closed (NC) timing contact. The operation of the pneumatic time-delay relay was covered in Chapter 4. In some cases the contacts are isolated. In other cases they have a common terminal. Sometimes time-delay relays have several NO and/or NC instantaneous contacts that operate immediately when the time-delay coil is energized.

When reference is made to *timers,* the time function may start on one or more of the contacts at energization, or at any time after energization during a preset time cycle. Likewise, the timing function may stop on one contact or more during the cycle after timing has been started on the particular contact(s).

In general, a timer opens or closes electrical circuits to selected operations according to a timed program. Instantaneous contacts are also available on some timers.

Timing functions will start from an electrical signal, initiated through any one of several components: a push-button switch, relay contact, temperature switch, pressure switch, limit switch, etc. Once an electrical signal is available to the timer, the timing function can, in general, be one of two types: ON delay (timing function after energizing) or OFF delay (timing function after de-energizing), as shown in Figure 10-2.

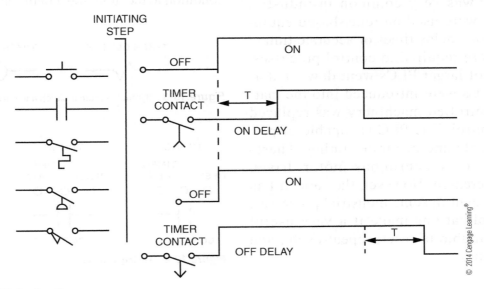

Figure 10-2 Basic Timing functions

Other forms of this basic timing function are found with the solid-state timer, including pulse and pulse-and-repeat operations. In some timers an inhibit function is available. This function is accomplished by applying a voltage to specific inhibit terminals. The timing cycle then stops and holds its outputs in the last state without resetting the circuit. Timing will continue and the outputs will be allowed to change according to their programmed operating modes when voltage is removed from the inhibit terminals.

10.2 Types of Timers

In grouping timers according to their method of timing, there are two types that are applied most frequently on industrial machines:

1. Synchronous motor driven
2. Solid state

Four other types of timers that have industrial applications are the dashpot, mechanical, electrochemical, and thermal. Since these have only limited use on industrial machines, they are not covered in this text.

10.3 Synchronous Motor-Driven Timers

At one time, the use of the *synchronous motor-driven timer* was very common in industry. These timers were used on relay-based equipment where long delay times or accurate timing sequences were required to control processes. As the price of larger PLCs went down and as lower end PLCs were introduced into the market, relay controlled machinery was replaced with PLC control. The PLC is capable of performing reliable and accurate timing. Therefore, the use of synchronous motor-driven timers has decreased. However, they are still an important element in some industrial processes. Its many applications make it a very useful tool. It is available in reset, repeat-cycle, and manual-set types.

10.3.1 Reset Timers
The *reset timer* depends on the clutch and the synchronous motor for its operation. The symbols for the clutch and synchronous motor are shown in Figure 10-3.

To provide an output for the timer to control electrical circuits, reset timers have a set of contacts to perform this function. In some timers, each contact can be adjusted to perform in the same way or in a different way, as required. Figure 10-4 shows that there are three conditions for each contact during a complete cycle:

1. Reset or de-energized—clutch and motor de-energized
2. Timing—clutch energized; motor may or may not be energized, depending on its circuit application
3. Timed out—motor de-energized; clutch may or may not be de-energized, depending on its circuit application

When the contact is open, the o symbol above the contact is shown. When the contact is closed, the symbol is shown as ×. For example, in a given timer contact arrangement, assume contact 1 is open in the reset condition, closed in the timing condition, and open in the timed-out condition. The complete symbol appearing above that contact would be o × o.

Remember that the contact changes from its reset condition to the timing condition when the clutch is energized. The contact changes from the timing condition to the timed-out condition when the motor

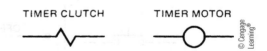

Figure 10-3 Symbols for timer clutch and motor.

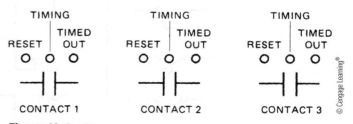

Figure 10-4 Timing symbols.

has run for the preset amount of time. The contact then changes back to the reset condition when the clutch is de-energized.

There are several arrangements available for each of the contacts. The possible configurations are shown in Figure 10-5.

In some reset timers, the motor connections are not brought out for external connection. In this case they are generally wired so that the motor is

energized from a normally open circuit contact that closes when the clutch is energized. On other timers, all connections are brought out. In this way, the motor as well as the clutch can be energized from an external source at any time during a cycle.

The circuit in Figure 10-6A illustrates the normal usage conditions in which the motor is energized directly from the clutch being energized. Figure 10-6B illustrates the motor energized at a later time, after the clutch is energized. Increased timing accuracy is possible with this arrangement if the timing period is relatively short in an overall long cycle. A third condition is shown in Figure 10-6C, in which the motor is energized before the clutch is energized. Increased accuracy can be obtained if the timing period is nearly that of the overall cycle.

Figure 10-7 shows an example of the reset timer.

Some of the simpler reset timers have only a snap-action switch that operates at the end of a preset time period. When the clutch is de-energized, the timer resets and the switch is released.

SEQUENCE NUMBER	RESET: Clutch deenergized; motor deenergized	TIMING: Clutch energized; motor energized	TIMED OUT: Clutch engaged until solenoid is deenergized; motor deenergized
1	O	O	X
2	O	X	O
3	X	X	O
4	X	O	X
5	O	X	X
6	X	O	O

O CONTACT OPEN   X CONTACT CLOSED

Figure 10-5 Time sequence chart.

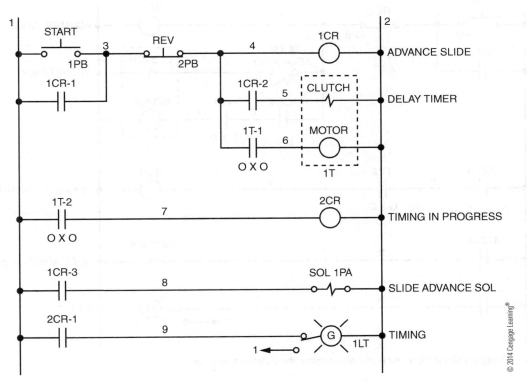

Figure 10-6A Timing circuit-timing motor is energized directly from the clutch being energized.

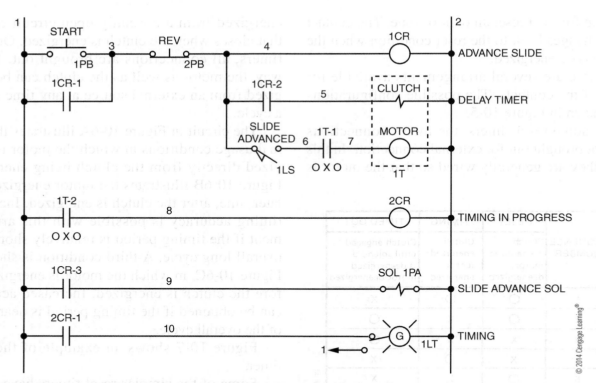

Figure 10-6B Timing circuit-timing motor is energized some time after the clutch is energized.

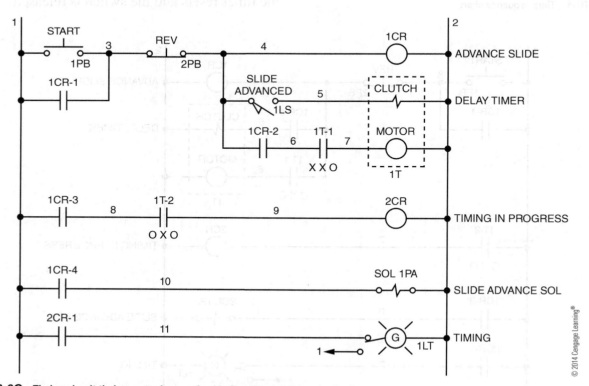

Figure 10-6C Timing circuit-timing motor is energized before the clutch is energized.

Figure 10-7 Reset timer. *(Courtesy of Eagle Signal Controls.)*

Figure 10-8 Multiple interval timer. *(Courtesy of Eagle Signal Controls.)*

Another form of the reset timer is often called a *multiple-interval timer.* It is used extensively in programming control. Multiple time periods can be controlled by the adjustment of each individual ON and OFF setting.

The timer displayed in Figure 10-8 consists of a series of contacts. Each contact has two fingers that ride on a cam plate. During the timing interval, a synchronous motor drives the plate downward through a solenoid-operated clutch. The contacts are closed and opened in the sequence set for the fingers to drop off the edge of the cam plate. When the clutch solenoid is de-energized to disengage the clutch, the cam plate is reset by a spring. Simultaneously, the contact fingers are lifted to their NO or NC RESET position.

10.3.2 Repeat-Cycle Timers

The *repeat-cycle timer* is used to control a number of electrical circuits in a predetermined sequence.

The timer shown in Figure 10-9 consists of a synchronous motor driving a cam shaft. The cam shaft rotates continuously as long as the motor is energized. Adjustable cams determine the point of closing and opening a switch during each cam-shaft revolution.

In the unique split-cam design, each side of the cam is separately screwdriver-adjustable in either direction. Either side determines the precise instant during the cycle when the switch will actuate; the other side determines how long the switch remains actuated. Adjustments are easy and precise. One quarter turn of the adjusting screw equals .5% of cycle time. A setting disc, calibrated in 1% increments, facilitates program setup and indicates cycle progress.

10.3.3 Manual-Set Timers

The *manual-set timer* requires manual operation of the timer to start operation. The timer then runs a selected time and stops automatically.

In the timers illustrated in Figures 10-10A and 10-10B, timing begins when the START switch is closed. At the same time, the timing LED goes on. A relaxation oscillator starts to run at a rate determined by the set point. The timer times out and the timing LED turns off when the oscillator count is to the level set by the range switch. At time out, the load relay is energized, transferring its contacts, and the timing circuit is automatically de-energized. Reset occurs when the START switch is opened or when power is interrupted.

Figure 10-9 Precision switch cam programmer. *(Courtesy of Automatic Timing and Controls Company, Inc.)*

Figure 10-10A Manual-set timer. *(Courtesy of Eagle Signal Controls.)*

Figure 10-10B Manual-set timers. *(Courtesy of Automatic Timing and Controls Company, Inc.)*

10.4 Solid-State Timers

A thorough review of the many types and variations of solid-state timers would more than fill this entire book. For our purposes, then, we limit our discussion to the following important design features:

- Microprocessor-based timers
- Digital set and optional digital readout
- Provision for external set

Examples of solid-state timers are pictured in Figures 10-11A and 10-11B.

The Eagle CX200 (Figure 10-11A) is a microprocessor-based timer/counter. Time or count operation, time range, and standard or reverse start operation are selected via seven miniature rocker switches located inside the unit housing. Time or count set points are entered into the unit using a sealed membrane keypad on the front of the unit. Each digit in the set point is individually increased

Figure 10-11A Microprocessor-based timer/counter. *(Courtesy of Eagle Signal Controls.)*

Figure 10-11B Microprocessor-based digital timer. *(Courtesy of Eagle Signal Controls.)*

or decreased by pressing an appropriate keypad switch. Time or count set point and progress are displayed on the front of the unit by a 4½-digit liquid crystal display with 0.5-inch digits. Operational mode annunciators also appear in the display area on the front of the unit. The mode annunciator flashes when the unit is timing or counting.

The outputs can be programmed to operate in one of four load sequences: ○○×, ○×○, ○○× with pulse output, and ○○× pulse output with repeat-cycle operation.

The setting accuracy for count is 100%. The setting accuracy for time is ±0.5% or 50 milliseconds (ms), whichever is larger. The repeat accuracy for count is 100%. For time it is .001% or 35 ms, whichever is larger.

The SX200 series (Figure 10-11B) is a microprocessor-based digital timer. The programmable features include eight time ranges and eight output operating modes. These operating modes and all other setup functions are programmed with miniature rocker switches located on the back of the housing.

This timer uses a nonvolatile RAM memory (see Chapter 15) to retain the set point, actual time values, and program parameters. The expected life of data in memory is ten years.

The front panel of the unit is a sealed membrane keypad that provides excellent protection for most industrial environments. The keypad includes a special surface just below the display on which the function of the timer can be marked with a pen or pencil.

The SET and ENT keys on the front panel provide access to the set point and to the front panel programmed functions. Programming changes are entered using the increment and decrement keys.

A keypad "lock" function is built into the software of the unit, which allows the set point to be viewed but does not allow unauthorized changes.

The timing cycle progress is shown on four 0.3-inch red LED displays for easy readability. The front panel also has a flashing LED at the right side of the display to indicate that the unit is in the timing cycle and an LED at the left side of the display that lights when the programmed contacts are energized.

10.5 Circuit Applications

Refer to Figure 8-10C. The ram advances until it reaches the work piece and builds pressure. On reaching a preset pressure, the normally closed pressure switch contact opens, and the ram reverses.

In Figure 10-12, a reset timer is added to the circuit shown in Figure 8-10C. The pressure switch contact is changed from NC to NO.

The sequence of operations is as follows:

1. Press the START push-button.
2. Relay coil 1CR is energized.

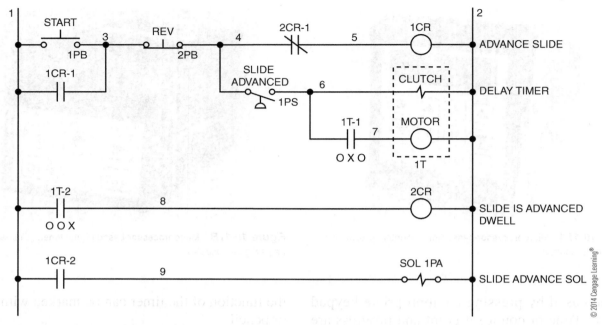

Figure 10-12A Timing circuit—advance, dwell, return.

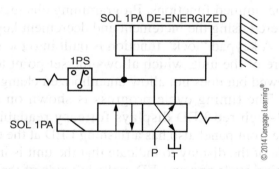

Figure 10-12B Pneumatic circuit-returned. Valve spool shifted to normal position (condition shown in drawings) P-pressure, T-tank.

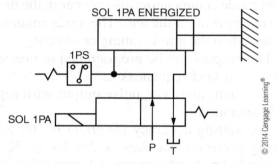

Figure 10-12C Pneumatic circuit-advanced. Valve spool shifted to right (not shown on drawings, for discussion only) P-pressure, T-tank.

a. Relay contact 1CR-1 closes, sealing around the START push-button.

b. Relay contact 1CR-2 closes, energizing solenoid 1PA. The valve spool shifts and the piston advances to the work piece and builds pressure.

3. The NO pressure switch contact closes, energizing the timer clutch.

a. Timer contact 1T-1 closes, energizing the timer motor.

4. When a preset time elapses:

a. timer contact 1T-1 opens, de-energizing the timer motor.

b. Timer contact 1T-2 closes, energizing relay 2CR.

5. Relay contact 2CR-1 opens, de-energizing relay coil 1CR.

a. Relay contact 1CR-1 opens, opening the seal circuit around the START push-button.

b. Relay contact 1CR-2 opens, de-energizing solenoid 1PA.

6. The valve spool shifts back, the piston returns, and pressure drops, opening pressure switch contact 1PS.

7. Timer clutch is de-energized.

8. Timer contact 1T-2 opens, de-energizing relay 2CR.

9. The NC relay contact 2CR-1 closes, setting up the circuit for the next cycle.

In the circuit shown in Figure 10-13, an additional cylinder-piston assembly is added with an additional pressure switch contact 2PS.

The function of this circuit is to start piston #2 with a time delay after piston #1 has built pressure to a preset level on the work piece. The sequence of operations proceeds as follows:

1. Press the START push-button.
2. As long as there is no pressure on piston #1 so that pressure switch 1PS is closed, relay coil 1CR is energized.
   a. Relay contact 1CR-1 closes, sealing around the START push-button.
   b. Relay contact 1CR-2 closes, energizing solenoid 1PA.

The valve spool shifts, piston #1 advances, and builds pressure on the work piece. When a preset pressure is reached, pressure switch contact 1PS actuates.

3. The NO pressure switch contact 1PS closes, energizing relay coil 2CR as long as there is no pressure on piston #2 so that pressure switch 2PS is closed.
   a. Relay contact 2CR-1 closes, sealing around the NO 1PS pressure switch contact.
4. The NC pressure switch contact 1PS opens, de-energizing relay coil 1CR.
   a. Relay contact 1CR-1 opens, opening the seal around the START push-button.
   b. Relay contact 1CR-2 opens, de-energizing solenoid 1PA.

The valve spool shifts back and piston #1 returns to its initial start position.

5. Timer clutch is energized.
   a. Timer contact 1T-1 closes, energizing the timer motor.

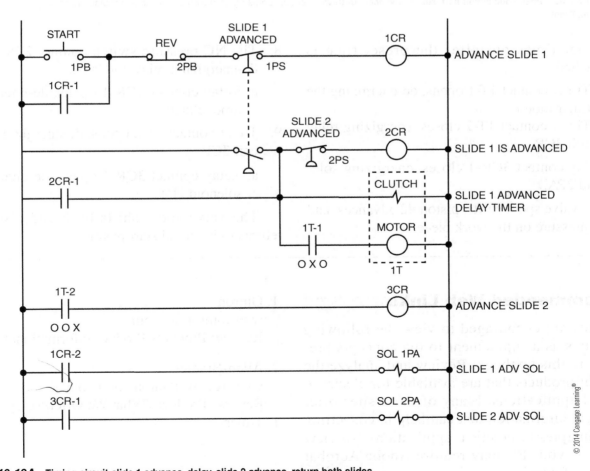

Figure 10-13A   Timing circuit-slide 1 advance, delay, slide 2 advance, return both slides.

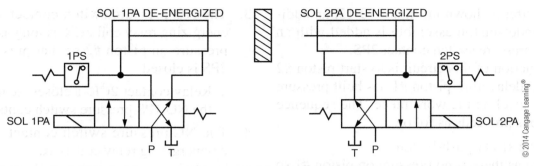

Figure 10-13B Pneumatic circuit-both slides returned. Valve spools shifted to normal position (condition shown in drawings) P-pressure, T-tank.

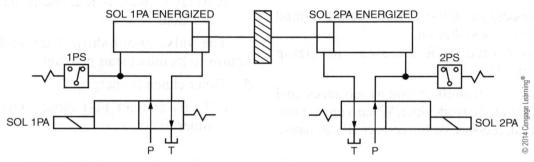

Figure 10-13C Pneumatic circuit-both slides advanced. Both valve spools shifted (not shown on drawings, for discussion only) P-pressure, T-tank.

6. Timer times out after the preset time is reached.

   a. Timer contact 1T-1 opens, de-energizing the timer motor.

   b. Timer contact 1T-2 closes, energizing relay coil 3CR.

7. Relay contact 3CR-1 closes, energizing solenoid 2PA.

   The valve spool shifts, piston #2 advances, and builds pressure on the work piece.

8. The NC pressure switch contact 2PS opens, de-energizing relay coil 2CR.

   a. Relay contact 2CR-1 opens, de-energizing timer clutch.

9. Timer contact 1T-2 opens, de-energizing relay coil 3CR.

   a. Relay contact 3CR-1 opens, de-energizing solenoid 2PA.

   The valve spool shifts back and piston #2 returns to its initial start position.

Recommended Web Links

Students are encouraged to view the following Web sites as a supplement to the concepts presented in this textbook. Review and analyze the array of products that are available for electrical control applications. Many of these sites offer technical information that can help in converting the principles to practical applications. To view catalogs, your PC may require Adobe Acrobat Reader software.

1. Omron
   www.omron-ap.com
   Review: Products/Product Information/Timers

2. Allen-Bradley
   www.rockwellautomation.com
   Review: Products/What We Offer/Relays & Timers

3. General Electric
   www.geindustrial.com
   Review: Products/Relays & Timers-Control/
   Electronic Timers

4. Automatic Timing and Controls
   www.flw.com
   Review: Distribution/ATC/Mechanical Timers

Achievement Review

1. Describe the differences between a time-delay relay and a timer.

2. List two types of timers that are applied in the industrial field.

3. What are the two major components responsible for the operation of the reset timer?

4. List the conditions of the clutch and the synchronous motor in a reset timer under the following conditions:
   a. Reset or de-energized
   b. Timing
   c. Timed out

5. In a control circuit using two push-button switches, two relays, a reset timer, and a solenoid, show how the clutch and motor of the reset timer can be energized at the same time. The cycle starts with the operation of a push-button switch.

6. What is the major use of the multiple-interval timer?

7. What controls the opening and closing of a switch in the repeat-cycle timer?

8. What four sequences can be programmed to operate in the Eagle CX200 timer/counter? Explain the $\times$ and O notation.

9. In the control circuit shown in Figure 10-12, which of the following describes when the timer clutch is energized?
   a. Relay 1CR is de-energized
   b. Relay 2CR is energized
   c. Pressure builds, operating pressure switch 1PS

10. In the control circuit shown in Figure 10-13, which of the following describes when solenoid 2PA is de-energized?
    a. Relay 1CR is energized
    b. The timer T1 times out
    c. Pressure builds, operating pressure switch 2PS

11. For the circuit in Figure 10-12, replace the pressure switch 1PS with a limit switch 1LS that will actuate when the slide is advanced. Draw the control circuit for the modified configuration. Will the revised circuit function the same?

12. For the circuit in Figure 10-12, add a returned limit switch 2LS that will actuate when the slide is returned. Draw the control circuit for the modified configuration showing 2LS as an initial condition to cycle the machine.

13. For the circuit in Figure 10-12, add a "cycled" memory circuit that will energize when the slide is advanced and resets when the START push-button is pressed. Draw the control circuit for the modified configuration.

14. For the circuit in Figure 10-13, add returned limit switches for each slide (1LS and 2LS). Draw the control circuit for the modified configuration showing 1LS and 2LS as initial conditions to cycle the machine.

11

CHAPTER

Count Control

After studying this chapter, you should be able to:

- Explain the basic difference between an electromechanical reset timer and an electromechanical reset counter.

- Discuss how the clutch and count motor operate through one complete cycle.

- Use the counter contact symbol in each of three conditions: de-energized, counting, and counted out.

- Follow through a typical control circuit using an electromechanical reset counter.

- Discuss some of the features of a solid-state counter.

11.1 Electromechanical Control Counters

The electromechanical control counter is similar to the electromechanical reset timer, except that the synchronous motor of the timer is replaced by a stepping motor. The stepping motor advances one step each time it is de-energized. After a preset number of electrical impulses have occurred, the counter reaches the "counted out" state and the contacts may transition states (depending on the configuration of the contact).

The electromechanical counter requires a minimum of approximately 0.5 seconds OFF time between input impulses to reset.

As in the electromechanical timer, there are several output contacts. In some counters, the contacts can be adjusted to perform in a specific manner. The conditions for each contact may be set up as shown in Figure 11-1.

CONTACT OPEN	RESET	COUNTING	COUNTED OUT
O	O	O	×
	O	×	O
CONTACT CLOSED	O	×	×
×	×	O	O
	×	O	×
	×	×	O

Figure 11-1 Counter contact conditions.

With the clutch de-energized, the contacts are in the reset condition. When the clutch is energized, the contacts go to the counting condition. When the count motor has received the same number of count impulses as set on the dial, the contacts go the counted-out condition. When the clutch is de-energized, the contacts return to the reset condition.

Figure 11-2 shows a typical electromechanical counter. These devices are generally available with analog set and readout dials.

© 2014 Cengage Learning®

158

Figure 11-2 Typical electromechanical counter. *(Courtesy of Eagle Signal Controls.)*

11.2 Circuit Applications

A control circuit utilizing a reset counter is shown in Figure 11-3. In this example, a single piston-cylinder assembly is used. The piston is to travel to the right as shown until it actuates limit switch 1LS at position P1. The piston returns to position P2, actuating limit switch 2LS. The piston now travels again to the right. This reciprocating motion continues until the preset number of counts has been reached. At this point, the counter is in the counted-out condition. The piston continues to the right (past 1LS) until a work piece is engaged and preset pressure builds to the setting of pressure switch 1PS. The piston now returns to the start position.

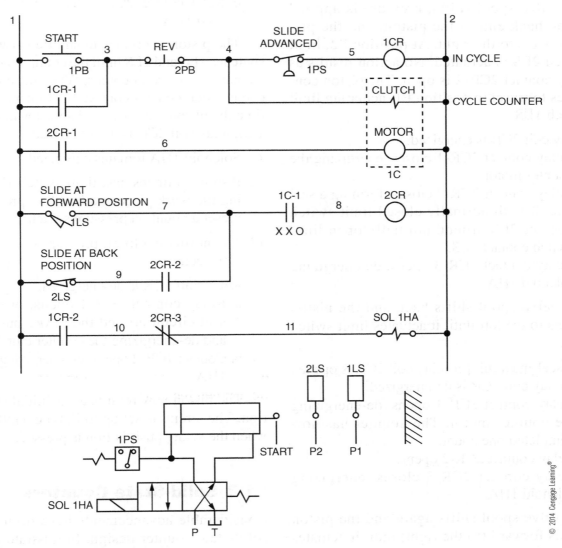

Figure 11-3 Control circuit using a reset counter.

© 2014 Cengage Learning®

A two-position, single-solenoid, spring return operating valve is used to supply fluid power to the cylinder. The circuit is arranged to return the piston to the start position at any time by pressing the REVERSE push-button.

The sequence of operation proceeds as follows:

1. Press the START push-button.

2. As long as pressure switch 1PS is not actuated, relay 1CR is energized.
   a. Relay contact 1CR-1 closes, sealing around the START push-button and energizing the counter clutch.
   b. Relay contact 1CR-2 closes, energizing solenoid 1HA.

3. The valve spool shifts, pressure is applied to the back end of the piston, and the piston travels to the right. At position P2, limit switch 2LS is actuated. No action results as relay contact 2CR-2 is open. The piston continues to travel to position P1, actuating limit switch 1LS.

4. Relay coil 2CR is energized.
   a. Relay contact 2CR-1 closes, energizing the counter motor.
   b. Relay contact 2CR-2 closes, forming a seal circuit with normally closed limit switch contact 2LS around normally open limit switch contact 1LS.
   c. Relay contact 2CR-3 opens, de-energizing solenoid 1HA.

5. The valve spool shifts back and the piston travels to the left until it actuates limit switch 2LS.

6. The seal-in circuit for relay coil 2CR is opened and relay coil 2CR is de-energized.
   a. Relay contact 2CR-1 opens, de-energizing the counter motor. The counter has now completed one count.
   b. Relay contact 2CR-2 opens.
   c. Relay contact 2CR-3 closes, energizing solenoid 1HA.

7. The valve spool shifts again and the piston travels forward (to the right) until it actuates limit switch 1LS.

8. Relay coil 2CR is energized.
   a. Relay contact 2CR-1 closes, energizing the counter motor.
   b. Relay contact 2CR-2 closes, sealing around limit switch 1LS.
   c. Relay contact 2CR-3 opens, de-energizing solenoid 1HA.

9. The valve spool shifts back and the piston travels to the left until it actuates limit switch 2LS.

10. The seal-in circuit for relay coil 2CR is opened and relay coil 2CR is de-energized.
    a. Relay contact 2CR-1 opens, de-energizing the counter motor. The counter has now completed two counts.
    b. Relay contact 2CR-2 opens.
    c. Relay contact 2CR-3 closes, energizing solenoid 1HA.

The piston continues to shuttle between limit switches 1LS and 2LS until the counter has counted the number of preset counts. When the counter has counted out, counter contact 1C-1 opens. The next time the piston advances and actuates limit switch 1LS, relay coil 2CR is not energized.

11. Solenoid 1HA remains energized.

12. Piston continues past limit switch 1LS, meeting the work piece and building pressure to a preset amount on pressure switch 1PS.

13. The normally closed pressure switch contact 1PS opens.

14. Relay coil 1CR is de-energized.
    a. Relay contact 1CR-1 opens, opening the seal circuit around the START push-button and de-energizing the counter clutch.
    b. Contact 1CR-2 opens, de-energizing solenoid 1HA.

The piston now returns to its initial start position. The circuit is set up to initiate another cycle when the START push-button is pressed.

11.3 Solid-State Counters

Considerable advancements have been made in solid-state counter design. Solid-state counters are available with digital set and digital readout.

High-speed pulse operation with 100% accuracy is also available.

Figure 11-4A shows a typical solid-state counter. It is a single preset counter with many programmable features. Two input channels (A and B) allow a variety of inputs. Three counting modes can be programmed (count directional, add/subtract, quadrature). This unit is able to count up from zero or count down from the set point. The counter can be programmed to give a pulse at count-out and automatically reset, or it can be programmed for single-cycle latch operation. The counter is programmed with six dip switches located on the side of the counter.

The counter shown in Figure 11-4B uses complimentary metal-oxide semiconductor (CMOS) integrated circuits for the counting function. Counter output action occurs when the count total indicated by the front-mounted thumbwheel switches is reached.

Referring to the terminal arrangements shown in Figure 11-4C, the counter sets to the selected thumbwheel setting when power is applied to

Figure 11-4A Typical solid-state counter. *(Courtesy of Eagle Signal Controls.)*

Figure 11-4B Solid-state counter with CMOS integrated circuits. *(Courtesy of Eagle Signal Controls.)*

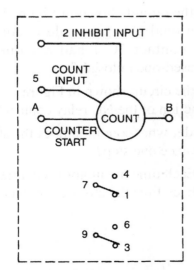

NOTE: RELAY OUTPUTS AND TRIAC OUTPUT ARE NOT AVAILABLE SIMULTANEOUSLY.
RELAY OUTPUTS—STANDARD
START INPUT—CLOSE TO START
             OPEN TO RESET
COUNT INPUT—COUNTS ON SWITCH
             OPENING
INHIBIT INPUT—CLOSE TO STOP
             COUNT PROGRESS
             OPEN TO RESTART
             COUNT PROGRESS

Figure 11-4C Wiring connections for counter shown in Figure 11-4B. *(Courtesy of Eagle Signal Controls.)*

terminals A and B. Counts are applied to a count input terminal, and each count is registered on contact opening. When registered counts equal the set point, the output changes state. The output remains in this state as long as the line voltage is applied to terminals A and B.

A count inhibit is available with this counter. When line voltage is applied to the count inhibit, incoming count pulses are not counted. The counter remembers count total at the time when the inhibit is applied and resumes counting from that point after the inhibit voltage is removed.

Referring back to Figure 11-1, this counter has four usable output sequences: ○○×, ××○, ○×○, and ×○×.

Recommended Web Links

Students are encouraged to view the following Web sites as a supplement to the concepts presented in this textbook. Review and analyze the array of products that are available for electrical control applications. Many of these sites offer technical information that can help in converting the principles to practical applications. To view catalogs, your PC may require Adobe Acrobat Reader software.

1. Eaton/Durant
   www.durant.com
   Review: Count Control/Products

2. Automatic Timing and Controls
   http://automatictiming.com
   Review: Counters

Achievement Review

1. How does the electromechanical reset counter differ from the electromechanical reset timer?

2. A given counter contact has been arranged to be opened in the de-energized condition of the counter, closed in the counting condition, and closed in the counted-out condition. Show the contact and symbol for these conditions.

3. What accuracy can you expect to receive when using a solid-state counter?

4. Given the following counter contact symbol, explain the condition of the contact in the de-energized condition of the counter, the counting condition, and the counted-out condition.

5. In the electromechanical counter, what is the condition of the contacts with the clutch de-energized?

6. What accuracy can be expected in modern solid-state counters?
   a. 75%
   b. 90%
   c. 100%

7. What are some of the features available in the solid-state timer-counters?

8. In the circuit shown in Figure 11-3, what operational change would occur if the counter contact 1C-1 had a sequence of ×○× (closed-open-closed)?

9. In the circuit shown in Figure 11-3, what is the function of the NC relay contact 2CR-3?

10. Under what condition does the stepping motor advance one step?
    a. Each time the motor is energized
    b. Each time the motor is de-energized

11. In a solid-state counter, when does the counter output action occur?

12. Explain the three output status conditions and the notation used.

13. For the circuit shown in Figure 11-3, draw the circuit to add a green pilot light that will turn on when the counter has counted out. The light will turn off when the counter is reset.

14. For the circuit shown in Figure 11-3, draw the circuit to add a "cycled" memory relay coil. The memory circuit will energize when the counter counts out and reset when the START push-button is pressed. Add a green pilot light that will turn on when the "cycled" memory circuit is energized.

15. Draw the circuit for counting the number of cars entering a parking garage. The circuit will be enabled when a two-position selector switch is rotated to the right position. A counter will increment when a vehicle passes over a surface that actuates a pressure switch. A red light will turn on when the number of cars entering the parking garage reaches the set point in the counter. The system will reset when the two-position selector switch is rotated to the left position.

12

CHAPTER

Control Circuits

After studying this chapter, you should be able to:

- Explain how all complete control circuits progress through three basic areas: information or input, decision or logic, and output or work.

- Draw a sequence bar chart for an electrical control circuit.

- Demonstrate how various electrical components are used to gather information from a machine or system.

- Demonstrate how electrical components are used to make a decision as to how the gathered information is to be used.

- Demonstrate how this decision affects the output or work the machine is to accomplish.

- Draw simple electrical control circuits from a given set of requirements.

- Describe how a seal circuit is used.

- Explain the function of the timer sequence symbol above each timer contact.

12.1 Placement of Components in a Control Circuit

Most of the components used in the electrical control of machines are discussed in Chapters 1 through 11 of this text. Figure 12-1 categorizes

some of these components based on their purpose in a control circuit. This methodology will be useful as the study of control circuits moves away from relay control systems into PLC based control systems. In this chapter, each component may be thought of as belonging to one of the three

INFORMATION OR INPUT	DECISION OR LOGIC	OUTPUT OR WORK
Push-buttons	Timers	Lights
Selector switches	Relays	Solenoids
Proximity switches	Contactors	Alarms
Limit switches	Motor starters	Heating elements
Pressure switches		Motors
Thermocouples		

Figure 12-1　Chart showing examples of division of control.

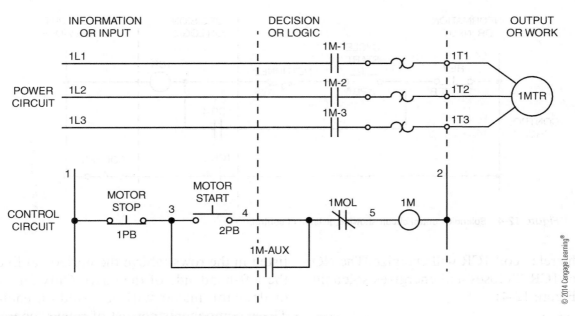

Figure 12-2 Motor starter control circuit showing division of control.

groups—input devices, decision makers, or output devices.

1. Input devices gather information from the operator, the machine, or the process.
2. The decision makers use the information obtained from the input devices (and other decision makers) to logically determine what should happen on the piece of equipment.
3. The output devices are used to perform the work that must be done as the machine sequences through its operations. This work may be in the form of motions, indicating lights, sounding horns, counting parts, and so on.

Here are three examples that show the procedure as the control moves from the *input* (information) to the *output* (work):

1. A motor START push-button is pressed, energizing a motor starter coil. The motor starter contacts then close, energizing the motor (Figure 12-2).
2. An NO thermostat contact closes, energizing the coil of a contactor. The contacts of the contactor then close, energizing the heating elements (Figure 12-3).
3. A CYCLE START push-button is pressed. If the NO limit switch 1LS is actuated (held closed),

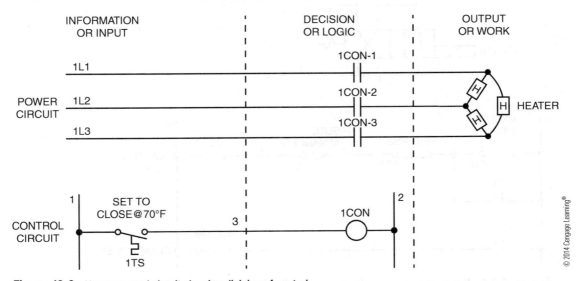

Figure 12-3 Heater control circuit showing division of control.

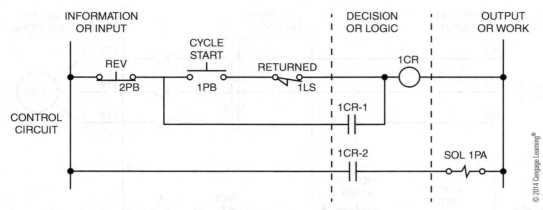

Figure 12-4 Solenoid control circuit showing division of control.

then the relay coil 1CR will energize. The NO contact 1CR-2 closes and energizes solenoid 1PA (Figure 12-4).

In the 12 circuit examples that follow, input devices are used to control output devices. For each example, a *sequence bar chart* or analysis of the circuit is shown. The bar chart is a pictorial representation that explains the functioning of the control components from input to output.

A sequence bar chart is created by drawing horizontal and vertical lines to form a matrix (or chart). The outputs and decision makers will be listed in the rows (above the horizontal lines) on the left-hand side of the chart. Only one output or decision maker will be listed on each row. These components consist of relays, contactors, motor starters, timers, and solenoids. The input devices are then listed on the vertical lines at the bottom of the chart. These devices are listed in the sequential order in which they are operated throughout the cycle. The input devices consist of switches being actuated, push-buttons being pressed, or timers timing out. These input devices changing states are the events that trigger

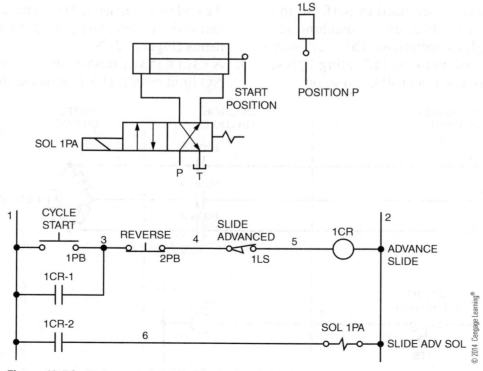

Figure 12-5A Motion control circuit with advanced limit switch.

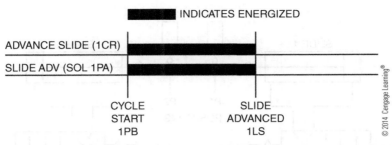

Figure 12-5B   Sequence bar chart for circuit in Figure 12-5A.

the outputs or decision makers to turn on or off. Each time an event happens causing an output or decision maker to be energized, a solid bar is drawn on the horizontal line opposite to the component being energized. When an event happens that causes an output or decision maker to be de-energized, the solid bar will no longer be drawn. Thus, at any time during a machine cycle, the energized or de-energized condition of the component may be readily observed. Checking along the vertical lines will clearly show the event that happened to cause the component to energize (or de-energize).

Figure 12-5A shows a control circuit for advancing a cylinder. The pneumatic drawing is included. At position P, limit switch 1LS is actuated, and both the coil of relay 1CR and solenoid 1PA are de-energized. Figure 12-5B is a sequence bar chart for this circuit.

12.2 Control Circuit Examples

Example #1—Figures 12-6A and 12-6B
A three-phase motor is used to drive a fluid power pump that provides fluid power pressure. A fused disconnect switch provides a disconnecting means and short-circuit protection. The control transformer has fuses in the primary winding circuit. The transformer secondary is grounded on the common side. The other side of the transformer secondary is fused. A selector switch is used for control disconnecting means. The operating cycle for the machine proceeds as follows.

The #1 piston moves from a start position to the right as shown. When the piston reaches a predetermined position P1, it stops and reverses to its start position. At the same time that #1 piston returns, #2 piston moves from a start position to the left as shown. When #2 piston reaches a predetermined position P2, it stops and returns to its start position. Pressure is available on the rod end of both pistons at the start positions. Provisions are made by pressing the REVERSE push-button to return either piston at any time during the cycle.

Example #2—Figures 12-7A and 12-7B
The same power and control circuit used for the motor and control transformer in Example #1 is used in this example. Therefore, in Example #2 only the ladder-type schematic circuit is shown.

As shown, the #1 piston moves from a start position to the right. The #2 piston moves from a start position to the left. The #3 piston moves from a start position up.

All three of the pistons will return to their start positions when any of the following conditions occurs:

- The #1 piston has engaged the work piece and built pressure to a predetermined setting
- A predetermined time setting on 1T has been reached
- The #3 piston has reached a predetermined position P3

Pressure is available on the rod end of the pistons in their start positions. Pressing the REVERSE push-button will return all pistons at any time during the cycle.

One practical advantage of this circuit is that any one of the three inputs can be adjusted

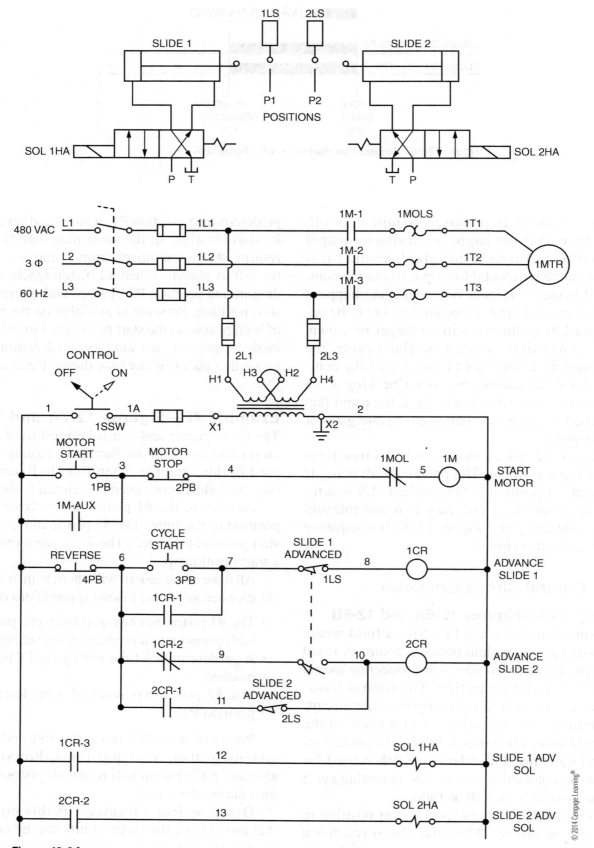

Figure 12-6A Motion control circuit for two slides with advanced limit switches.

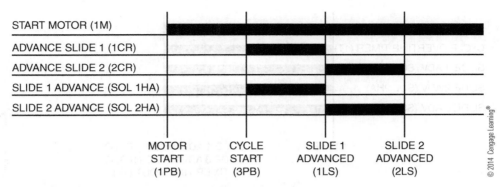

Figure 12-6B Sequence bar chart for circuit in Figure 12-6A.

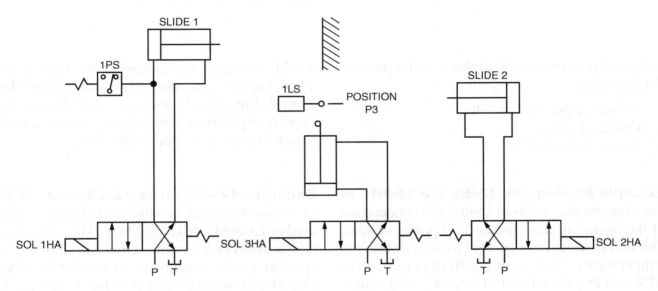

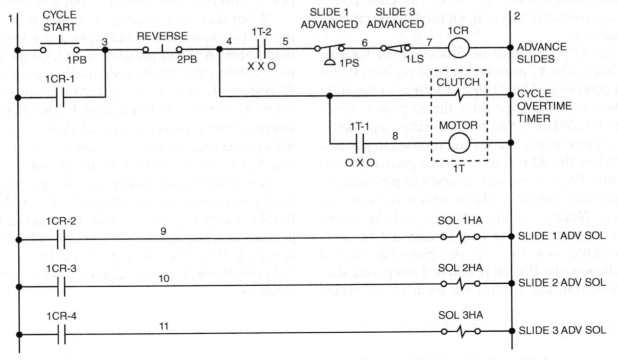

Figure 12-7A Motion control circuit for three slides, advanced limit switch, advanced pressure switch, and cycle overtime timer.

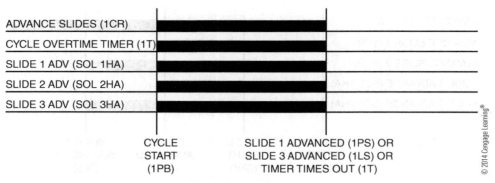

Figure 12-7B Sequence bar chart for circuit in Figure 12-7A.

independently to control all three of the pistons. For example:

1. Adjust the pressure switch.
2. Change the timing.
3. Move the limit switch.

Example #3—Figures 12-8A and 12-8B In

this example the goal is to control the temperature of the operating fluid in the fluid power system. The motor should not be able to start unless the temperature of the operating fluid is more than 70°F and less than 100°F. The motor will stop if the temperature exceeds 100°F. The same power and control transformer circuit used in Example #1 is used here.

The #1 piston must be at position P1 for start condition. The #2 piston must be at position P4 for start condition. The #1 piston moves to the right as shown. At the same time, the #2 piston moves to the left. When the #1 piston reaches a predetermined position P2, it stops and reverses to position P1. When the #2 piston reaches a predetermined position P3, it stops and reverses to position P4. No pressure can be on the pistons at the start positions. Notice that the advance and the return coils for each cylinder may or may not be energized at the same time. The two pistons are started simultaneously. If each piston had a separate start push-button, the control for each of the slides

would be completely independent from one another. This form of control is called *independent* control. Pressing the REVERSE push-button will cause both pistons to return (if they are not already returned) at any time during the cycle.

Example #4—Figures 12-9A and 12-9B

The same motor and control transformer circuit used in Example #1 is used here. Therefore, only the ladder-type schematic circuit is shown. Two cylinder-piston assemblies are used: #1 and #2. The #1 piston must be at position 1, and the #2 piston must be at position 2 for start conditions.

When the CYCLE START push-button is pressed, the #1 piston moves to the right and engages a work piece. A preset pressure is built on the work piece. When the preset pressure is reached, the #2 piston advances to the left. At the same time, the #1 piston returns to position 1. The #2 piston advances for a preset period of time. When the preset time expires, the #2 piston returns to position 2. A reset timer is used in this circuit.

Notice that these pistons are not independent from one another as was shown in Example #3. In this example, piston #2 will not advance until piston #1 has completed its motion and 1PS has actuated. This form of control is called *sequential* control since the operations occur in a given sequence.

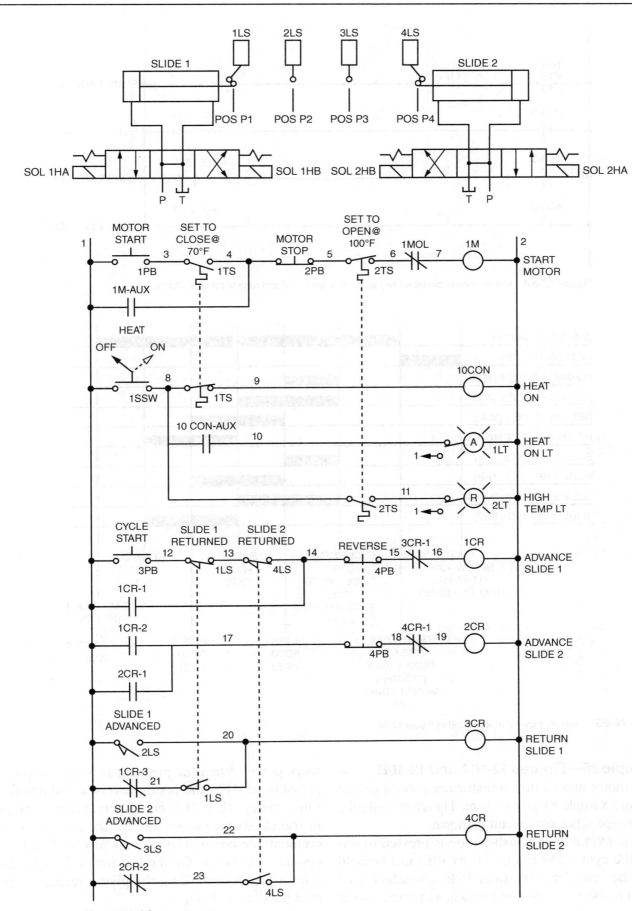

Figure 12-8A *(continued)*

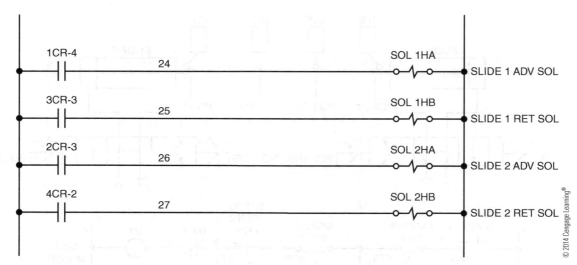

Figure 12-8A Motion control circuit for two slides with advanced and returned limit switches.

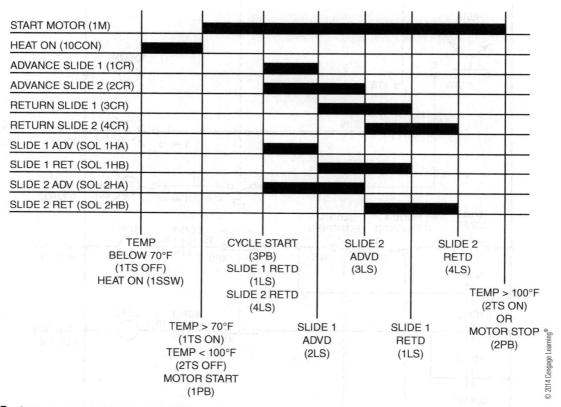

Figure 12-8B Sequence bar chart for circuit in Figure 12-8A.

Example #5—Figures 12-10A and 12-10B The
same motor and control transformer control circuit used in Example #1 is used here. Therefore, only the ladder-type schematic circuit is shown.

The CYCLE START push-button is pressed to initiate the cycle. The CYCLE START PB must be held until the time delay on timer 1TR is reached. One piston is used. The piston extends and contacts the work piece. When the pressure reaches the preset pressure on 1PS, the piston reverses and actuates limit switch 1LS. 1LS being actuated causes the piston to advance again and for the counter to increment one count. The cycle is repeated until the counter counts out. On the last forward stroke and after a preset time (2T), the piston returns to the start position and stops.

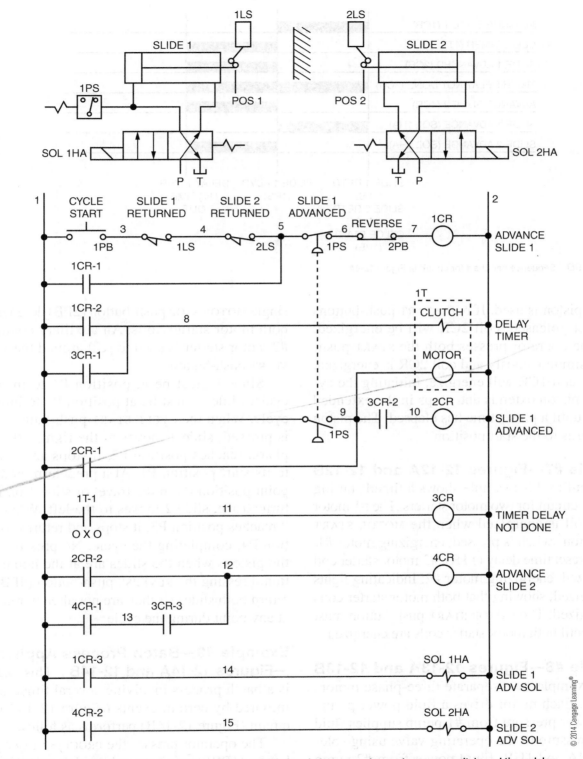

Figure 12-9A Motion control circuit for two slides with returned limit switches, one advanced pressure switch, and timer delay.

© 2014 Cengage Learning®

Example #6—Figures 12-11A and 12-11B

The same motor and control transformer circuit used in Example #1 is used here. Therefore, only the ladder-type schematic circuit is shown. This circuit is known as an *anti-tie down*; that is, it is required that both start push-buttons be operated to start a cycle. Both switches must be also released to the deactuated condition before the next cycle can be initiated.

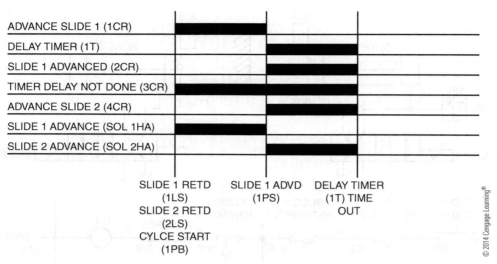

ADVANCE SLIDE 1 (1CR)
DELAY TIMER (1T)
SLIDE 1 ADVANCED (2CR)
TIMER DELAY NOT DONE (3CR)
ADVANCE SLIDE 2 (4CR)
SLIDE 1 ADVANCE (SOL 1HA)
SLIDE 2 ADVANCE (SOL 2HA)

SLIDE 1 RETD
(1LS)
SLIDE 2 RETD
(2LS)
CYLCE START
(1PB)

SLIDE 1 ADVD
(1PS)

DELAY TIMER
(1T) TIME
OUT

© 2014 Cengage Learning®

Figure 12-9B Sequence bar chart for circuit in Figure 12-9A.

One piston is used. If both START push-buttons have been released, coil 2CR will be energized. When the operator presses both the START push-buttons simultaneously and coil 2CR is energized, the relay coil 1CR will energize, initiating the cycle. The piston extends and stops in the extended position until a preset time has elapsed. The piston then returns to the start position.

Example #7—Figures 12-12A and 12-12B

The circuit in this example shows a timed starting sequence circuit for two motor starters. The #1 motor starter coil is energized when the MOTOR START push-button switch is pressed, energizing motor #1. After a preset time delay (1TR), #2 motor starter coil is energized, energizing motor #2. Indicating lights are energized, showing that both motor starter coils are energized. The MOTOR START push-button must be held until both motor starter coils are energized.

Example #8—Figures 12-13A and 12-13B

In this example, two separate three-phase motors are used. Each motor drives a fluid power pump. Fluid power pressure from #1 pump supplies fluid power pressure to the operating valve using solenoids 1HA and 1HB. Fluid power from #2 pump supplies fluid power pressure to the operating valve using solenoids 2HA and 2HB.

The motor start circuit uses only one push-button (1PB) for starting. The #2 motor starter coil is energized through the closing of an NO auxiliary contact on #1 motor starter. The operation of a single MOTOR STOP push-button (2PB) de-energizes both motor starter coils. An auxiliary contact on #2 motor starter is used to seal around the MOTOR START push-button.

Slide 1 must be at position P1 to initiate a cycle. Slide 2 must be at position P4 to initiate a cycle. When the CYCLE START push-button (3PB) is pressed, slide 1 moves to the right. When this piston reaches position P2, it stops and reverses to its start position P1. At a predetermined mid-point position P5 in the travel of slide 1 on its return stroke, slide 2 moves to the left. When slide 2 reaches position P3, it stops and returns to position P4, completing the cycle. No pressure is on the pistons when the slides are at the home position. Pressing the REVERSE push-button (4PB) will return both slides (if they are not already returned) at any point during the cycle.

Example #9—Batch Process Application —Figures 12-14A and 12-14B

This example is a batch process involving several timed events initiated by certain events (Figure 12-14A). The circuit (Figure 12-14B) performs as follows.

The operator presses the PROCESS START push-button (1PB) to activate solenoid A, which allows product A (a liquid) to enter the tank. As the liquid level in the tank rises, the float switch 1FS activates, de-energizing solenoid A and starting 1TR to time and start the mixer motor. At the end of the (1TR) timing cycle, the mixer motor will stop, timer 2TR will start timing, and solenoid B will

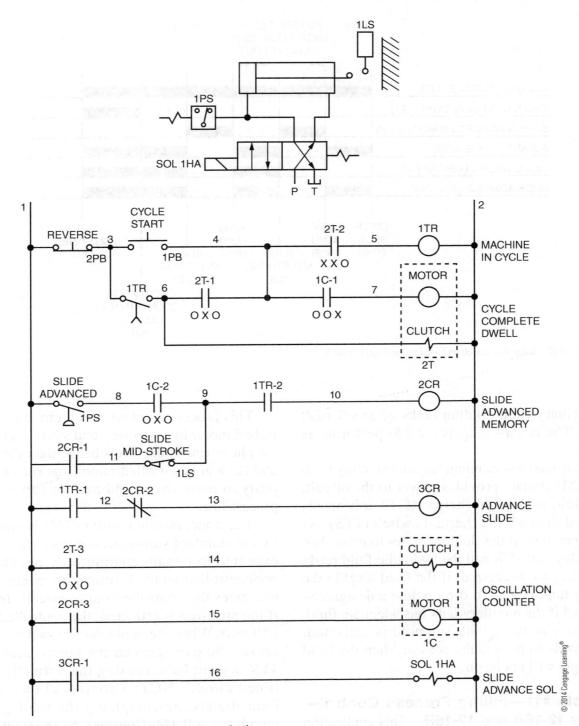

Figure 12-10A Motion control circuit using a counter and a timer.

energize allowing the product to leave the tank. When timer 2TR times out, relay coil 1CR will turn off causing the cycle to be complete and the circuit to be set up for another cycle to be initiated.

Example #10—Fluid Control in an Off-Road Vehicle—Figure 12-15 This example is an application of monitoring hydraulic fluid level and maintaining fluid temperature in both the primary

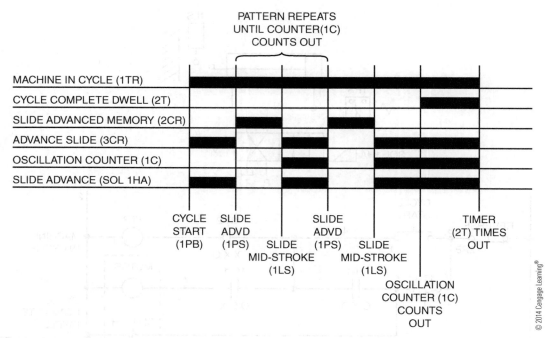

Figure 12-10B Sequence bar chart for circuit in Figure 12-10A.

and auxiliary fluid holding tanks of an off-road tractor. The circuit (Figure 12-15) performs as follows.

As the motor is running, the control relay master (CRM) contact provides power to the circuit. If the fluid is too cold (below 45°F), a heater is activated through the thermal switch (1TS). As the temperature of the fluid increases to more than 45°F, relay coil 1CR and the hydraulic fluid ready light (1LT) are energized. If the fluid level in the primary tank begins to drop below a designated level, and if the auxiliary tank has adequate fluid, the bypass primary tank solenoid is activated. If the levels in both tanks are low, then the fluid ready light will not be on.

Example #11—Filling Process Control—Figures 12-16A and 12-16B
This application illustrates the control of a filling process found in many material packaging operations (Figure 12-16A). The circuit (Figure 12-16B) performs as follows.

This process employs two motors: (1) a controlled motor that will stop and start in relation to the placement of the boxes below the filling chute and (2) a noncontrolled motor that runs continuously to move the filled boxes further down the process line.

The noncontrolled motor (2M) is controlled by the standard stop–start control. The motor is expected to remain continuously on. Once the noncontrolled motor is turned on, contact 2M-2 energizes the controlled motor circuit. Initially, if the push-rod is retracted, the controlled motor will start. When the photoelectric switch 1PED-1 senses the gap between the boxes, timing coil 1TR is energized, causing the normally closed timed closed (NCTC) contact (1TR) to open immediately, de-energizing the top leg of the circuit for coil 1M. However, NO contact 1TR-1 will keep the circuit sealed in and the motor running until the edge of a box is detected.

When the motor stops, 1M-AUX2 and 1CR-1 will both be closed, thus causing the push-rod to be

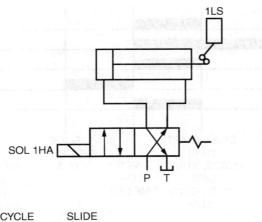

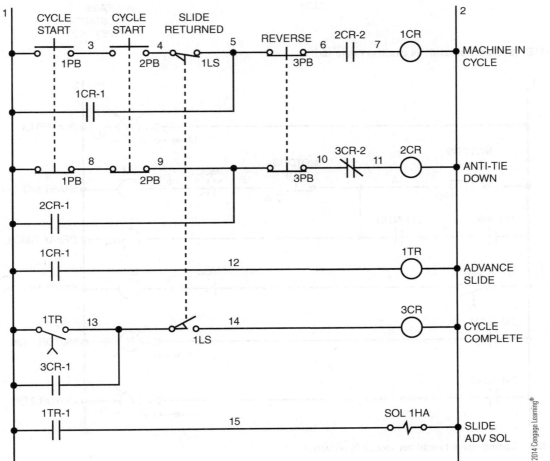

Figure 12-11A Motion control circuit using anti-tie down circuit.

© 2014 Cengage Learning®

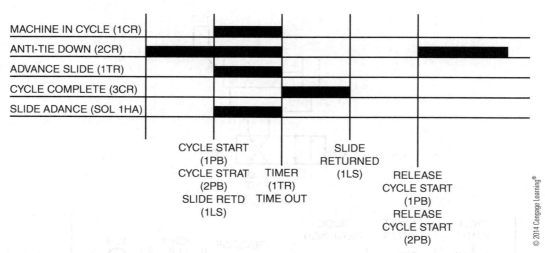

Figure 12-11B Sequence bar chart for circuit in Figure 12-11A.

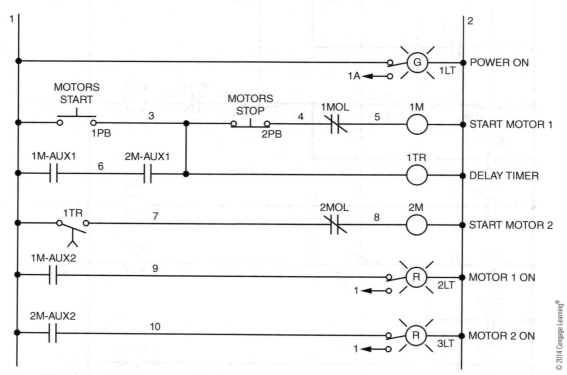

Figure 12-12A Control circuit to start two motors in sequence.

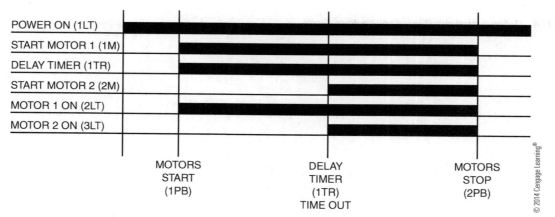

Figure 12-12B Sequence bar chart for circuit in Figure 12-12A.

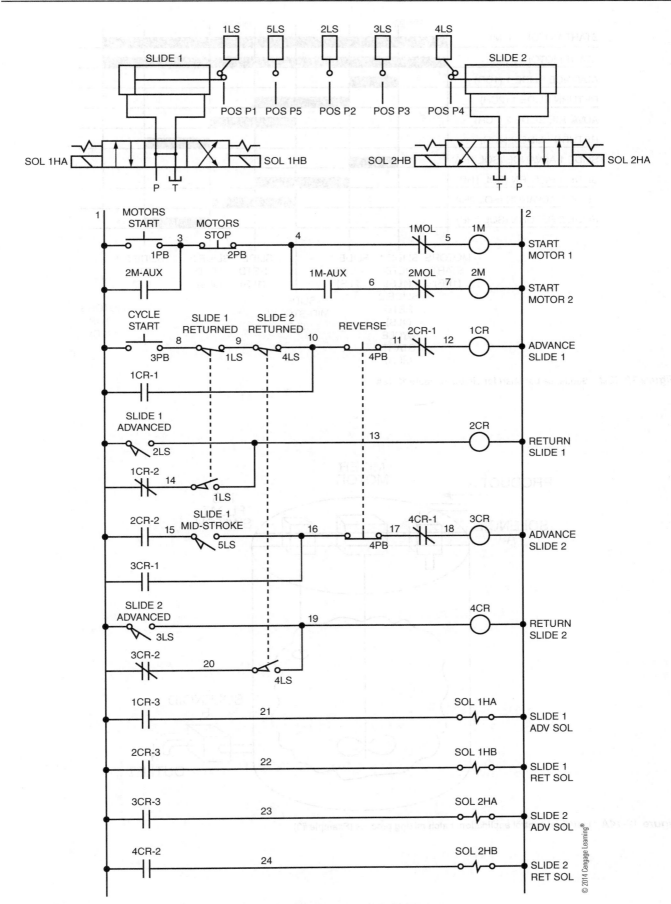

Figure 12-13A Motion control circuit with two motors, two slides, and five limit switches.

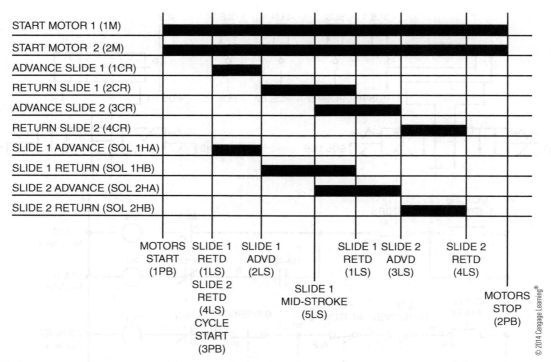

Figure 12-13B Sequence bar chart for circuit in Figure 12-13A.

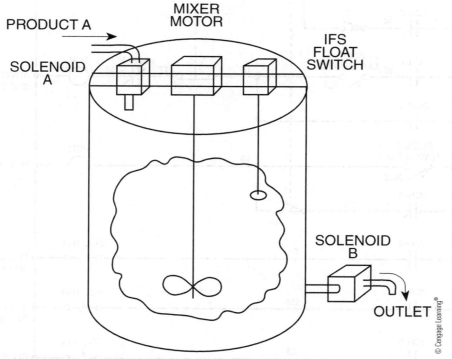

Figure 12-14A Industrial control application: batch mixing process (Example #9).

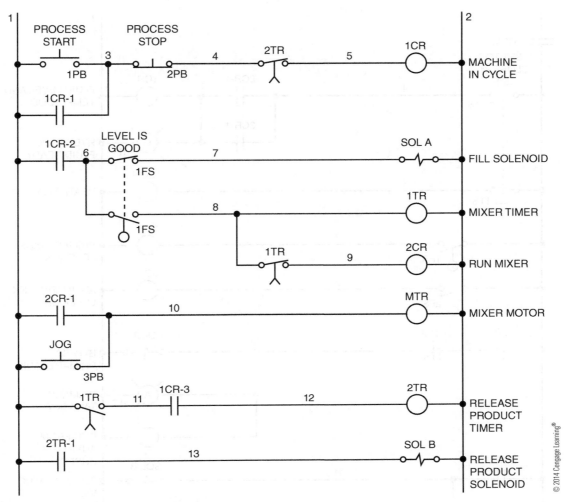

Figure 12-14B   Control circuit for batch mixing process.

extended into the box. Once the push-rod has fully extended into the box, limit switch 1LS will be actuated and the fill gate will open, causing the material to begin filling the box. As the material rises limit switch 2LS is actuated, the fill gate (SOL2PA) turns off, relay coil 1CR energizes, and the push-rod returns. When the push-rod is fully returned, limit switch 3LS is actuated, the controlled motor (1M) starts up again, and the sequence is repeated.

Example #12—Controlling the Rate of Speed and Position of an Electrohydraulic Proportional Valve
This application example illustrates the use of control circuits to switch other electronic components. In this case, the circuit will switch potentiometers that control the acceleration and deceleration ramp rates for a proportional valve. The circuit functions as follows.

Figure 12-17A shows the block diagram of the overall system. The function of the ramp module is to provide the hydraulic valve with a varying DC voltage. This voltage will cause the valve's solenoid to adjust the spool to a desired fluid flow rate. The module will accept a voltage developed from a resistance circuit, which creates a ramp rate (voltage change in time). The spool will, in turn, control the fluid flow to a cylinder. The cylinder's rod is the end-effector of this circuit.

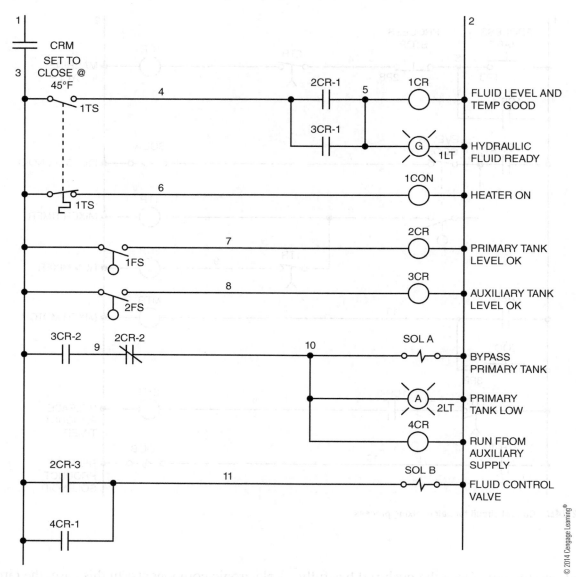

Figure 12-15 Control circuit for Example #10: fluid control in an off-road vehicle.

© 2014 Cengage Learning®

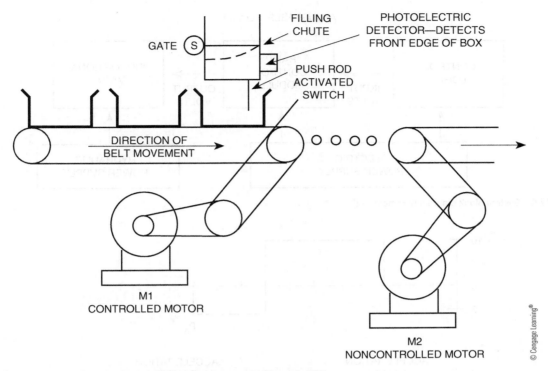

Figure 12-16A **Industrial control application: Filling process control (Example #11).**

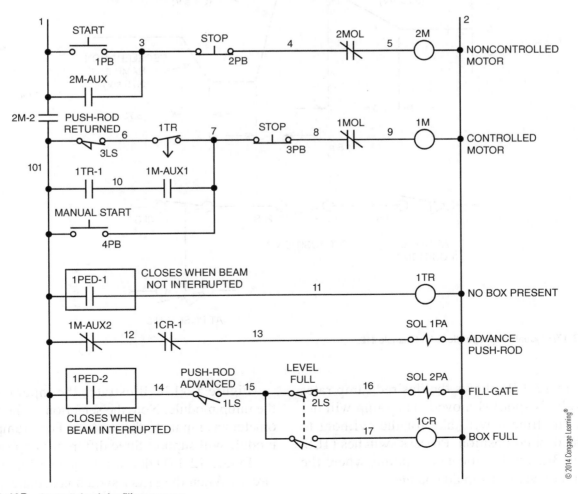

Figure 12-16B **Control circuit for filing process.**

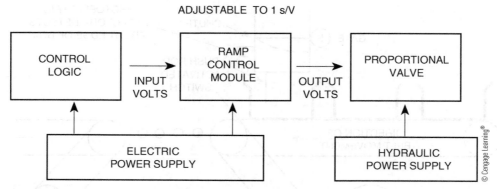

Figure 12-17A **System block diagram (Example #12).**

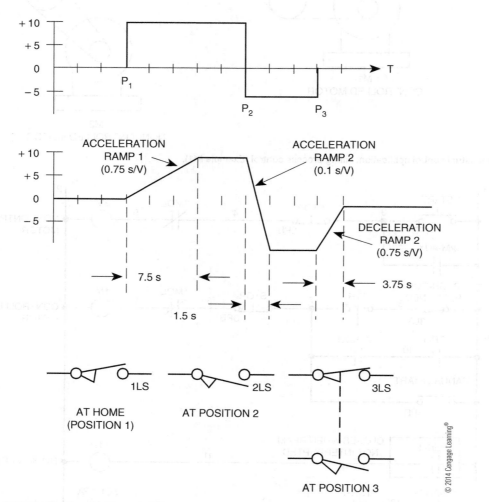

Figure 12-17B **Control specifications for Example #12.**

Figure 12-17B shows the desired ramp rates as the valve's spool is moved. The ramp will determine the time it will take for the cylinder to reach the desired position. Process switches (1LS, 2LS, and 3LS) will indicate the points where the cylinder's movement should change.

Figure 12-17C illustrates the input circuit to the ramp module. Note the selection of the potentiometers as inputs to the module. For example, this module will support three different ramp rates.

Figure 12-17D defines the ladder logic control circuit. When the START switch is actuated, 1CR-4

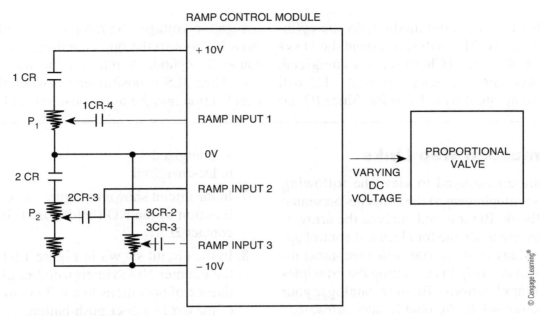

Figure 12-17C Control circuit for Example #12.

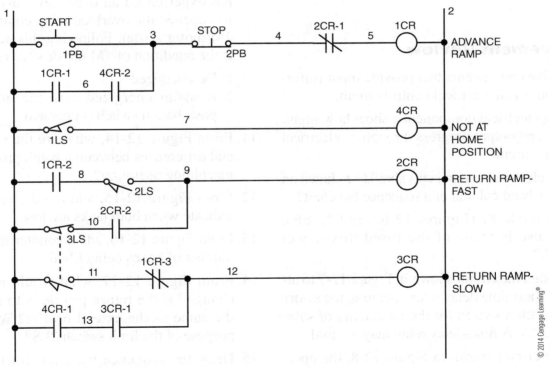

Figure 12-17D Control circuit for Example #12.

will switch P1 to the ramp module. As the cylinder begins to move, 1LS will seal around the START push-button allowing 1CR to remain energized. When the cylinder reaches position 2, 2LS will switch the ramp input from P1 to P2. Since P2 has a negative voltage, the ramp output will shift the valve's spool to the opposite direction. This shift causes the cylinder to retract at the designated ramp rate. Then 3LS is positioned to switch the ramp to deceleration into the home position, and stop.

Recommended Web Links

Students are encouraged to view the following Web site as a supplement to the concepts presented in this textbook. Review and analyze the array of products that are available for electrical control applications. Many of these sites offer technical information that can help in converting the principles to practical applications. To view catalogs, your PC may require Adobe Acrobat Reader software.

1. Entertron Industries, Inc.
   www.entertron.com
   Review: The information about ladder logic software

Achievement Review

1. List five components that provide input (information) to an electrical control circuit.

2. Using electrical components, show how input, logic, and output progress through an electrical control circuit.

3. What electrical components should be shown on the left-hand column of a sequence bar chart?

4. In Example #1 (Figures 12-6A and 12-6B), what use is made of the fused disconnect switch?

5. Redesign the circuit shown in Figure 12-7 to allow a short time delay after operating the START push-button switch for the energizing of solenoid 3HA. A time-delay relay may be used.

6. In the circuit shown in Figure 12-8, the operating fluid temperature being sensed by temperature switch 1TS is at 80°F. The heat off/on switch has been turned on. What is the condition now of contactor coil 10 CON?

a. Energized
b. De-energized

7. In the circuit shown in Figure 12-9, what is the function of the NO (held closed) limit switch contact 2LS?

8. In the circuit shown in Figure 12-10, when is the counter clutch energized? Explain the sequence of operations that follows the operation of the CYCLE START push-button.

9. In the circuit shown in Figure 12-11, what is the function of relay contact 2CR-2?

10. In the circuit shown in Figure 12-12, 2M motor has experienced an overload. This condition has opened the overload relay contacts on the 2M motor starter. Following this action, what is the condition of 1M motor starter?

a. De-energized
b. Remains energized until the MOTOR STOP push-button switch is operated

11. From Figure 12-14, what are the similarities and differences between a batch process and a machining operation?

12. From Figure 12-15, add a red alarm light to indicate when both tanks are low.

13. From Figure 12-16, add a counter to count the number of boxes being filled.

14. From Figure 12-17, what would need to be changed if the return profile is to be exactly the same as the extend profile? What is the purpose of the limit switch 2LS?

15. Draw the sequence bar chart for the circuit shown in Figure 12-14B.

16. Draw the sequence bar chart for the circuit shown in Figure 12-15.

17. Draw the sequence bar chart for the circuit shown in Figure 12-16B.

18. Draw the sequence bar chart for the circuit shown in Figure 12-17D.

19. Design a control circuit to energize three lights (red, amber, and green). The red light will turn on as soon as a START push-button is pressed.

Two seconds later, the amber light will turn on. The green light will turn on two seconds after the amber light. All lights will remain on until a STOP push-button is pressed.

20. Draw the sequence bar chart for the circuit described in problem 19.

Motors and Drives

After studying this chapter, you should be able to:

- Name the important components of a DC motor
- Explain the principles of operation for a DC motor
- Know the basic types of DC motors
- Explain the characteristics of each type of DC motor
- Describe the basic operation of a DC drive
- Explain the principles of operation for an AC motor
- Name the important components of an induction motor.

- Explain *slip* and its importance in the operation of a three-phase induction motor.
- Describe the basic operation of a variable frequency drive (VFD)
- Explain the major difference between a polyphase motor and a single-phase motor.
- List four different types of single-phase motors.
- Explain why the shaded-pole single-phase motor is different from the other single-phase motors.
- Become acquainted with the operation of a brushless DC motor

13.1 DC Motors—Principles of Operation

Motors produce work by using electrical energy to cause motion. When electricity is supplied to a motor, a shaft begins to rotate. If the shaft is coupled to a mechanical load, work will be performed by moving the load. Motors operate based on the principle of *electromagnetism*. That is, when a current is passed through a conductor a magnetic field is created. When the current flow is stopped, the magnetic field will collapse. The strength of the magnetic field will be increased if the conductor is looped into a coil configuration, if the amount of

current is increased, if the number of turns or coils is increased, or an iron core is placed in the center of the loop. Motors utilize all of these factors to increase the magnetic field and to control the motor.

Consider the diagram, shown in Figure 13-1, to illustrate the basic principle of operation for a DC motor. In this example, a single loop of wire is attached to a commutator to form a simplified armature. The *commutator* is two half circles that are constructed from copper. The *brushes* are spring loaded and sit against the commutator as it rotates. The contact between the brushes and the commutator allow a path for current to flow from

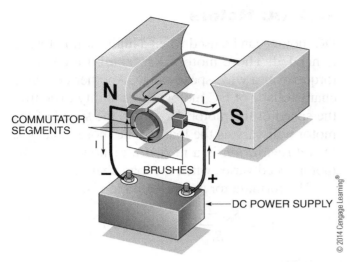

Figure 13-1   Single loop armature.

the power source through the armature. The *stator* is represented by the two magnetic poles.

Since the system shown is a closed circuit, a current will flow from the power supply through the brushes, commutator, and loop assembly. This flow of current through the loop assembly will create a magnetic field around the wires as shown in Figure 13-2. The repelling forces between the magnetic field generated around the wires and the magnetic field produced by the stator poles will generate a force that causes the armature to rotate. This rotating effect created by the interaction between the magnetic fields is called *torque*.

As the wire loop rotates, the commutator will rotate also. For every 180 degrees of rotation, each brush will switch the commutator segment that it was riding against. This switching of

commutator segments causes the direction of current flow to remain the same as the wire loop rotates within the field. Therefore, the magnetic flux created around the wire loop will also remain in the same direction. This will cause the rotating force to continually be applied to the armature in the same direction and the motor to continue to rotate in the same direction. More torque can be created by increasing the amount of current in the wire or by increasing the magnetic field of the stator. More torque can also be created by increasing the number of turns of wire per loop and by adding more loops to the armature assembly as shown in Figure 13-3. Increasing the number of loops also creates a more consistent interaction between the magnetic fields allowing the motor to continue its rotation. Finally, the armature windings are wrapped around slots in an iron core to increase the magnetic capabilities of the armature. An armature assembly will consist of the iron core, the armature windings, and the commutator.

The magnetic field in the stator of a DC motor is created in the pole pieces. The pole pieces on a permanent magnet motor are constructed using permanent magnets. Since the magnetic field in this type of motor is fixed, the only method of controlling this motor is through making adjustments to the armature.

Other types of DC motors use coils of wires wrapped around the pole pieces to create the magnetic field in the stator. These coils of wire are

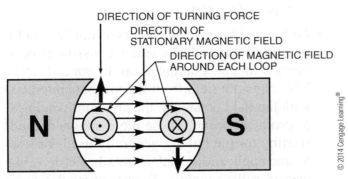

Figure 13-2   Flux lines in the same direction repel each other, and flux lines in the opposite direction attract each other.

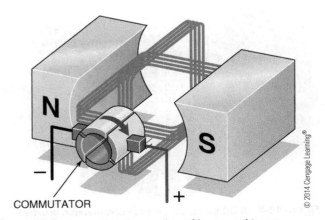

Figure 13-3   Increasing the number of loops and turns increases the torque.

called the *field windings*. The field windings can be self-excited or they can be excited from an external source. As electrical current flows through the field windings, a magnetic field is produced that will interact with the armature field to produce rotation of the motor.

DC drives control a DC motor by modifying the signal that is sent to the armature and to the field windings. For permanent magnet DC motors, a signal may only be sent to the armature. Reversing of the motor may be accomplished by reversing the polarity of the signal sent to the armature or the polarity of the signal sent to the field, but not both signals. When one of the signals is reversed in polarity, the direction of current flow will be in the opposite direction, resulting in a reversed magnetic field, and a reversed direction of rotation for the shaft of the motor.

A DC motor will generally require more maintenance than an AC motor. The brushes ride against the commutator to provide current to the armature. Springs push against the brushes so they maintain contact with the commutator as it rotates. The brushes are made of a softer material than the commutator (typically carbon or copper), so they will be the wear point in the system. Adjusting the location of the brushes, adjusting the spring tension on the brushes, and replacing the brushes are common issues when working with DC motors.

See Figure 13-4 for a cutaway of a typical DC motor.

Figure 13-4 **A DC motor with visible commutator, brushes, armature windings, and field windings.** *(Courtesy of Baldor Electric Company.)*

13.2 DC Motors

DC motors can be used when a high starting torque is needed. These motors also exhibit a constant torque over a wide speed range. Another desirable characteristic of DC motors is the ability to control the speed of the motor. However, the speed of the motor will vary as the load on the motor changes. *Speed regulation* is a measure of how much the motor speed varies when the load changes.

The formula for speed regulation is:

$$\%R = \frac{S_{NL} - S_{FL}}{S_{FL}} \times 100$$

where:

S_{NL} = no load speed (in rpm)
S_{FL} = rated full load speed (in rpm)

The principle of electromagnetic induction states that any time a wire moves through a magnetic field a voltage will be induced in the wire. When the motor rotates, the armature windings move through the magnetic field generated by the field poles. Therefore, a voltage is induced in the armature windings. This induced voltage is called the counter-electromotive force (CEMF). From Lenz's Law, we know that the polarity of the CEMF will be in the opposite direction of the polarity of the voltage source. Therefore, the current flow through the armature will be limited.

13.2.1 DC Motor Classifications
The speed and torque characteristics of a DC motor are determined by the way the armature and the field are connected. The basic types of DC motors are permanent magnet, series wound, shunt wound, and compound wound.

- *Permanent magnet* DC motors do not have field windings. Instead, the stator consists of permanent magnets mounted on the stator frame. The magnetic field from the magnets interacts with the field created in the armature windings to produce torque. These motors have a good starting torque but are generally used for low torque applications such as small electric vehicles or trolling motors. The speed regulation is not very good but may be improved upon with modifications to the design. Since the field for this motor is fixed, the motor is controlled by varying the voltage and current to the armature. The direction of rotation may be reversed by switching the leads to the armature. These motors generally experience lower maintenance problems than other types of DC motors.

• *Series wound* DC motors are wired so the field windings are connected in series with the armature windings as shown in Figure 13-5A. The field coils are wound using large diameter wire with only a few turns. Therefore, the resistance of the field windings is low. This motor has very high starting torque characteristics and is used for applications such as trains, cranes, and hoists. This motor is not used to handle light loads since the speed of the motor increases significantly as the load is reduced. Series wound DC motors are very fast at no load condition and slow down significantly at high load conditions. Large motors can rotate so fast that if they lose their load they can self-destruct. Therefore, they must be mechanically coupled to the load without the use of belts or chains.

• *Shunt wound* DC motors are wired so the field windings are connected in parallel with the armature windings as shown in Figure 13-5B. The field coils are wound using small diameter wire with many turns. The small wire cannot handle a lot of current. However, the large number of turns allow for a higher magnetic field to be produced. A majority of the current flows through the armature. This motor does not have very good starting torque so it must be used for lighter loads such as belt drives or adjustable speed machine tools. The speed does not vary much between full load and no load so shunt wound DC motors are sometimes referred to as constant speed motors.

• *Compound wound* DC motors have two sets of field windings. One set is connected in series with the armature and one set is wired in parallel with the armature as shown in Figure 13-5C. This configuration utilizes the speed characteristics of the shunt wound motor and exhibits better starting torque from the addition of the series field.

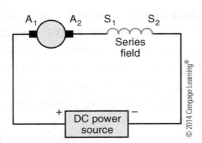

Figure 13-5A Series motor connections.

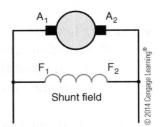

Figure 13-5B Shunt motor connections.

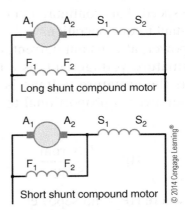

Figure 13-5C Compound motor connections.

13.2.2 DC Motor Speed Control The speed of a DC motor may be controlled by varying the voltage applied to the armature windings or to the field windings. The *base speed* of a DC motor refers to the speed at which the motor will rotate if full voltage is applied to both the armature and the field. Armature control is used at speeds below the base speed and field weakening is used to control the motor rpm at speeds higher than the base speed.

13.2.2.1 Armature Control Armature control is used to vary the motor rpm for speeds below the base speed of the motor. A constant maximum voltage is applied to the field windings throughout this mode of operation. As the voltage applied to the armature increases, the speed will also increase.

The output torque of a DC motor is proportional to the product of the main pole flux, armature current, and a machine constant that is a function of the armature winding. Thus,

$$T = K\varphi I_a$$

where T = torque
K = machine constant
φ = main pole flux
I_a = armature current

Therefore, with constant shunt field excitation, the torque is dependent on armature current only. A DC motor operated in this mode will develop rated torque at rated armature current independent of the speed. This configuration is commonly called constant torque operation.

Horsepower, at constant current, depends on applied armature voltage and, as a result, will increase as the armature voltage is increased. Since horsepower is proportional to torque and speed,

$$\text{HP} = \frac{T \times \text{rpm}}{5252}$$

where HP = horsepower
rpm = revolutions per minute
T = torque in lb-ft

From this equation, it may be seen that under constant torque conditions, the horsepower will vary directly with the speed of the motor.

In summary, increasing the armature voltage while maintaining a constant voltage to the field, will result in an increase in both the speed of the motor and the horsepower of the motor. Refer to the top graph in Figure 13-6 for an illustration

of the constant torque mode of operation for a DC motor.

13.2.2.2 Field Weakening Field weakening is used to vary the motor rpm for speeds above the base speed of the motor. Armature control is used to operate the motor below base speed. Once the base speed of a motor is obtained, a further increase in the armature voltage will not result in an increase in rpm of the motor. The speed of the motor may be increased by reducing the amount of excitation applied to the field windings. This process is commonly referred to as field weakening. Reducing the shunt field voltage decreases the field current, which, in turn, reduces the main pole flux. Since the stator field is weakened the amount of CEMF produced by the armature is reduced. This reduction in CEMF allows the motor to rotate at higher speeds. However, there is a trade-off in torque as the speed is increased above the base speed. As was indicated previously, torque is proportional to the product of flux, armature current, and the machine constant by the equation

$T = K\varphi I_a$. Therefore, at constant armature current, the torque decreases as the flux is reduced to obtain higher speeds.

The motor horsepower output at constant or rated armature current is essentially constant as the field voltage or flux is varied. Therefore, this configuration is commonly called constant horsepower operation.

In summary, decreasing the field voltage will result in an increase in the speed of the motor and a decrease in the torque of the motor. Refer to the bottom graph in Figure 13-6 for an illustration of the constant horsepower mode of operation for a DC motor.

13.2.3 DC Drives Adjusting the speed of a DC motor may be accomplished through various methods. A simple method is to use an adjustable resistance in either the armature or field windings. Armature control or field weakening may be accomplished by varying the amount of resistance

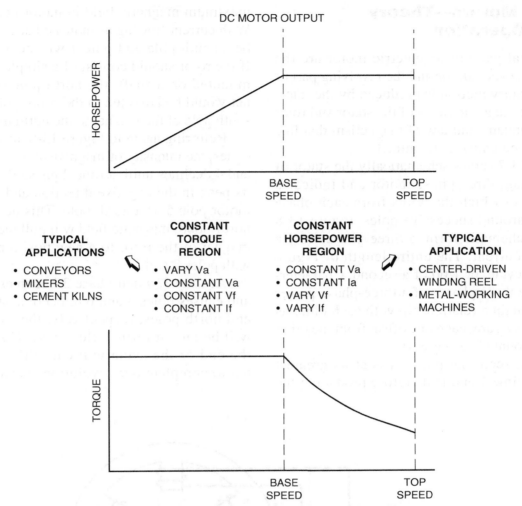

Figure 13-6 Torque and hp characteristics of constant horsepower (top), constant torque load (bottom). *(Courtesy of General Electric Company, General Purpose Control, Bloomington, IL.)*

in the motor circuit. DC drives may be used to provide accurate control of a DC motor.

Armature control is accomplished by maintaining a constant voltage to the field. The armature voltage to the motor is controlled through the use of silicon control rectifiers (SCRs). Basically, the drive will rectify the incoming AC signal to generate a pulsating DC signal. When the SCRs are fired, the output will be applied to the motor. The drive will control the firing of the SCRs to maintain the voltage to the armature at the correct level to control the speed of the motor. Feedback devices may be used to monitor the speed of the motor. The drive will use this feedback information to modify the firing of the SCRs to accurately control the motor. If field weakening is utilized, the drive will also use this feedback information to control the output voltage applied to the field windings. Several adjustments may be made to properly set-up a DC motor drive. Typically, switch settings or jumpers may be used to properly configure the drive for the application. Potentiometers may be adjusted to set specific parameters for the application such as acceleration/deceleration rates, maximum speed, maximum current, and so on.

13.3 AC Motors—Theory of Operation*

The principal parts of an electric motor are the stationary frame or *stator* and the revolving part or *rotor*. The rotary motion is produced by the interaction of the magnetic fields of the stator and rotor under the fundamental law of magnetism that like poles repel and unlike poles attract.

Figure 13-7 shows schematically the stator of an alternating, three-phase motor and indicates the manner in which the wires from each phase are wound around successive poles. Figure 13-8 shows, in schematic form, a three-phase, 60-Hz alternating current. The entire length of Figure 13-8 is one cycle, or 1/60 of a second. Characteristically, the three phases of a three-phase power system maintain a relationship with each other as shown in this figure, each one offset from the other an equal amount (120 degrees).

It can be seen that phase A is at its greatest intensity at time 1 and is therefore producing the maximum magnetic field in motor poles 1 and 4. With current flowing as indicated at I_A, pole 1 will be a south pole and pole 4 will be a north pole. If the rotor should consist of a simple bar magnet mounted on a shaft, the north pole of the magnet would be attracted to the south pole 1 and the south pole of the magnet to the north motor pole 4.

Referring again to Figure 13-8, as phase A declines, the intensity of magnetism in motor poles 1 and 4 declines until at time 2 phase B has reached its peak in the negative direction and has excited motor pole 5 as a north pole. This new dominant north–south magnetic field will pull the simple bar magnet of the rotor to a new position to line up with poles 2 and 5.

At time 3, when phase 3 becomes maximum and causes poles 3 and 6 to become strong south and north poles, respectively, the rotor magnet will be pulled further clockwise. Continuing to the end of the cycle will cause the rotor magnet to complete one revolution. Since one cycle

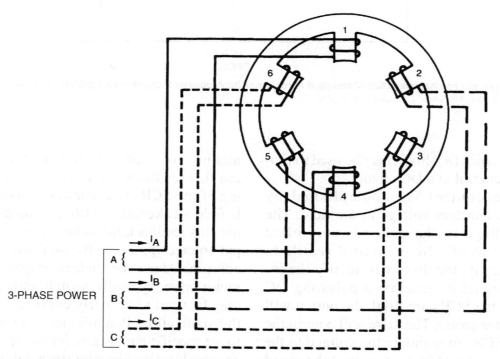

Figure 13-7 **Three-phase AC rotating field.** *(Courtesy of TECO-Westinghouse Electric Corporation.)*

*Information courtesy of Westinghouse Electric Company.

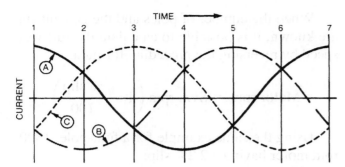

Figure 13-8 One complete three-phase (A, B, and C) AC cycle. *(Courtesy of TECO-Westinghouse Electric Corporation.)*

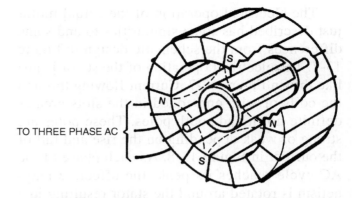

TO THREE PHASE AC

Figure 13-9 Schematic diagram of polyphase squirrel-cage induction motor. *(Courtesy of TECO-Westinghouse Electric Corporation.)*

and one revolution took 1/60 second, the speed of this motor is evidently 60 × 60 or 3600 rpm. Each phase of this motor has two poles, so it is known as a "two-pole" motor. The formula for speed is:

$$S = \frac{120f}{p}$$

where S is the synchronous speed in revolutions per minute, f is the frequency in hertz, and p is the number of poles. Therefore:

$$S = \frac{120 \times 60}{2}$$
$$= 3600 \text{ rpm}$$

13.4 Polyphase Squirrel-Cage Induction Motors

The most common AC motor is the three-phase squirrel-cage induction motor shown schematically in Figure 13-9. The stator does not have solid poles as in Figure 13-7. Solid iron poles would overheat and give poor electrical efficiency. In actual practice the stator consists of a number of flat sheets called *laminations,* held together in a package and insulated from each other so as to reduce the circulating or eddy currents.

A typical stator lamination has a circular periphery and a round hole in the center. Notches or teeth are cut around the inner edge of the center hole. When the laminations are stacked in a

package, these notches are lined up to form continuous channels called *slots.*

Insulated copper wires are coiled and laid in these slots in such a way that the ends of the wires can be connected to a three-phase circuit to bring about an electrical and magnetic arrangement similar to the schematic as shown in Figure 13-7.

The small discs that were punched out of the center of the stator laminations are now available to become the rotor. These discs are notched around the periphery, insulated, and then pressed on the shaft in a package with the notches lined up to form slots. Copper or aluminum is shoved or cast in these slots, and the ends of the bars at each end of the rotor package are connected by fastening them to a ring of the same metal. The bars and rings give the appearance of a squirrel cage, hence the name of this motor.

The basic mechanical part is the frame into which the stator is tightly pressed. The frame has feet or other devices for firmly fastening the motor in place. Covers at each end of the frame carry the bearings that support the shaft and hold the rotor in proper relationship with the stator so as to allow smooth rotation. These covers are usually called *brackets* or *end bells.* Protection from mechanical damage, dirt, moisture, and so on, is provided by various configurations of the frame and end bells while at the same time allowing necessary air circulation for cooling.

The electrical operation of the actual motor just described has some similarities to and some differences from the schematic design in Figure 13-7. The electrical operation of the stator is just like that in Figure 13-7. Current flowing through the coils of wire embedded in the slots creates definite north and south poles. These poles are strong or weak depending on the rise and fall of the current in Figure 13-8. As each phase of the AC cycle reaches its peak, the effective magnetism is rotated around the stator resulting in a rotating magnetic field.

The electrical operation of the rotor is somewhat different. The description of Figure 13-7 refers to a simple bar magnet fastened to a shaft. Actually, the soft iron laminations of the real rotor have no permanent magnetism. It is the rotating magnetic field in the stator that induces a voltage in the squirrel cage of the rotor. This voltage causes a current to flow in the cage, which sets up effective north and south poles or magnetic fields in the rotor. It is the interaction of the stator and rotor fields that produces torque in the motor.

Continuous voltage will be induced in the rotor, provided the rotor's conductors continue to cut across the magnetic flux lines of the stator field. To bring this about, it is necessary for the rotor to travel at a speed somewhat less than the synchronous speed of the stator's rotating field. The difference between synchronous speed and actual rotor or motor speed is referred to as *slip* and is normally expressed as a percent of synchronous speed:

$$\% \text{ slip} = \frac{\text{Syn. rpm} - \text{Motor rpm}}{\text{Syn. rpm}} \times 100$$

For example, a four-pole motor might have a full-load speed of 1760 rpm, which indicates a slip of 40 rpm below synchronous speed of 1800 rpm. Using the formula, the calculation results in a slip of 2.2%.

When the number of poles and the percent slip are known, it is possible to calculate the full-load speed by reversing the preceding formula:

$$\text{Motor rpm} = \text{Syn. rpm} \left(1 - \frac{\% \text{ slip}}{100} \right)$$

Using the same example for a four-pole, 1800-rpm motor having a 2.2% slip:

$$\begin{aligned}
\text{Motor rpm} &= 1800 \left(1 - \frac{2.2}{100} \right) \\
&= 1800(1 - 0.022) \\
&= 1800(0.978) \\
&= 1760
\end{aligned}$$

The magnitude of the torque produced by the interacting magnetic fields depends on several factors, but for any given motor the torque is mainly a function of the slip. Remember that any induction motor must have slip to produce torque.

13.5 Variable Frequency Drives*

Motor starters can be used to start and stop three-phase AC motors. Many examples of circuit designs using motor starters are provided in Chapter 14. Motor starters work well for simple ON/OFF type of motor control. However, the use of a variable frequency drive (VFD) will allow for accurate speed control, energy cost savings, acceleration/deceleration ramp adjustments, monitoring motor information, and displaying faults. In addition, many drives may be networked to other devices, such as PLCs, to transfer information. For example, the PLC may send speed commands to the drive and the drive may send the motor speed back to the PLC. AC drives are typically programmed using an interface module that communicates directly with the drive. The user may configure the drive with information such as motor nameplate data, selected

*Courtesy of Allen-Bradley, a Rockwell International Company.

speed commands, acceleration/deceleration rates, maximum allowable current, I/O configuration, or stop commands. The parameter names and methods for entering information will vary between manufacturers. However, the same functionality will usually be available on most drives.

13.5.1 Basic Operations

A variable frequency drive converts the incoming AC power to DC (rectifier circuit and DC bus), chops up the DC signal to create three-phase AC that is used to control the motor (inverter), and monitors the output of the motor to make adjustments to the firing circuit (control). The principle of operation is based on the fact that varying the frequency of the AC signal to the motor will result in a corresponding change in the speed of the motor.

13.5.1.1 Rectifier Circuit

The rectifier circuit converts the incoming AC power to DC. The incoming AC power is typically three-phase 480 VAC. However, some drives may require other voltage levels for proper operation. In this case, an isolation transformer may be used to provide the proper voltage level for the drive and to isolate the drive circuitry from variations to the incoming power. The rectified signal will contain ripple that must be filtered by the DC bus supply circuit (Figure 13-10).

13.5.1.2 DC Bus Supply Circuit

The DC bus supply circuit filters the rectified DC voltage from the rectifier circuit. The resultant output signal will be a filtered DC signal. The purpose of the DC bus circuit is to maintain the DC bus voltage at the proper level for use in the inverter circuit. The magnitude of the DC bus voltage will be greater than the magnitude of the incoming AC power to the drive. For example, a three-phase 230 VAC input may result in a DC bus voltage of over 320 VDC. Capacitors are used to generate the DC bus voltage. Therefore, a lethal voltage may exist on the DC bus for a significant amount of time after the system is shut down. Always consult the drive literature for critical safety precautions when working on drive systems.

13.5.1.3 Inverter Circuit

The inverter circuit uses the DC bus voltage to chop up the signal and create an AC waveform that is used to control the motor. The inverter will control the voltage level, the frequency, and the current sent to the motor. These parameters will control the torque and the speed of the motor.

Pulse width modulation (PWM) is a process of controlling the amount of voltage that is sent to a motor. PWM is accomplished by using high speed switching of the DC bus voltage through the firing of insulated gate bipolar transistors (IGBTs). When an IGBT is off, no voltage is supplied. The switching ON of an IGBT generates an output voltage to the motor.

Figure 13-11 shows how the switching of IGBTs may be used to control the voltage being sent to a motor. The longer the duration of the IGBT being switched on, the wider the pulse, and the higher the DC voltage supplied to the motor. In summary, pulse width modulation varies the width of the output signal to control the amount of voltage supplied to the motor.

The inverter circuit of a VFD uses the principle of PWM to control a motor. In addition, another set of IGBTs are used to generate a negative (or inverted) output voltage. Pulse width modulation

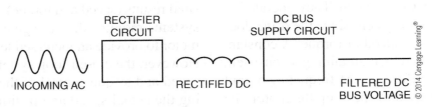

Figure 13-10 Signal conversion from AC to rectified DC to filtered DC bus voltage.

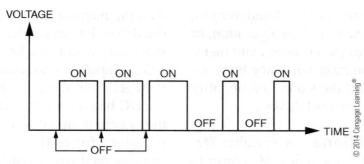

Figure 13-11 Pulse with modulation.

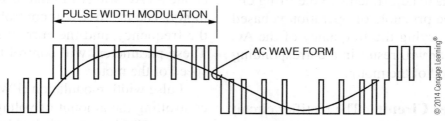

Figure 13-12 Inverter output (one phase shown).

is used to simulate an AC waveform to the motor (Figure 13-12).

Three identical circuits are used to create a three-phase output (offset by 120 degrees) to operate the motor. The frequency of the simulated AC waveform is adjusted by the inverter circuit to control the speed of the motor.

13.5.1.4 Control Circuitry
The control circuitry monitors the speed and torque of the motor and sends commands to the inverter circuit to adjust the output signal to the motor. The control circuitry also monitors the entire system and will generate a fault and potentially shut down the system if a fault condition occurs.

13.5.2 Volts/Hertz Drives
Volts/Hz drives are the simplest type of VFD system. They operate basically as an open loop system since the drive does not monitor the motor speed or torque. A constant torque is accomplished by maintaining a consistent Volts/Hz ratio. Therefore, as the frequency output of the drive is increased to speed up the motor, the voltage is also increased to maintain the Volts/Hz ratio (Figure 13-13).

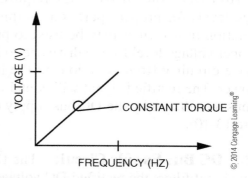

Figure 13-13 Volts/Hz drive characteristics.

13.5.3 Vector Drives
Vector drives provide better speed and torque control than the general purpose Volts/Hz drives. These drives monitor the output of the motor and will calculate the exact vector required (voltage and frequency) to obtain the desired results (speed and torque). The open loop drive systems monitor the voltage and current from the motor to provide an increased level of speed control. However, the closed-loop systems provide accurate speed and torque control of the motor by monitoring the actual speed and position of the motor. As the features and level of control of a drive system increase, so does the cost and complexity of the drive.

13.5.4 Servo Drives

Servo drives provide accurate motion control. A level of speed and torque control identical to the closed-loop vector drive is supplied on a servo drive system. Accurate position control of the motor is also obtained through the use of a servo drive. A servo drive is the most complex type of drive system available. Accurate speed, torque, and position control may be obtained. Servo systems may be used for sophisticated pick and place types of operation, for accurate control of a machining process, or for the precise control required in a test system application where many variables must be controlled simultaneously. The servo drive is the most costly type of VFD system. Typically, many parameters must be programmed into the drive to properly set up the drive for the given application. Also, servo drives must be programmed to execute a sequence of commands required to perform a specific sequence of operations. The programming language and the method for configuring the drive will vary based on the manufacturer of the drive system.

13.6 Single-Phase Motors

Most industrial applications utilize three-phase AC motors. However, a three-phase power source is not available to residential customers. Therefore, single-phase motors are primarily used for applications outside of industry. For example, the motor used to run a pump or a fan in your home will most likely be a single-phase motor.

The power supplied to industries consists of three voltages that are identical in magnitude and are 120 degrees out of phase with one another. For a three-phase motor, the 120 degree phase relationship creates a rotating magnetic field in the stator. With a single source of AC voltage connected to a single winding, a flux field is created that pulsates in strength as the AC voltage varies. However, this field does not rotate. Consequently, if a stationary rotor is placed in this pulsating stator field, it will not rotate. If the rotor is spun by hand, artificially creating relative motion between the rotor winding and the stator field, it will pick up speed and run. A single-phase, single-winding motor will run in either direction if started by hand, but it will not develop any starting torque (Figure 13-14).

Therefore, a rotating magnetic field is required to start the single-phase motor. A second winding will be required to accomplish this task. Recall from the analysis of basic electrical circuits that the current through two paths with different impedances will be out of phase with one another. This phase difference may produce a rotating field between the two windings that is adequate to start the motor.

Various combinations of resistance, capacitance, or inductance may be used to create the different impedances in the two windings. The components used and the value of the impedances will impact the characteristics of the motor. Therefore, many designs are available to optimize the performance for a given application.

One of the windings may be opened once the motor has started. This winding is referred to as the start coil. Several techniques may be used to switch the start coil out of the circuit once the motor has reached an adequate speed. Common methods utilize a centrifugal switch or starting relays.

The other winding is called the run coil. This coil is in operation at all times. It is used

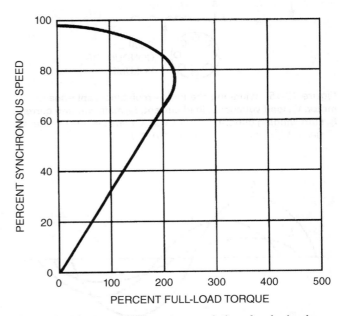

Figure 13-14 **Speed–torque characteristics of a single-phase, single-winding motor.** *(Courtesy of General Electric Company, General Purpose Control, Bloomington, IL.)*

to create a rotating field for starting and to run the motor once the start coil is switched out of the circuit.

13.7 Resistance Split-Phase Motors

The winding arrangement of a typical distributed winding resistance, split-phase motor is shown schematically in Figure 13-15.

If the main winding only is energized and the main winding current is recorded, the relationship between the supply voltage (V_L) and the main winding current (I_M) would be shown as in Figure 13-16. The main winding current lags behind the line voltage because the coils embedded in the steel stator naturally build up a strong magnetic field that slows the build-up of current in the winding.

The start winding is not wound exactly like the main winding, but its coils contain fewer turns of a

much smaller diameter wire than that of the main winding coils. This difference in windings is required to reduce the amount the start winding current lags behind the voltage if it is connected to the line.

When both windings are connected in parallel across the line, the main and start winding currents will then be out of time phase by about 30 degrees. Being out of phase produces a weak rotating flux field that is sufficient to provide a moderate amount of torque at standstill and start the motor.

The total current that the motor draws while starting is the vector sum of the main and start winding currents. Because of the small angle between these two vectors, the line current drawn during starting (inrush current) of a split-phase motor is quite high. Also, the small-diameter wire in the start winding carries a high current density, so it heats up very rapidly. A centrifugal switch and mechanism or relay must be provided to remove the start winding from the circuit once the motor has reached an adequate speed to allow running on the main winding only. The speed–torque relationship of a typical split-phase motor on both running and start connection is shown in Figure 13-17. The efficiency of the split-phase motor is moderate (50–65%). The starting torque is moderate with a high inrush current during starting.

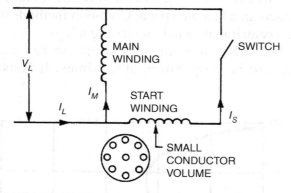

Figure 13-15 Winding schematic—resistance, split-phase motor. V_L, load voltage; I_L, load current; I_M, main winding current; I_S, start winding current. *(Courtesy of General Electric Company, General Purpose Control, Bloomington, IL.)*

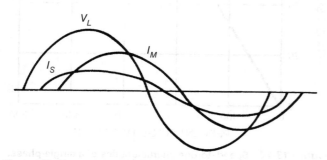

Figure 13-16 Phase relationships of Figure 13-5. *(Courtesy of General Electric Company, General Purpose Control, Bloomington, IL.)*

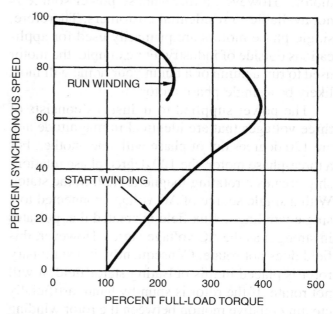

Figure 13-17 General performance characteristics of the motor in Figure 13-15. *(Courtesy of General Electric Company, General Purpose Control, Bloomington, IL.)*

13.8 Capacitor Start Motors

The capacitor start motor utilizes the same winding arrangement as the split-phase motor but adds a short time-rated capacitor in series with the start winding (Figure 13-18). The effect of the addition of this capacitor is shown in Figure 13-19.

The main winding current (I_M) remains the same as in the split-phase case, but the start winding current is much different. Because of the addition of the capacitor, it now leads the line voltage rather than lagging as does the main winding. The start winding itself is also different. It contains slightly more turns in its coils than the main windings. It also utilizes wire diameters only slightly smaller than those of the main winding.

The net result is a time-phase shift that is much closer to 90 electrical degrees than with the split-phase motor. A stronger rotating field is therefore created, and the starting torque is higher than with the split-phase design.

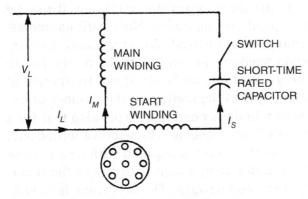

Figure 13-18 Winding schematic for capacitor start motor. *(Courtesy of General Electric Company, General Purpose Control, Bloomington, IL.)*

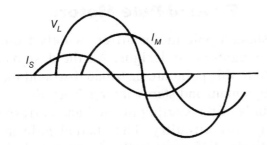

Figure 13-19 Phase relationships of Figure 13-18. *(Courtesy of General Electric Company, General Purpose Control, Bloomington, IL.)*

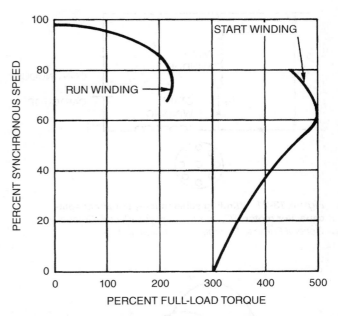

Figure 13-20 General performance characteristics of the motor in Figure 13-18. *(Courtesy of General Electric Company, General Purpose Control, Bloomington, IL.)*

The start and running speed–torque characteristics of a typical capacitor start motor are shown in Figure 13-20. Again, a centrifugal switch and mechanism or relay must be used to protect both the start winding and capacitor from overheating. When running, the capacitor start motor performs identically to the split-phase motor. The efficiency of the capacitor start motor is moderate (50–65%). The starting torque is moderate to high. The capacitor controls inrush current.

13.9 Permanent Split-Capacitor Motors

The windings of the permanent split-capacitor motor are arranged like those of the split-phase and capacitor start designs (Figure 13-21).

The capacitor is capable of running continuously and replaces the intermittent-duty capacitor of the capacitor start motor. The centrifugal switch is also removed. Once again, the main winding current remains similar to the previous designs, lagging the line voltage (Figure 13-22). The start winding current and the start winding itself, however, are somewhat different from the capacitor start design.

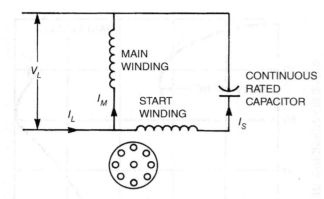

Figure 13-21 Winding schematic—permanent split-capacitor motor. *(Courtesy of General Electric Company, General Purpose Control, Bloomington, IL.)*

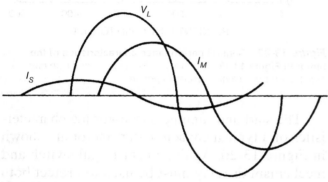

Figure 13-22 Phase relationships of Figure 13-21. *(Courtesy of General Electric Company, General Purpose Control, Bloomington, IL.)*

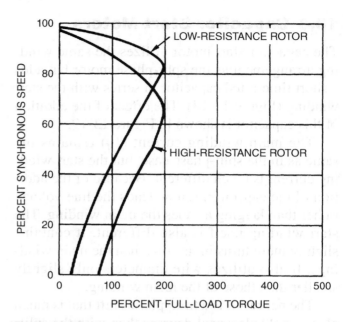

Figure 13-23 General performance characteristics of the motor in Figure 13-21. *(Courtesy of General Electric Company, General Purpose Control, Bloomington, IL.)*

Continuously rated capacitors are normally provided in small microfarad, high-voltage ratings. Therefore, the start winding is altered to boost the capacitor voltage to the correct level by adding significantly more turns to its coils than are in the main winding coils. Start winding wire size remains somewhat smaller than that of the main winding. The smaller microfarad rating of the capacitor produces more of a leading phase shift and less total start winding current. Starting torques will be considerably lower than with the capacitor start design.

The real strength of the permanent split-capacitor design lies in the start winding and the capacitor remaining in the circuit at all times. Therefore, the split-capacitor design results in better efficiency, better power factor, and lower 120-Hz torque pulsations than in the equivalent capacitor start and split-phase designs.

Figure 13-23 illustrates typical speed–torque curves for permanent split-capacitor motors. Note that varying the rotor resistance changes the shape of the speed–torque curve. Now various operating rpms can be utilized. With the addition of extra main windings in series with the original main windings, motors can be designed to operate at different speeds depending on the number of extra main windings energized. Note also that for a given full-load torque, less breakdown torque and, thus, a smaller motor is required with a permanent split-capacitor design than with any of the previously discussed designs. The efficiency is moderate to high (50–70%), and the starting torque is low to moderate (varies with rotor resistance).

13.10 Shaded-Pole Motors

The shaded-pole motor differs widely from the other single-phase designs. All the designs discussed to this point are distributed-wound motors having a main and start winding. They differ only in details of the starting method and corresponding starting circuitry. The shaded-pole motor is entirely different, both in construction and operation.

The shaded-pole motor is the most simple in construction and hence the least expensive of the single-phase designs. It consists of a run winding only plus shading coils that take the place of the conventional start winding. The construction of a typical four-pole shaded-pole motor is shown in Figure 13-24. The stator is of salient pole construction, having one large coil per pole wound directly in a single large slot. The shading coils are short-circuited copper straps wrapped around one pole tip of each pole.

The shaded-pole motor produces a very crude approximation of a rotating stator field through magnetic coupling between the shading coils and the stator winding. The placement and resistance of the shading coil is chosen so that as the stator magnetic field increases from zero at the beginning of the AC cycle to some positive value, current is induced in the shading coil. As previously noted, this current will create its own magnetic field opposing the original field. The net effect is that the shaded portion of the pole is weakened and the magnetic center of the entire pole is located at point A. As the flux magnitude becomes nearly constant across the entire pole tip at the top of the positive half-cycle, the effect of the shading coil is negligible and the magnetic center of the pole shifts to point B. As slight as this shift is, it is sufficient to generate torque and start the motor.

In Figures 13-25 and 13-26 the effect of the salient pole winding configuration on motor speed–torque output is shown. A sinusoidal distribution of winding turns produces the best efficiency and lowest noise and vibration levels. The single coil winding is the crudest possible approximation of such a distribution. Thus, shaded-pole efficiency suffers greatly in the presence of winding harmonic content, particularly the third harmonic. This drop in efficiency produces a dip in the speed–torque curve at approximately 1/3 synchronous speed (Figure 13-27). In addition, there are losses present in the shading coils. These factors combine to make the shaded-pole motor the least efficient and noisiest of the single-phase designs. The efficiency is low (20–40%) and the starting torque low (plus there is the third harmonic dip).

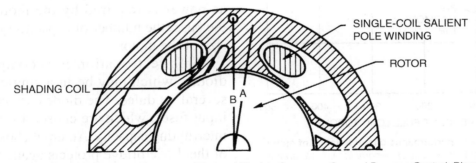

Figure 13-24 **Four-pole shaded-pole motor.** *(Courtesy of General Electric Company, General Purpose Control, Bloomington, IL.)*

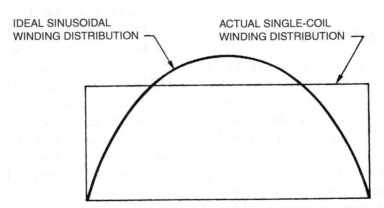

Figure 13-25 **Harmonic content resulting from single-coil winding distribution—ideal versus actual.** *(Courtesy of General Electric Company, General Purpose Control, Bloomington, IL.)*

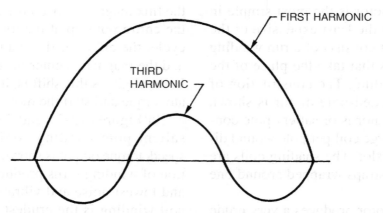

Figure 13-26 Harmonic content resulting from single-coil winding distribution showing first and third harmonic. *(Courtesy of General Electric Company, General Purpose Control, Bloomington, IL.)*

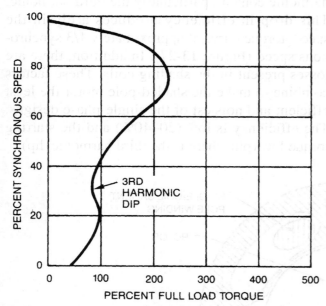

Figure 13-27 General performance characteristics of motor shown in Figure 13-24. *(Courtesy of General Electric Company, General Purpose Control, Bloomington, IL.)*

13.11 Brushless DC Motors

Brushless DC motors may be used in a wide range of applications such as computers or industrial robots. These motors do not contain a commutator or a brush assembly. Therefore, they have less maintenance issues and typically last longer than brushed DC motors.

The rotor is constructed using permanent magnets. The stator consists of windings similar to an AC induction motor. When a rotating field is created in the stator the permanent magnets in the rotor are attracted to the magnetic field. Therefore, the rotor will turn in the same direction and at the same speed as the rotating field. The brushless DC motor requires a controller to operate. Three-phase AC plant power is converted to DC by the input side of a brushless DC motor control to charge up a bank of storage capacitors (called the *buss*), whose function is to store energy and supply DC power to the power transistors in the output bridge as power is required by the motor. The size of the buss (the number of capacitors) varies with the size of the motor.

This rectification is accomplished by six diodes, which may be in a single package or in several modules. The diodes are protected by the input fuses, which are chosen for their speed and interrupting capacity. An input choke in the DC leg of the diode bridge protects against line transients and limits the rate at which input current may increase or decrease (Figure 13-28).

There are three power carrying wires going to the motor. Each of these wires has to be, at synchronized times, connected to either side of the DC buss. This is accomplished with a six transister bridge configuration (Figure 13-29).

Applying power to the motor requires turning on one transistor connected to the positive side of the buss and another transistor connected to the negative side of the buss, but never the two transistors in the same leg of the output. When the two transistors are turned on, the entire buss voltage is applied to the windings of the motor through

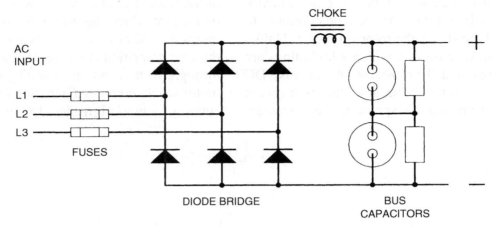

Figure 13-28 **Simplified drawing of input power section of a brushless DC motor control.** *(Courtesy of POWERTEC Industrial Motors.)*

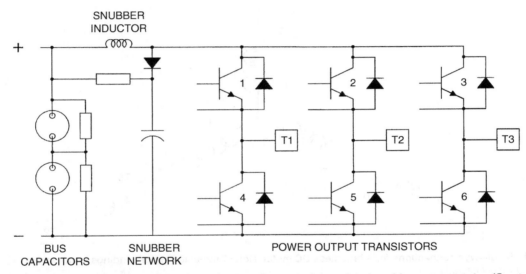

Figure 13-29 **The power output bridge consists of six power transistors and associated snubber components.** *(Courtesy of POWERTEC Industrial Motors.)*

the two wires connected to those transistors and current will flow (if the CEMF of the motor is not greater than the buss) until the transistors are turned off. Because of the inductive nature of the motor windings, the current will not cease immediately when the transistors are turned off. It will decay quickly, but voltages in the bridge would rise dangerously if the snubber network were not present to prevent this from happening.

If two transistors were turned on and left on for any length of time, the current would build to very high levels too quickly, so the transistors are turned on for only brief intervals at a time. If the motor is lightly loaded, it will not take a lot

of current (torque) to get it going. If the motor is heavily loaded, each turn-on interval will cause the current to build up until there is sufficient torque to turn the load. Once the motor has started turning, the current supplied will be, in either case, just enough to keep the motor and load turning. If a heavy load is taken off the motor, the current will quickly drop to the new level and an applied heavy load will be quickly picked up. Quick response to load changes accounts for the high efficiency of brushless DC.

Figure 13-30 is a schematic representation of a typical winding of the brushless DC motor. The connection shown in the drawing is a single-wye.

The transistors are in fired in a specific sequence to cause the field to rotate. As the field continues to move around the stator, the rotor follows it. Unlike the induction AC motor, in which all windings are continuously excited, the brushless DC stator is a DC excited field, and it moves because the windings are continuously switched in response to the movement of the rotor. The non-excited windings are carrying no current, since the windings in a brushless DC motor are either on or off. They do not go through the slow transitions that the AC motor windings go through. If the winding switching were to stop, the motor would come to a halt. See Figure 13-31 for a cutaway view of a brushless DC motor.

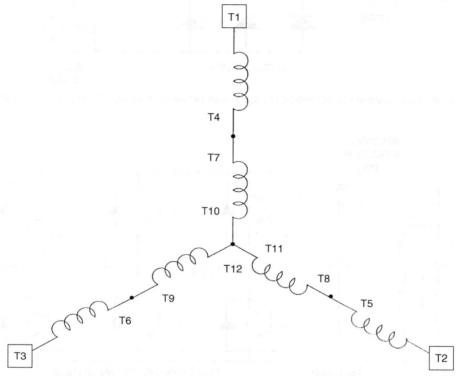

Figure 13-30 Single-wye connections for a brushless DC motor. Note: T indicates stator windings terminations. *(Courtesy of POWERTEC Industrial Motors.)*

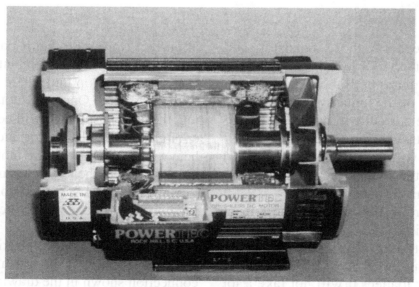

Figure 13-31 Cutaway view of a brushless DC motor. *(Courtesy of POWERTEC Industrial Motors.)*

Recommended Web Links

Students are encouraged to view the following Web sites as a supplement to the concepts presented in this textbook. Review and analyze the array of products that are available for electrical control applications. Many of these sites offer technical information that can help in converting the principles to practical applications. To view catalogs, your PC may require Adobe Acrobat Reader software.

Motors, General

1. General Electric
   www.geindustrial.com
   Review: Products/Motors

2. TECO-Westinghouse
   www.tecowestinghouse.com
   Review: Stock Motors

3. US Motors
   www.usmotors.com
   Review: Our Products/Motors

Brushless DC Motors

1. POWERTEC Industrial Motors
   powertecmotors.com
   Review: Brushless Motors

Achievement Review

1. How was the name *squirrel cage,* as applied to a three-phase induction motor, derived?

2. In a two-pole, three-phase, 60-hertz squirrel cage induction motor, what is the synchronous speed?

3. A four-pole, three-phase induction motor has a full-load speed of 1740 rpm. Calculate the percent slip from the synchronous speed.

4. Why does a single-phase motor have no starting torque if only a single winding is used?

5. Why is a centrifugal switch required in many single-phase motors?

6. How does the addition of a capacitor in a single-phase motor create a rotating field?

7. What is the difference between a capacitor start motor and a permanent split-capacitor motor?

8. Why is the efficiency of a shaded-pole motor low?

9. How are DC motors generally classified?

10. Why should a series motor be loaded at all times?

11. What is the relationship between the output torque and main pole flux, armature current, and a machine constant in a DC motor?

12. What is the relationship between hp, torque, and rpm in a DC motor?

13. How is speed regulation determined in a DC motor?

14. In the description of a brushless DC motor in this text, how is the rectification of three-phase AC power accomplished?

15. What is the difference in the rotor in a brushless DC motor and the AC induction motor?

16. What are the Hall-effect switches used for on a brushless DC motor, and how are they used on a four-pole motor?

17. In the brushless DC motor, what is speed regulation in terms of rpm difference between set-speed and actual speed?

18. If the no load speed of a DC shunt motor is 3550 rpm and the full load speed is 3450 rpm, what is the speed regulation?

19. What is the horsepower of a DC shunt motor running at 1750 rpm and having a torque of 25 ft lbs?

20. A squirrel cage induction motor with four poles per phase is connected to a 60 Hz line. Determine the synchronous speed.

21. State which type of drive (volts/Hz, vector, or servo) you would use for the following applications and why:

    a. Positioning of a slide to take measurements on a part. The part must be moved in a forward direction while test measurements are taken on the part. The speed must be accurately controlled as a function of the position of the slide so that accurate data may be captured.

    b. Operation of an exhaust fan.

    c. Drill spindle requiring accurate speed control.

14 CHAPTER

Motor Starters

OBJECTIVES

After studying this chapter, you should be able to:

- Discuss the difference between a contactor and a motor starter.

- Explain why overload relays are used on motor starters.

- Describe how the normally closed contact in the overload relay is connected into the motor starter control circuit.

- Explain how the ambient compensated overload relay operates.

- Draw the basic power and control circuit for a magnetic full-voltage motor starter.

- Explain why both mechanical and electrical interlocks are used on the reversing motor starter.

- Describe the relationship between applied voltage on a motor and the resulting torque.

- Draw the basic power and control circuit for a resistor-type, reduced-voltage motor starter and explain how it operates.

- Explain why a reduced-voltage motor starter should be used in some cases rather than a full-voltage motor starter.

14.1 Contacts and Overload Relays

The motor starter is similar to the contactor in design and operation. Both have one important feature in common: contacts operate when the coil is energized. The important difference is the use of overload relays on the motor starter.

Manual motor starters are available in which the contacts are closed by manual operation. The end results are energizing and protecting a motor. However, most machines today require additional control, safety, and convenience of remote control on one motor or multiple motors. Thus, very limited use is made of the manual motor starter on industrial machines.

The *magnetic motor starter* has three main contacts in the form used most frequently. These three contacts are normally open (NO). This arrangement is used in the starting of three-phase motors. Most industrial plants in the United States have three-phase power available as standard. Sometimes two-phase, four-wire power is used. In such cases, four NO contacts are supplied on the starter.

Like the power contactor, magnetic motor starters are available in many sizes. They start at approximately the relay size and extend up in capacity. Figure 14-1 lists motor starters, showing size, rated horsepower, and rated voltage.

The important concern with motor starters is the closing of contacts that connect the motor to the source of electrical power. Unfortunately, it is not always possible to control the amount of work

NEMA SIZE	MOTOR STARTERS HORSEPOWER		VOLTS
	SINGLE PHASE	TWO AND THREE PHASE	
00	1	1 1/2 2 2	200–230 460 575
0	2	3 5 5	200–230 460 575
1	3	7 1/2 10 10	200–230 460 575
2	7 1/2	15 25 25	200–230 460 575
3	15	30 50 50	200–230 460 575
4		50 100 100	200–230 460 575
5		100 200 200	200–230 460 575
6		200 400 400	200–230 460 575
7		300 600 600	200–230 460 575
8		450 900 900	200–230 460 575

© Cengage Learning®

Figure 14-1 **Chart of motor starter horsepower.**

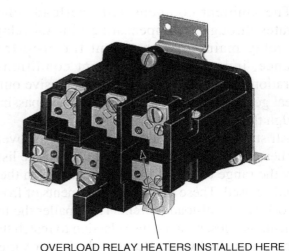

OVERLOAD RELAY HEATERS INSTALLED HERE

Figure 14-2 **Typical thermal overload relay.** *(Courtesy of TECO-Westinghouse Electric Corporation.)*

applied to the motor. Therefore, the motor may be overloaded, resulting in serious damage. For this reason, overload relays are added to the motor starter. Figure 14-2 shows a typical thermal overload relay.

The goal is to protect the motor from overheating. The current drawn by the motor is a reasonably accurate measure of the load on the motor and thus of its heating.

Thermal protected overload relays today use a thermally responsive element. That is, the same current that goes to the motor coils (causing the motor to heat) also passes through the thermal elements of the overload relays.

The thermal element is connected mechanically to an normally closed (NC) contact. When excessive current flows through the thermal element for a long enough time period, the contact is tripped open. This contact is connected in series with the control coil of the starter. When the contact opens,

the starter coil is de-energized. In turn, the starter power contacts disconnect the motor from the line.

A motor can operate on a slight overload for a long period of time or at a higher overload for a shorter period of time. Overheating of the motor will not result in either case. Therefore, the overload heater element should be designed to have heat-storage characteristics similar to those of the motor. However, they should be more sensitive enough so that the relay will trip the NC relay contact before excessive heating occurs in the motor.

The *ambient* (surrounding air) temperature in the location of the motor and starter also has some effect. It is necessary to specify the rating of a given temperature base plus the allowable temperature rise caused by the load current. For example, an open motor rating is generally based on 40°C (104°F). The motor nameplate will specify the allowable temperature rise from this base.

The motor and starter are usually located in the same general ambient temperature. Thus, the overload heater elements are affected by the same temperature conditions. The overloads will open the motor starter control circuit either through excessive motor current, a high ambient temperature, or a combination of both.

Ambient compensated overload relays are used when the control is located in a varying ambient temperature and the motor that it protects is in a constant ambient temperature.

The ambient compensated overload relay operates through a compensating bimetal relay. The relay maintains a constant travel-to-trip distance, independent of ambient conditions. Operation of this bimetal relay is responsive only to heat generated by the motor overcurrent passing through the heater element.

All starter manufacturers list the size of overload heaters for a given starter application. The lists show the range of motor currents with which they should be used. These may be in increments of from 3% to 15% of full-load current. The smaller the increment, the closer can be the selection to match the motor to its actual work. Because the heater varies with the enclosure, this information is also included.

All overload relays should be *trip-free*; that is, it should be impossible for anyone to hold down or block a RESET button, resulting in damage to the motor. After a relay has tripped, the cause of the overload should be investigated. The problem should be solved before the RESET button is depressed to put the starter back into operation.

There is one point about overload relays that is often misunderstood. Overload relays are not intended to protect against short-circuit currents. Short-circuit protection is the function of fuses and circuit breakers.

Another feature of the motor starter is the auxiliary contact normally used in the starter control circuit. In some small starters, an additional load contact can be used. It is not generally used in large starters.

Auxiliary contacts may be NO or NC and have a 10-ampere rating. In most cases, one NO contact is supplied as standard. Additional contacts (up to four) can be obtained. These can be purchased with the original starter or can be added later. A kit is available from the manufacturer containing all hardware necessary to make the installation.

14.2 Across-the-Line (Full-Voltage) Starters

The *across-the-line (full-voltage) starter* is often referred to as the "simple" type. This starter has one set of contacts that close when the starter coil is energized. When the contacts close, the motor is connected directly to the line voltage. This type of motor starter has control that is relatively simple, inexpensive, and easy to maintain.

Figure 14-3A shows a typical across-the-line motor starter. Such items as the line terminals, load terminals, motor starter coil, overload relays, and auxiliary contact can be seen. Figure 14-3B is a complete circuit diagram of a typical full-voltage starter. In this circuit, the control voltage is shown as 120 VAC. Motor starters may be purchased with different operating coil voltages. Options are available for various AC voltages (at 50 or 60 Hz) or for 24 VDC.

Notice the use of an auxiliary contact in this circuit to provide a seal around the MOTOR START push-button. In the event of an overload condition, the overload contacts will open, causing the motor starter coil to de-energize. When the fault condition is resolved and the RESET button is pressed, the motor must be intentionally restarted by pressing the MOTOR START push-button.

Figure 14-3C shows an across-the-line motor starter assembled in a NEMA 12 enclosure with a control transformer and disconnecting and protective devices. This arrangement is referred to as a *combination starter*. The NEMA 12 enclosure is designed to be oil-tight and dust-tight. (Enclosures are discussed in more detail in Appendix E.) Note that the disconnecting device is mounted in the dead front of the enclosure to allow it to be locked in the open position with the door open or closed.

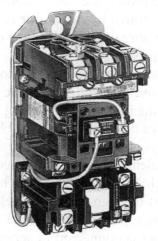

Figure 14-3A **Full-voltage starter.** *(Courtesy of Rockwell Automation, Inc.)*

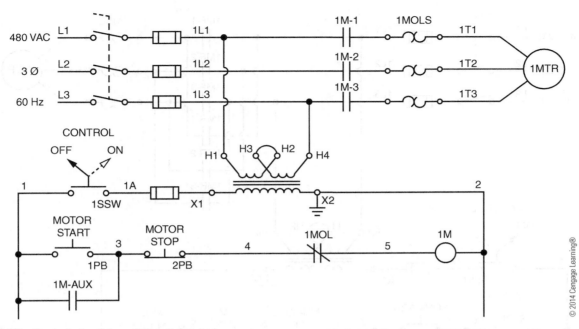

Figure 14-3B Control circuit for a full-voltage starter.

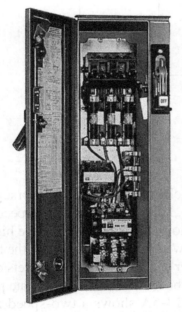

Figure 14-3C Size 1, NEMA 12 combination motor starter **(fusible).** *(Courtesy of Eaton Corporation.)*

14.3 Reversing Motor Starters

Reversing and multispeed motor starters may be considered as special applications of across-the-line starters.

In the *reversing starter,* there are two starters of equal size for a given horsepower motor application. The reversing of a three-phase, squirrel-cage induction motor is accomplished by interchanging any two line connections to the motor. The concern is to properly connect the two starters to the motor so that the line feed from one starter is different from the other. Both mechanical and electrical interlocks are used to prevent both starters from closing their line contacts at the same time. Only one set of overloads is required as the same load current is available for both directions of rotation.

A reversing starter is shown in Figure 14-4A. Figure 14-4B is the circuit diagram for a typical reversing starter. Note the phase reversal of the connections between the forward and reverse contacts in lines 1L2 and 1L3.

Figure 14-4A **Reversing starter.** *(Courtesy of Rockwell Automation, Inc.)*

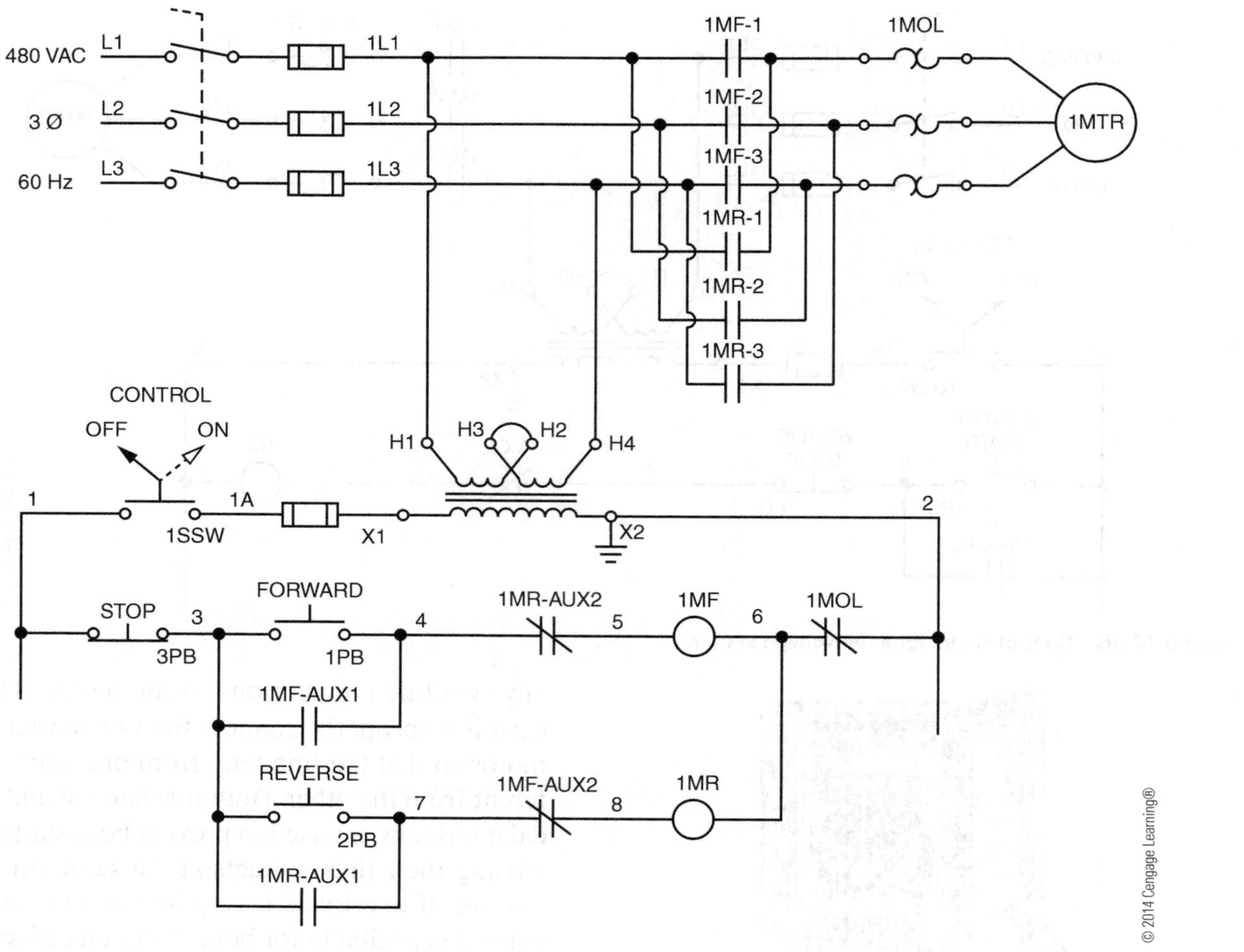

Figure 14-4B **Control circuit for a reversing circuit.**

14.4 Multispeed Motor Starters

Many industrial applications require the use of more than one speed in their normal operation. Although many methods for adjustment of speed are available, the *pole changing* method is still used. For example, in a squirrel-cage induction motor, the speed depends on the number of poles. Pole changing is obtained by the design of the stator winding. At 60 Hz, which is standard in the United States, a two-pole motor operates at approximately 3600 rpm. A four-pole operates at 1800 rpm, a six-pole at 1200 rpm, an eight-pole at 900 rpm, and a ten-pole at 720 rpm. By having two or more sets of winding leads brought out into the terminal connection box, the number of effective poles can be changed. Also, one winding can be connected in more than one way.

With a change in speed, the horsepower will also change. For example, in a two-speed motor for which the slower speed is one-half the higher speed, the horsepower will also be one-half the horsepower of the higher speed. Therefore, two sets of overload relays must be used to provide adequate protection.

Figure 14-5A shows a two-speed reconnectable multispeed starter. The circuit of a typical two-speed motor and starter is shown in Figure 14-5B.

Notice the use of two auxiliary contacts for each motor starter coil. The NO contact is used as a seal around the initiating push-button. The NC contact is used as an interlock to prevent the simultaneous operation of both starter coils. The overload contacts will open the power wiring to the motor and the control wiring to the starter coil. In the event of an overload condition, the motor will need to be intentionally restarted.

Figure 14-5A **Two-speed reconnectable multispeed starter.** *(Courtesy of Eaton Corporation.)*

14.5 Additional Across-the-Line Starter Circuits

Additional across-the-line starter circuits are shown in Figures 14-6 through 14-12. To start this group of circuits, it is important to become acquainted with two more expressions in the electrical field: *no-voltage or low-voltage release* and *no-voltage or low-voltage protection.*

No-voltage or low-voltage release means that if there is a voltage failure, the starter will open its contacts or "drop out." However, the contacts will close again, or "pick up," as soon as the voltage returns. A typical wiring diagram showing no-voltage or low-voltage release is shown in Figure 14-6A. Note that this is a wiring diagram, not a schematic or elementary diagram. The schematic diagram is shown in Figure 14-6B.

In the event of an overload condition, it will open the power wiring to the motor and the control

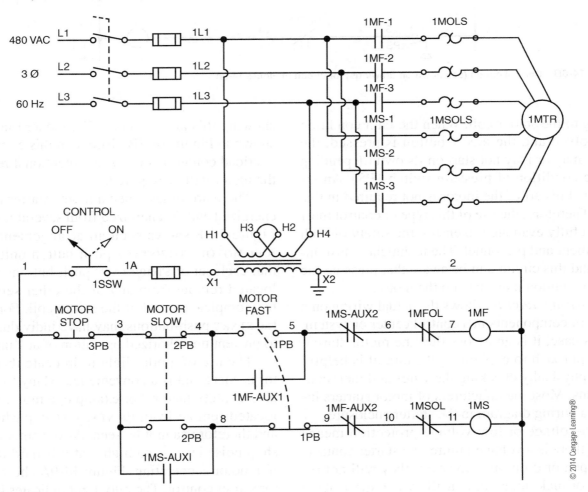

Figure 14-5B **Control circuit for two-speed starter.**

© 2014 Cengage Learning®

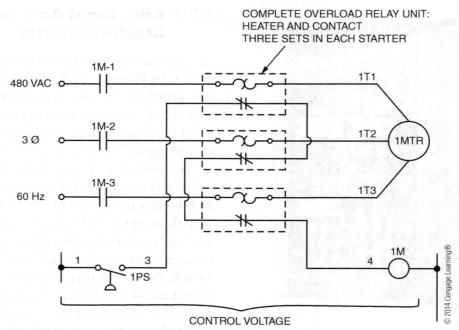

Figure 14-6A   Circuit incorporating no- or low-voltage release. Actual wiring to devices.

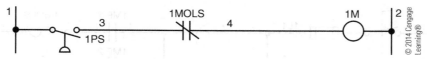

Figure 14-6B   Control circuit (as shown on drawings) for circuit in Figure 14-6A.

wiring to the starter coil. When the fault condition is resolved and the RESET button is pressed, the motor may or may not start on its own (depending on the condition of pressure switch 1PS). An intentional restart of the motor is not required in this case. Therefore, the use of this type of control must be carefully evaluated to ensure the safety of both equipment and personnel. The technician must understand this circuit and be aware that pressing the RESTART button may start-up the motor.

A *wiring diagram* shows the actual wiring on a group of components in a control center or system. In this case, it is an across-the-line motor starter. This approach to explaining the circuit is helpful when physically checking the wires and their connections. Most manufacturers of motor starters include a wiring diagram in the starter enclosure.

No-voltage or low-voltage protection means that if there is a voltage failure, the starter contacts will open or drop out. However, they will not re-close, or pick up, automatically when the voltage returns. Figure 14-7A is a typical wiring diagram

showing this arrangement. The ladder diagram is shown in Figure 14-7B. Notice in this circuit if an overload condition occurs, an intentional restart of the motor will be required.

There are times when a motor starter must be energized and de-energized from several locations. Figure 14-8 shows such an arrangement, using four MOTOR STOP/START push-button units. Note that each set of STOP/START push-buttons may be located remote from any of the other sets. Each has complete control of the starter coil. For safety, the START push-buttons may have individual locks to prevent unauthorized operation of any units.

The use of a pilot light to indicate that a motor is energized is a convenience. Many times it is also a safety factor. For example, a motor may be located remote from its START/STOP push-button and therefore cannot be seen. Also, sometimes the shop noise level is so high that it is hard to know if a motor is operating. Figure 14-9A shows a motor starter control. The pilot light indicates its energized condition (Figure 14-9A). The circuit shown

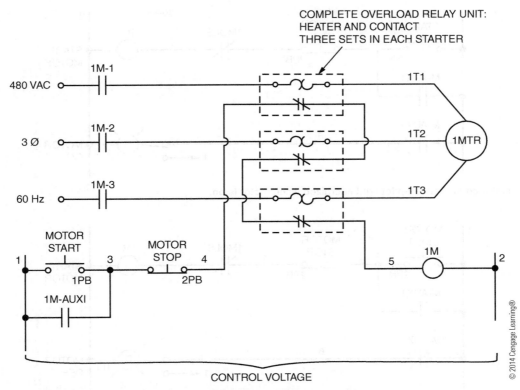

Figure 14-7A   Circuit incorporating no- or low-voltage release. Actual wiring to devices.

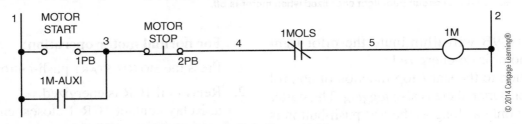

Figure 14-7B   Control circuit (as shown on drawings) for circuit in Figure 14-7A.

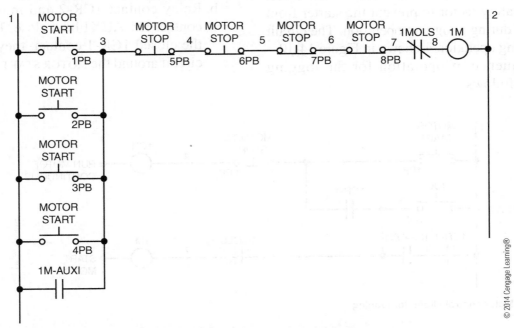

Figure 14-8   Control circuit utilizing four START/STOP push-buttons to allow operation from several locations.

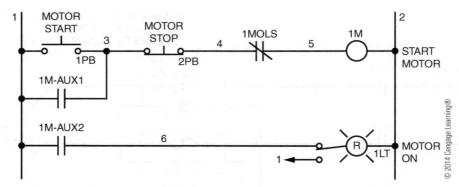

Figure 14-9A Motor control circuit-pilot light energized when motor is on.

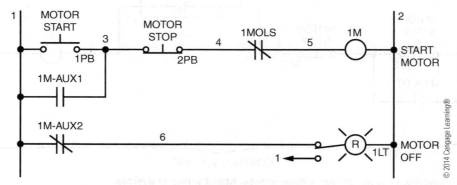

Figure 14-9B Motor control circuit-pilot light energized when motor is off.

in Figure 14-9B will illuminate the pilot light when the motor is de-energized.

In addition to the start/stop function of control for the motor starter, there is also *jogging*. The starter is energized only as long as the JOG push-button is pressed and held. The use of the relay in this circuit provides a safety factor to prevent the starter from "sealing in" during a jogging operation. The circuit for the jogging operation is shown in Figure 14-10.

The sequence of operation for the jogging circuit is as follows:

For normal motor operation:

1. Press the MOTOR START push-button.

2. Relay coil 1CR is energized.
   a. Relay contact 1CR-1 closes, energizing the motor starter coil.
   b. Relay contact 1CR-2 and motor auxiliary contact 1M-AUX1 both close. Together with the closed 1CR-1 contact, they form a seal circuit around the MOTOR START push-button.

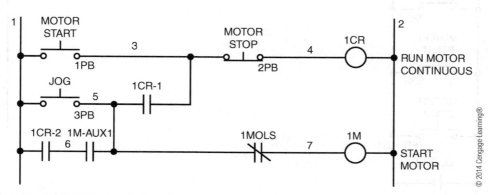

Figure 14-10 Motor control circuit for jogging.

3. Pressing the MOTOR STOP push-button de-energizes the motor starter coil 1M and the relay coil 1CR.

For jogging operation:

1. Press and hold the JOG push-button.

2. The motor starter coil 1M will remain energized only as long as the JOG push-button switch is held operated. Note that relay coil 1CR is not energized in this operation.

Many machines have multiple motors. Because of the high inrush current when starting squirrel cage motors and depending on power line conditions, it may not be advisable to start all the motors at the same time.

An arrangement of three motors starting from a single START push-button but energizing at different times is shown in Figure 14-11. Two time-delay relays are used. They can be set to start #2 and #3 motors at predetermined time intervals after #1 motor is energized. The STOP push-button will de-energize all three motors. Additional motors can be started from this arrangement by adding time-delay relays for each motor starter. They should be

added in the same way that #1 and #2 time-delay relays are used.

The sequence of operations for the circuit shown in Figure 14-11 is as follows:

1. Press the MOTOR START push-button.
2. Motor starter coil 1M is energized.
   a. Time-delay relay coils 1TR and 2TR are energized.
3. Motor starter auxiliary NO contact 1M-AUX1 closes, sealing around the MOTOR START push-button.
4. After a preset time delay as set on timing relay 1TR, timing relay contact 1TR closes.
5. Motor starter coil 2M is energized.
   a. Motor starter auxiliary NO contact 2M-AUX1 closes, sealing around the 1TR timing relay contact.
6. After a longer preset time as set on timing relay 2TR, timing relay contact 2TR closes.
7. Motor starter coil 3M is energized.
   a. Motor starter auxiliary NO contact 3M-AUX1 closes, sealing around the 2TR timing relay contact.
8. All motors are now energized.

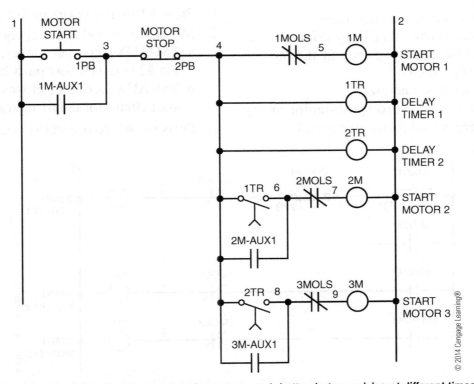

Figure 14-11 Motor control circuit. Starting three motors from same push-button but energizing at different times.

9. Operation of the MOTOR STOP push-button de-energizes all motor starter coils and the timing relay coils at any point during the cycle.

A number of motors can be sequenced into operation using the auxiliary contact of the preceding motor starter. This method gives a short, fixed time delay of only the starter closing time. The opening of any of the motor starters through an overload condition (opening overload contacts) will de-energize all the motors in sequence. In addition, all the motors will be de-energized through operation of the one common STOP push-button. This circuit is shown in Figure 14-12. Note that the START push-button must be pressed and held until #3 motor starter coil is energized and the 3M-AUX1 contact closes.

The sequence of operations for the circuit shown in Figure 14-12 is as follows:

1. Press the MOTOR START push-button.
2. Motor starter coil 1M is energized.
   a. 1M motor starter auxiliary contact 1M-AUX1 closes.
3. Motor starter coil 2M is energized.
   a. 2M motor starter auxiliary contact 2M-AUX1 closes.
4. Motor starter coil 3M is energized.
   a. 3M motor starter auxiliary contact 3M-AUX1 closes, sealing around the MOTOR START push-button.
5. All motors are now energized.
6. Operating the MOTOR STOP push-button at any time de-energizes all motor starter coils.

Additional motors can be added to this sequence circuit. The auxiliary contact on the last motor starter is generally used to seal around the START push-button.

Occasionally, it is necessary to prevent one motor from stopping until some time has elapsed after the stopping of another motor. This arrangement is used in fluid power applications in which a larger motor is driving a main pressure pump. If the motor driving the main pressure pump and a small motor driving a pilot pressure pump were de-energized at the same time, the small motor would come to rest before the main pressure pump motor. This sequence would cause a loss of control pressure on the large main pressure pump during the time the large motor was coming to rest.

Figure 14-13 shows a circuit using two motor starters with an OFF delay timer. This arrangement prevents #2 motor starter from being de-energized until after a preset time has elapsed from the time #1 motor starter is de-energized. The sequence of operations for the circuit shown in Figure 14-13 uses two motor starters with a time delay. Motor starter coil 2M must be energized first.

The sequence proceeds as follows:

1. Press #2 MOTOR START push-button.
2. Motor starter coil 2M is energized.
   a. 2M-AUX1 contact closes, sealing around the #2 MOTOR START push-button.
   b. 2M-AUX2 contact closes, setting up the start circuit for the #1 motor.
3. Press the #1 MOTOR START push-button.

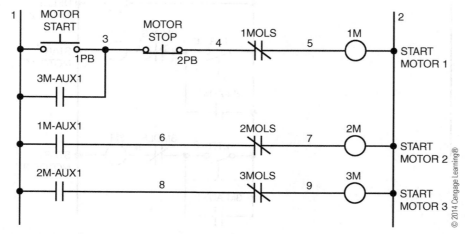

Figure 14-12 **Motor control circuit for starting three motors in sequence.**

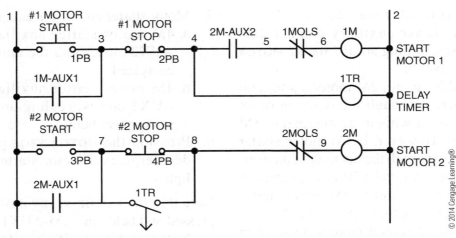

Figure 14-13   **Motor control circuit for controlling two motors. Delay circuit for de-energizing motor 2.**

4. Motor starter coil 1M is energized.
5. Time-delay relay coil 1TR is energized.
6. OFF delay timer contact 1TR closes, sealing around the #2 MOTOR STOP push-button.

Motor starter coil 1M must be de-energized before coil 2M. Pressing the #2 MOTOR STOP push-button at this point will not cause anything to happen in the circuit.

7. Press the #1 MOTOR STOP push-button.
   a. Motor starter coil 1M and time-delay relay coil 1TR are de-energized.
8. After a preset time delay as set on 1TR, the OFF time-delay contact opens.
9. Press the #2 MOTOR STOP push-button.
10. Motor starter coil 2M is de-energized.

The circuit shown in Figure 14-14 is similar to that shown in Figure 14-11. The main difference is in the operation of the MOTOR START push-button to initially energize the #2 motor starter coil 2M. In this circuit, the 2M motor starter coil cannot be energized until the preset time on timing relay has expired.

The sequence proceeds as follows:

1. Press the #1 MOTOR START push-button.
2. Motor starter coil 1M and timing relay coil 1TR are energized.
   a. 1M motor starter auxiliary contact 1M-AUX1 closes, sealing around the #1 MOTOR START push-button.
3. After a time delay set on timing relay 1TR has expired, timing contact 1TR closes.
4. Press the #2 MOTOR START push-button.

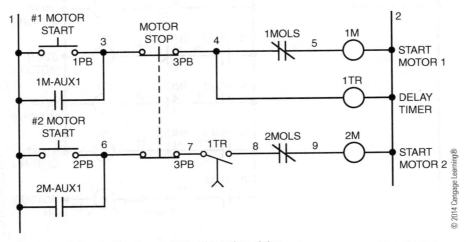

Figure 14-14   **Motor control circuit for starting two motors with a time delay.**

5. Motor starter coil 2M is energized.
   a. 2M motor starter auxiliary contact 2M-AUX1 closes, sealing around the #2 MOTOR START push-button.

Note that in this circuit, 2M motor starter coil can be de-energized through the opening of an overload relay contact without de-energizing 1M motor starter coil. However, if 1M motor starter coil is de-energized through the opening of an overload contact, timing relay coil 1TR is de-energized. This action opens the circuit to 2M motor starter coil, de-energizing the motor starter coil 2M.

The sequencing of several motors is shown in Figure 14-15.

The sequence of operations for that circuit is as follows:

1. Press the MOTOR START push-button.
2. Motor starter coil 1M is energized.
   a. Green pilot light is energized.
   b. 1M motor starter auxiliary contact 1M-AUX1 closes.
3. Motor starter coil 2M is energized.
   a. 2M motor starter auxiliary contact 2M-AUX1 closes.
4. Motor starter coil 3M is energized.
   a. 3M motor starter auxiliary contact 3M-AUX1 closes.

5. Motor starter coil 4M is energized.
   a. 4M motor starter auxiliary contact 4M-AUX2 closes and the amber pilot light is energized.
   b. 4M motor starter auxiliary contact 4M-AUX1 closes, sealing around the MOTOR START push-button.
6. Pressing the MOTOR STOP push-button will de-energize all motor starter coils and pilot lights.

Note that the MOTOR START push-button must be pressed and held until 4M-AUX1 closes.

Note that this circuit is similar to that shown in Figure 14-12. The advantage of the arrangement in Figure 14-15 is that the STOP push-button switch has complete control in opening the circuit (de-energizing) to all the starter coils. In the case of the circuit in Figure 14-12, the starters drop out (de-energize) in sequence from the preceding starter. The disadvantage of the circuit in Figure 14-15 is that it may limit the number and/or size of starters used. The current used to energize the starter coils comes through the push-button switches. Therefore, the start circuit is controlled to a degree by the current rating of the push-buttons.

The use of a MASTER MOTOR STOP push-button switch can be arranged with each motor starter

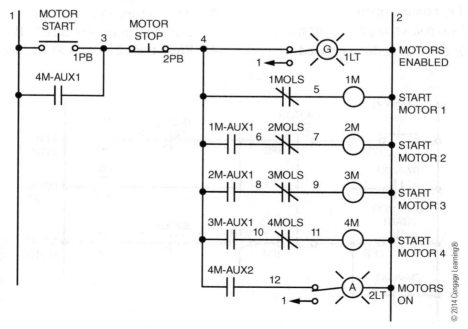

Figure 14-15 **Motor control circuit for sequencing the starting of four motors.**

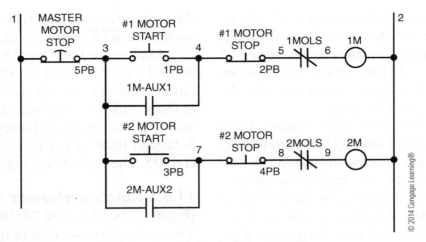

Figure 14-16 Motor control circuit for two motors. Each motor has individual START and STOP control with one MASTER STOP PB.

circuit having an individual STOP push-button. The MASTER MOTOR STOP push-button switch can be used as a convenience or as emergency safety. This switch is generally supplied with a mushroom head operator. This circuit is shown in Figure 14-16.

As a safety precaution, it is often necessary to provide an overstroke limit switch to de-energize a motor starter coil. For example, a machine part such as a moving table may travel a given distance under safe operating conditions. However, if this distance is exceeded, damage to equipment or injury to personnel may result. The limit switch is located to actuate if the proper travel distance is exceeded. When the limit switch is actuated, the motor starter coil is de-energized. The motor must then be energized on an emergency basis to back the operating dog off the limit switch. This action is accomplished by switching the power leads on the motor. The motor is reversed and moved off the limit switch by inching (momentarily pressing) the MOTOR START push-button.

The power leads to the motor are switched to resume normal operation. This circuit is shown in Figure 14-17.

14.6 Reduced-Voltage Motor Starters

Reduced-voltage motor starters can be divided into several different designs:

- Autotransformer (or compensator)
- Primary resistor
- Wye (or star) delta
- Part winding

In connecting a motor directly across the line, the resulting current at start condition may be on the order of 4 to 10 times the full-load rating of the motor. This current is sometimes referred to as *locked rotor current*. The high current often causes line disturbances if the power supply is inadequate or the motor is large.

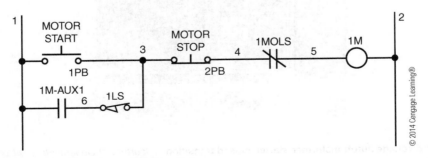

Figure 14-17 Motor control circuit with overtravel limit switch.

The basic principle of the reduced-voltage starter is to apply a percentage of the total voltage to start. After the motor starts to rotate, switching is provided to apply full line voltage. For example, in the autotransformer type, starting steps may be 50%, 65%, or 80% of full voltage.

A timer is provided in the starter control circuit so that the time between starting and running may be adjusted to suit actual operating conditions.

It should be pointed out that at reduced voltage, the torque available from the motor will be reduced. This reduction may cause concern when the motor is starting under load. Torque varies as the square of the impressed voltage. For example, starting on the

- 50% voltage tap, torque will be 25% of normal.
- 65% voltage tap, torque will be 42% of normal.
- 80% voltage tap, torque will be 64% of normal.

Each type of reduced-voltage motor starter is best explained by referring to its circuit diagram. Therefore, a circuit diagram and photo accompany the description of each type of reduced-voltage starter on the following pages.

Detail control may vary between manufacturers and with the use of the starter. Therefore, the circuit diagrams are not necessarily those of the devices shown in the photographs. The manufacturer lists the maximum horsepower for a range of 5 through 1000 for voltages of 200, 230, 380, 460, and 575. The NEMA size is given for each rating.

14.6.1 Autotransformer or Compensator (Figures 14-18A and 14-18B)

Two contactors are used in the control. One is a five-pole called the *start contactor* (SC). The second contactor is a three-pole contactor, the *run contactor* (RC). A typical starter is shown in Figure 14-18A.

An autotransformer of sufficient size to handle the horsepower rating of the motor is shown connected to the line leads and motor through the open contacts. Taps are available for a starting voltage of 80%, 65%, or 50% of full line voltage.

Figure 14-18A **Reduced-voltage autotransformer starter, closed transition.** *(Courtesy of General Electric Company, General Purpose Control, Bloomington, IL.)*

On pressing the MOTOR START push-button, the start contactor is energized and the SC contacts close. This action connects the motor to the line through the selected autotransformer tap.

When the time set on time 1TR runs out, the start contactor is de-energized and the run contactor is energized. This condition connects the motor to full line voltage.

This type of starting has one disadvantage. The motor is momentarily disconnected from the line during the changing from the start contactor to the run contactor. This occurrence is known as *open transition*. In some localities or under some conditions open transition may be objectionable. If in doubt, consult the local power company.

The circuit for open transition is shown in Figure 14-18B.

A more practical solution is to use a *closed-transition* autotransformer-type motor starter. The circuit for the closed-transition type is shown in Figure 14-18C. With closed transition, the motor is not disconnected from the line when changing from the start condition to the run condition.

14.6.2 Primary Resistor (Figures 14-19A and 14-19B)

Two contactors, each three-pole, are used in the control circuit. One is a *line contactor* (LC). The other is the *accelerating contactor* (AC). Three banks of resistors are used. These are shown connected to the line and the motor through the contacts. The resistors must be of sufficient capacity to handle the horsepower rating of the connected motor. This type of starter is shown in Figure 14-19A.

On pressing the MOTOR START push-button, the line contactor is energized, closing the LC contacts and connecting the motor to the line through the resistors. After a preset time delay, the AC contacts are closed, connecting the motor directly to the line.

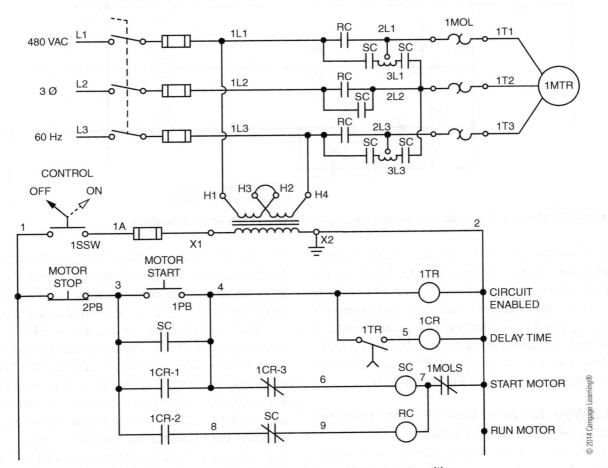

Figure 14-18B **Motor control circuit for reduced-voltage autotransformer starter-open transition.**

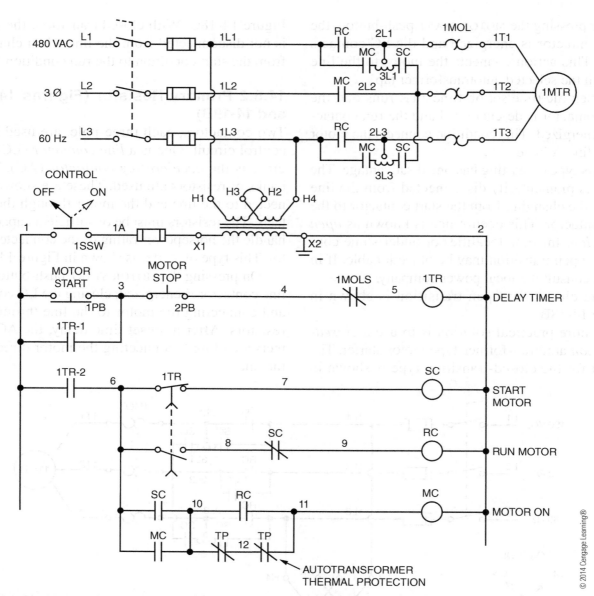

Figure 14-18C **Motor control circuit for reduced-voltage autotransformer starter-closed transition.**

One advantage of the primary resistor-type starter is the smooth acceleration. As the motor accelerates, the current decreases. Thus, the voltage drop across the resistors is reduced, increasing the voltage applied to the motor. The disadvantage is inefficient operation, as the power dissipated in the resistors is comparatively high. The circuit for this type of starter is shown in Figure 14-19B.

14.6.3 Wye- (or Star) Delta Starter (Figures 14-20A, 14-20B, and 14-20C)

A wye-delta reduced-voltage starter is shown in Figure 14-20A. The circuit diagram in Figure 14-20B shows that to use this type of starting, all leads from each three-phase stator winding must be brought out for connecting to the starter. This design provides 33% of the normal starting torque.

Three contactors are used: the LC, SC, and *run contactor* (RC). Pressing the MOTOR START push-button energizes the LC and the SC. The start contacts connect the motor in wye or star, putting approximately 58% of the line voltage across each motor phase. After a preset time delay, the start contacts open and the run contacts close, connecting the motor in delta and putting full voltage on the motor.

Figure 14-19A Primary resistor-type reduced-voltage starter. *(Courtesy of Square D/Schneider Electric.)*

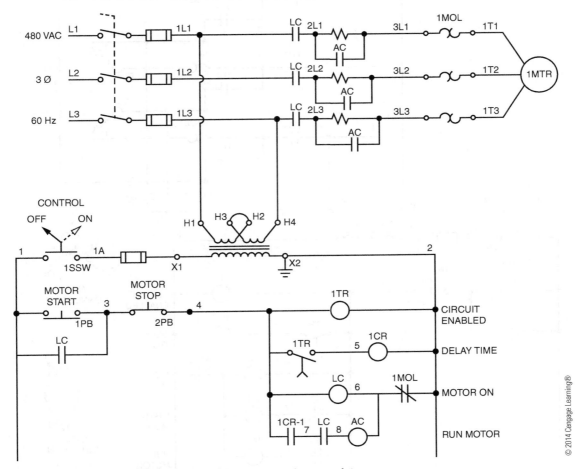

Figure 14-19B Motor control circuit for reduced-voltage starter-primary resistor.

Figure 14-20A **Wye-delta reduced-voltage starter.** *(Courtesy of Square D/Schneider Electric.)*

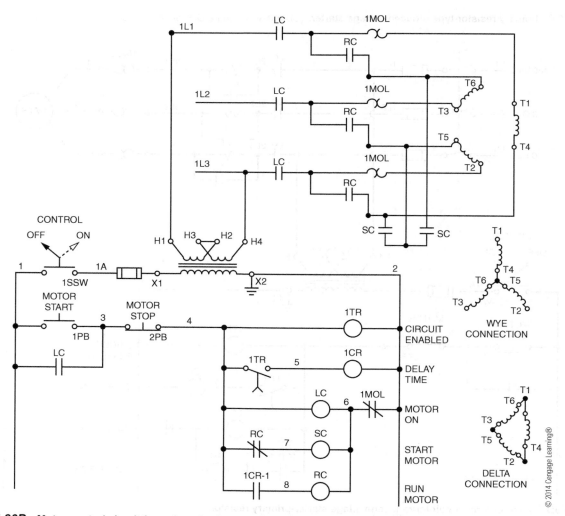

Figure 14-20B **Motor control circuit for reduced-voltage starter-wye-delta (open transition).**

The cost of the control components for this type of starting is less than for the autotransformer and primary-resistor types. However, the motor must be supplied with all the leads brought out for this type of starter. Thus, wye-delta starters can be used only with wye-delta motors.

Closed transition versions are available that keep the motor windings energized for the few cycles required to transfer the motor windings from a wye connection to a delta connection. Such starters are provided with one additional contactor, plus a resistor bank. Figure 14-20C shows the circuit of a wye-delta closed transition motor starter.

14.6.4 Part Winding (Figures 14-21A, 14-21B, and 14-21C)

Like the wye-delta starter, the part-winding starter depends on the motor having the proper leads brought out for connection to the starter. The part-winding starter supplies 48% normal starting torque. Thus, whereas part-winding starting is not truly a reduced-voltage means, it is usually so classified because of the resulting reduced current and torque.

If the starter windings of the motor are divided into two or more equal parts, with all terminals for each section available for external connection,

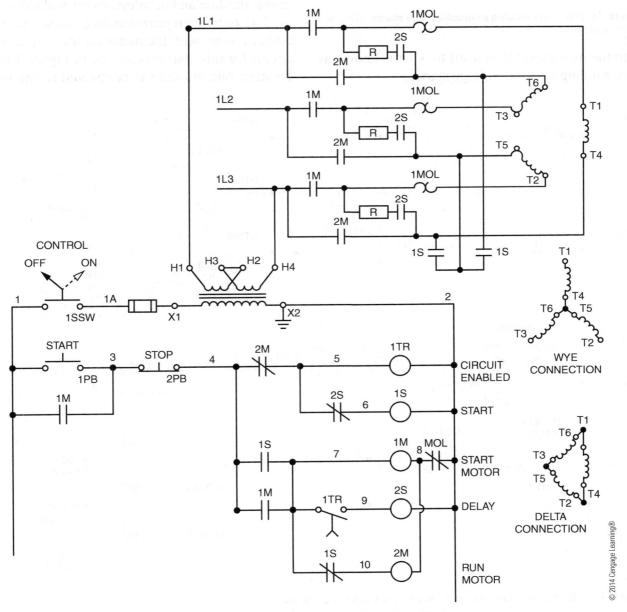

Figure 14-20C Motor control circuit for reduced-voltage wye-delta starter (closed transition).

Figure 14-21A Part winding reduced-voltage starter. *(Courtesy of Square D/Schneider Electric.)*

then the motor may lend itself to what is known as part-winding reduced-voltage starting.

Part-winding starters are designed for use with squirrel-cage induction motors generally having two or three windings. A photo of a typical part-winding starter is shown in Figure 14-21A.

The circuit for the *two-step part-winding starter* is shown in Figure 14-21B. This starter is designed so that when the control circuit is energized, one winding of the motor is connected directly to the line. The winding draws about 60% of normal locked rotor current. It develops approximately 45% of normal motor torque. After about four seconds, the second winding is connected in parallel with the first. The motor is then fully across-the-line and develops its normal torque.

The *three-step part-winding starter* has resistance in series with the motor on the first step. The circuit for this starter is shown in Figure 14-21C. To start, one winding is connected to the line in

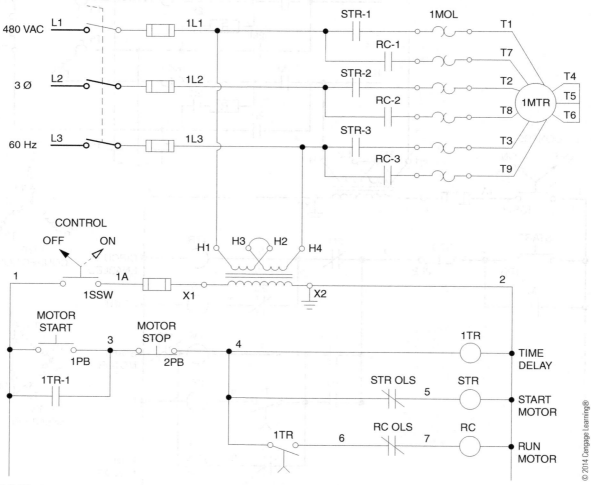

Figure 14-21B Motor control circuit for two-step part-winding starter.

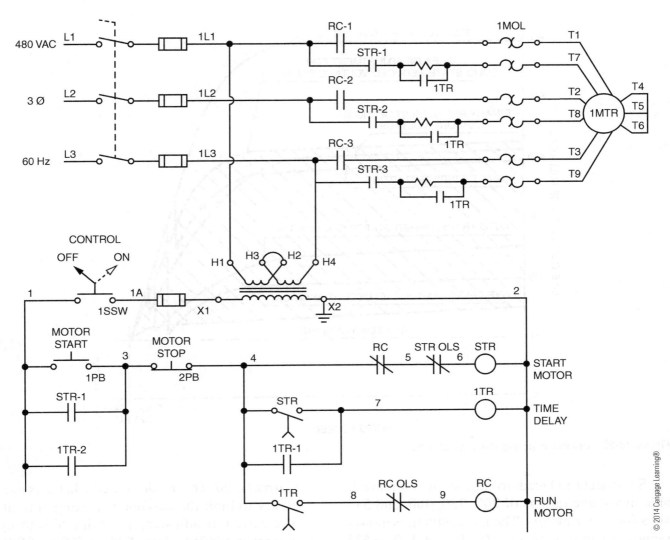

Figure 14-21C **Motor control circuit for three-step part-winding starter.**

series with the resistor. After a short interval, approximately two seconds, the resistor is shorted out. The first winding is connected directly to the line. About two seconds later the second winding is connected in parallel with the first winding. Part-winding starters provide closed transition starting.

14.7 Solid-State Motor Starters

Many applications require the lower starting torque and smooth acceleration offered by solid-state systems. Typical of such applications are conveyor systems, pumps, and compressors. Solid-state reduced-voltage controllers provide smooth, stepless acceleration of a motor through the use

of silicon-controlled rectifiers. By controlling the conduction of the silicon-controlled rectifiers, voltage is gradually applied to the motor. This operation is sometimes referred to as providing a "soft start" to the motor. An adjustable current-limit feature limits current to 25–70% and starting torque to 6–49% of full voltage values.

Figure 14-22 shows the relationships among various types of reduced-voltage starters for the line current drawn from zero to full-load speed. It is also compared to full-voltage starting. Note that the entire range of most reduced-voltage starting is covered by solid-state reduced-voltage starters. This feature gives the solid-state type more flexibility over other types of reduced-voltage starting.

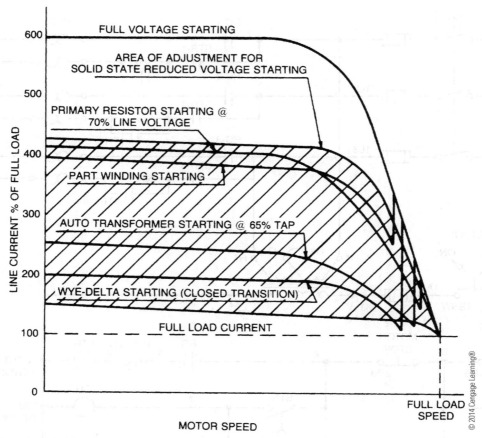

Figure 14-22 Graph comparing starting currents.

Solid-state reduced-voltage starters are available in voltage ratings of 200, 230, 460, and 575 volts for 60 Hz motors. The maximum horsepower ranges are approximately 10–1000 at 460 or 575 V. The maximum range for 200, 230 V drops to approximately 10–300 hp.

The SMC PLUS™ SMART motor controller shown in Figure 14-23 provides microcomputer-controlled starting for standard squirrel-cage induction motors. There are three standard modes of operation:

1. *Soft start with selectable kickstart.* This mode has the most general application. The motor voltage gradually increases during the acceleration ramp period, which can be adjusted from 2 seconds to 30 seconds. This setting is made for the best starting performance over the required load range. Kickstart is for high friction applications and is selectable from 0.4 second to 2 seconds. It provides a pulse of 500% of full-load amperage.

2. *Current limit.* This mode is used when it is necessary to limit the maximum starting current. The current is adjusted, according to starting current restrictions, from 50% to 500% of full-load amperage.

Figure 14-23 Solid-state motor starter. *(Courtesy of Rockwell Automation, Inc.)*

3. *Full voltage.* For applications requiring a full-voltage start, the acceleration ramp time is set to minimum ($\frac{1}{4}$ second), in effect allowing the controller to start the load across-the-line.

The microcomputer-based controller is self-calibrating. It adjusts itself for any line voltage level and frequency within its rating and any current value at or below the continuous rating of the controller.

14.8 Starting Sequence

A circuit showing the sequencing of two main drive motors using magnetic reduced-voltage starters is shown in Figure 14-24. The reason for the sequence starting is to reduce the line inrush current.

Associated with the two main drive motors are two auxiliary motors. They are used for pilot pressure and cooling or lubrication. These motors must be started first and must drop out (de-energize) the main motors in case either auxiliary motor de-energizes through an overload.

When the auxiliary motor starters are energized, the #1 and #2 auxiliary contacts close.

The sequencing of the two main drive motor starters proceeds as follows:

1. Press the MOTORS START push-button.

   a. The energizing of the control section of the #1 motor starter is now initiated.

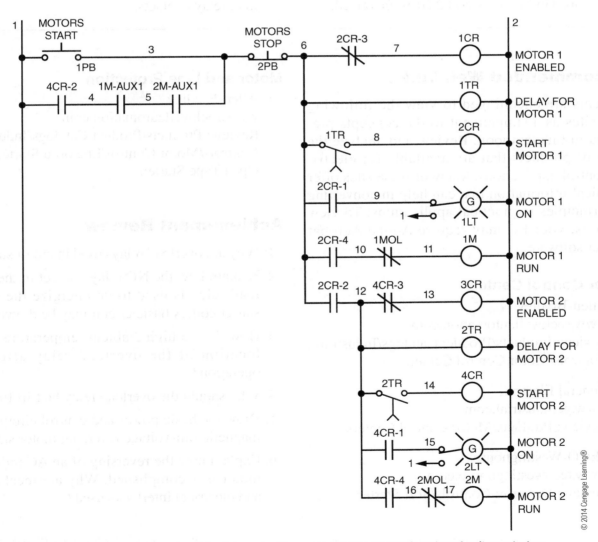

Figure 14-24 Motor control circuit for sequencing two main drive motors using magnetic reduced-voltage starters.

1a. Relay coil 1CR is energized.

1b. Timing relay coil 1TR is energized.

1c. Timing relay 1TR times out, closing the NOTC 1TR contact.

1d. Relay coil 2CR is energized.

1e. Relay contact 2CR-1 closes, energizing the green pilot light.

1f. Relay contact 2CR-3 opens, de-energizing relay coil 1CR.

2. The sequence for energizing the control section for #1 motor starter is now complete.

3. Relay contact 2CR-2 closes.

   a. The energizing of the control section of the #2 motor starter is now initiated.

   3a. Relay coil 3CR is energized.

   3b. Timing relay coil 2TR is energized.

3c. Timing relay 2TR times out, closing the NOTC 2TR contact.

3d. Relay coil 4CR is energized.

3e. Relay contact 4CR-1 closes, energizing the green pilot light.

3f. Relay contact 4CR-3 opens, de-energizing relay coil 3CR.

4. Relay contact 4CR-2 closes, sealing around the MOTORS START push-button.

5. The sequencing for the energizing of motor starter #2 is now complete.

6. Both motor starters are energized.

7. Both motor starters are de-energized at any time by the operating of the MOTORS STOP push-button or the opening of any of the overload relay contacts.

Recommended Web Links

Students are encouraged to view the following Web sites as a supplement to the concepts presented in this textbook. Review and analyze the array of products that are available for electrical control applications. Many of these sites offer technical information that can help in converting the principles to practical applications. To view catalogs, your PC may require Adobe Acrobat Reader software.

Motor Control Centers

1. Allen-Bradley
   www.rockwellautomation.com
   Review: Products/Product Catalogs/Industrial Controls/Motor Control Centers

2. General Electric
   www.geindustrial.com
   Review: Products/Motor Control Centers

3. TECO-Westinghouse
   www.tecowestinghouse.com
   Review: Drives & Controls/Soft Starter

Motor and Line Protection

1. Allen-Bradley
   www.rockwellautomation.com
   Review: Products/Product Catalogs/Industrial Controls/Motor Control/Enclosed Starters & Open Type Starters

Achievement Review

1. Why are overload relays used in motor starters?

2. Explain how the NC relay contact in the overload relay is used to de-energize the motor starter coil. A basic circuit may be drawn.

3. How does a high ambient temperature in the location of the overload relay affect its operation?

4. Why should the overload relay be trip-free?

5. Draw the basic power and control circuit for a magnetic full-voltage reversing motor starter.

6. Explain how the reversing of an AC induction motor is accomplished. Why are mechanical and electrical interlocks used?

7. A reduced-voltage motor starter is set to apply 65% of full voltage to the motor for starting. What torque, in percent of full voltage, will be developed in the motor under these starting conditions?

8. Explain how the resistor-type, reduced-voltage motor starter reduces the voltage to the motor on starting. A basic power and control circuit should be drawn.

9. What is the difference between no-voltage or low-voltage release and no-voltage or low-voltage protection?

10. What is meant by *jogging* in motor starter control?

11. What protection is provided in the solid-state reduced-voltage starter to prevent a high-temperature condition owing to overcurrent?

   a. Special fuses
   b. Current meter
   c. Thermal sensor

12. How is the effective voltage applied to a motor using a solid-state reduced-voltage starter controlled?

   a. By means of a timer
   b. Through controlling the conduction of the silicon-controlled rectifiers
   c. By the current drawn by the motor

13. What is the difference between open transition and closed transition in a reduced-voltage motor starter?

14. For the circuit shown in Figure 14-10, replace the NO JOG push-button and the NO MOTOR START push-button with a three-position maintained selector switch. Does the circuit function the same as before?

15. For the circuit shown in Figure 14-11, can ON delay timer 2TR be located somewhere else in the circuit and still have the same results?

16. For the circuit shown in Figure 14-14, can ON delay timer 1TR be located somewhere else in the circuit and still have the same results?

15

CHAPTER

Introduction to Programmable Control

OBJECTIVES

After studying this chapter, you should be able to:

- Explain the three basic logic control functions.
- Describe electrical control devices that provide information or input to a control system.
- Describe electrical control devices that provide decision making or logical sequencing of a control system.
- Describe electrical control devices that provide work or output from a control system.
- Understand the programming language of a programmable logic controller (PLC).
- List the advantages of a PLC over hard-wired relay logic circuits.
- Explain the relationship of the PLC processor to input/output (I/O).
- Identify the input and output symbols of a PLC ladder logic diagram.

- Draw a simple motor starter circuit using a PLC ladder logic.
- List the different types of devices to "condition" the input signals to a PLC.
- Explain the relationship of PLC memory to I/O.
- Explain the types of computer memory used in PLCs.
- Describe the logical path of an input device through the input and output tables of a PLC.
- Draw a simple PLC control circuit.
- Convert an existing relay logic circuit to a PLC controlled circuit.
- Understand the role of the data communication highway.
- Describe the function of a complete PLC system utilized in a typical industrial application.

15.1 Primary Concepts of Relay Control

Before we begin discussion on the programmable logic controller (PLC), it may be helpful to review some fundamental concepts of relay control. The concepts that we considered with electromechanical control are similarly applicable to the PLC. In fact, the PLC was designed to replace relay control. Therefore, there are many similarities between

the two systems. For example, when the relay coil is energized the normally open (NO) contact will transition and pass power, while the normally closed (NC) contact will open and de-energize the path in the circuit. This concept applies to the PLC also. The difference is that the PLC is a computer. The programming logic uses basic elements that look like the relay coil, the NO contact, and the NC contact. These logic elements function in the same manner as the relay circuit devices. However, in the PLC the devices are used to solve logic code whereas in the relay circuit the devices are used to pass electricity through the physical wires in the circuit. Further discussion on this topic will follow throughout this chapter.

Another concept worthy of review is the classification of devices into one of three categories based on their role in the relay circuit. Each component may be classified as an input device, a decision maker, or an output device. Figure 15-1A duplicates the information in Chapter 12 for review purposes.

The input devices gather information from the operator, the machine, or the process. The decision makers use the information obtained from the input devices to logically determine what should happen on the piece of equipment. The output devices perform the work that must occur as the machine sequences through its operations.

For the PLC system, the input devices are wired to input terminals on the PLC. Similarly, the output devices are wired to output terminals on the PLC. All the decisions are made by the logic program in the PLC. Therefore, the only devices in the decision maker category for the PLC system are the virtual relay coils that reside in the PLC logic program. Timing and counting functions may also be performed in the PLC logic program. Electromechanical relays may be used in the PLC system to switch output loads whose current draw exceeds the limitations of the PLC output cards. In this case, the relay coil is connected to a PLC output and the PLC logic will determine when the coil will be energized. The contacts of the relay will be used to switch the high current load. If the current draw for the load exceeds the current rating for the relay contacts, then a contactor coil may be connected to the PLC output card. A motor starter coil may also be connected to a PLC output if the PLC logic is used to control the motor. Figure 15-1B shows the modified classification chart for a PLC system.

The PLC gathers information from the input devices. The PLC logic program uses this information to make decisions while the program is solved. The decisions that are made by the PLC logic are used to control the appropriate output devices as the machine sequences through its operations.

15.2 Introduction to Programmable Logic Controllers

Electromechanical control is covered in Chapters 1 through 12. Throughout these chapters, different devices were used and circuits were modified as the requirements of the system changed. These basic circuits required only a handful of relays, timers, or counters to perform the required operations. Large equipment requires a significant number of relays and control devices to operate the machinery. The control panels used to house the devices are fairly large in size and troubleshooting the system can be difficult with the components far away from one another. In addition, implementing significant design changes to relay systems presents a challenge to production since the equipment will

INFORMATION OR INPUT	DECISION OR LOGIC	OUTPUT OR WORK
Push-buttons	Timers	Pilot lights
Selector switches	Relays	Solenoids
Proximity switches	Contactors	Alarms
Limit switches	Motor starters	Heaters
Pressure switches	Counters	Motors
Thermocouples		

Figure 15-1A Classification of devices in relay control.

© 2014 Cengage Learning®

INPUTS	LOGIC	OUTPUTS
Push-buttons	Virtual relay coils	Pilot lights
Selector switches	Programmed timers	Solenoids
Proximity switches	Programmed counters	Alarms
Limit switches		Relays
Pressure switches		Contactors
Thermocouples		Motor starters

Figure 15-1B Classification of devices in PLC control.

© 2014 Cengage Learning®

RELAY LOGIC	PROGRAMMABLE CONTROLLER	BENEFITS OF PROGRAMMABLE CONTROLLERS OVER RELAY LOGIC
Relays	Solid-state components	High reliability
Hard wiring	Programming	Easy to implement
		Easy to change
Large size	Small size	Save floor and panel space
Timer/counters	Programmed functions	Easy to change presets
	Microcomputer based	Greater functionality/Faster speed
	I/O interface	Adjust to most field devices
	Diagnostic indicators	Simplify troubleshooting

© 2014 Cengage Learning®

Figure 15-2 **Chart of benefits of programmable controllers versus relay logic.**

not function until the changes are completed and the system is verified to function correctly. The invention of the PLC alleviated many of these issues for industry. The utilization of computer-based control has also proven to be less expensive to implement than a hard-wired relay logic circuit. Traditionally, the rule of thumb has been that a control system requiring more than six relays would be less expensive to implement with a PLC. Today, lower-end PLCs are available that cost under $100 (US dollars). In addition, a PLC system offers many advantages over a relay-based control system. Some of these advantages are identified in Figure 15-2.

15.3 Programmable Logic Controller Concepts

The National Electrical Manufacturers Association (NEMA) standard ICS 3–1988, Part ICS 3–304.01 defines a programmable controller as a digital operating electronic apparatus that uses a programmable memory for the internal storage of instructions for implementing specific functions such as logic, sequencing, timing, counting and arithmetic to control, through digital or analog input/output modules, various types of machines or processes. A digital computer that is used to perform the function of a programmable controller is considered to be within this scope. Excluded are drum and similar mechanical type sequencing controllers.

To avoid any confusion in abbreviating programmable controller, we use the term *programmable logic controller* (PLC) to differentiate it from personal computer (PC).

As stated earlier, the input devices from the machine are connected to terminals on the PLC input cards. The output devices that perform the work are connected to terminals on the PLC output cards. The logic resides in the *central processing unit* (CPU) of the PLC. A block diagram showing this structure is shown in Figure 15-3.

The most common programming language used in industry is called *ladder logic*. This programming language is intended to look very similar to a conventional relay circuit. There are vertical power rails on the left and right sides of the programming screen. The different programming elements that are available in the programming language are called *instructions*. The instructions are programmed between the

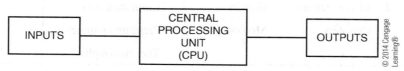

Figure 15-3 **Block diagram of a PLC system.**

vertical power rails in a horizontal orientation. The compilation of all available instructions for a given PLC is called the *instruction set*. The instruction set will vary between PLCs even if the manufacturer is the same. The appearance of an instruction used in the program may vary drastically between different manufacturers. However, the functionality is usually very similar. Regardless, the basic programming elements (Figure 15-4) are the same in any PLC, both in appearance and functionality.

These instructions solve in logic as follows:

- The relay coil will solve TRUE (ON) if any path through the logic connected to the coil solves TRUE. Notice that the symbol for the logic coil is very similar to the symbol for an electromechanical relay except that the circle is not completely closed for the logic coil.
- The NO contact will solve TRUE (allow logic continuity) if the address assigned to the contact solves TRUE (or is ON). The address may refer to an input or to a coil that is used elsewhere in the program. Otherwise, the NO contact solves FALSE and does not allow continuity of logic past this element.
- The NC contact will solve TRUE (allow logic continuity) if the address assigned to the contact is FALSE (or is OFF). The address may refer to an input or to a coil that is used elsewhere in the program. If the referenced address is ON, the NC contact solves FALSE, and does not allow continuity of logic past this element.

Notice that these instructions solve in the same manner that contacts and coils operate in an electromechanical relay circuit. To associate a contact to a specific relay coil in PLC logic, the same address must be used for both the relay coil and the contact. The addressing scheme for instructions varies drastically between manufacturers and between different families of PLCs from the same manufacturer. It would be very difficult to cover all of the different addressing schemes that are used throughout industry in this text. Therefore, only the basics for a few configuration schemes are explained here.

Some systems use an addressing scheme where inputs are addressed with a one (1) followed by a unique number for that input. Outputs are designated with a zero (0) followed by a unique number for that output. The mapping of the physical *input/output* (I/O) points to the addressing scheme is accomplished through the programming of a configuration set up file.

Some systems use an addressing scheme that is based on the physical wiring location of the input and the output devices. An input is named with the letter I followed by the slot number and terminal wiring connection to which the input is wired. Similarly, an output is addressed with the letter O followed by the slot number and terminal wiring connection to which the output is wired. For example, an input that is wired to the card in slot 5, terminal 3, would be designated as I:5/3. Also, an output wired to a card located in slot 6 terminal 4 would be addressed as O:6/4. Internal relays are coils that are used in the logic program that are not connected to a physical output device. Internal relays are addressed with a B3 designation, for example, B3:2/5. Internal relays function in the same way as programmed relay coils that are mapped (addressed) to outputs. Contacts from the internal relays may be used as logical conditions in the PLC program. The difference is the relay coil does not map to real-world output devices. For simplicity and consistency purposes, this addressing scheme will be used on examples throughout this chapter.

Some systems use a tag-based addressing scheme. For these PLCs, a unique name is given to the coil, for example, "Machine in Cycle." This same name must then be given to all contacts associated with this coil. Inputs and outputs are also given a unique name. The PLC is set up to link all tag names to the physical input/output devices wired to the different terminals on the PLC system.

In Figure 15-5, a simple motor starter circuit is shown using conventional relay control. The motor overload contacts are not shown in the drawing.

In Figure 15-6, the input and output devices are shown wired to PLC I/O cards. It may help

NC CONTACT　　NO CONTACT　　RELAY COIL

© 2014 Cengage Learning®

Figure 15-4 **Basic ladder logic programming instructions.**

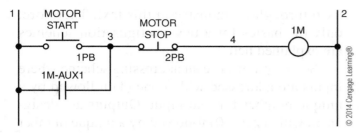

Figure 15-5 Motor control circuit-relay control.

to make this transition to the PLC by asking the question—what are the inputs and what are the outputs in the circuit shown in Figure 15-5? Finally, Figure 15-7 shows the PLC logic that is required to perform the task of starting the motor. Notice that an NO contact is used for the MOTOR STOP contact (I:5/1) in logic. This is because the PLC input is ON when the MOTOR STOP push-button is NOT actuated, causing the NO contact in logic to solve true.

15.4 PLC Input/Output (I/O)

As we have seen in reviewing electromechanical relay control circuits, all control must depend on initial and continuing information from the machine and/or the process. This information comes from discrete inputs such as pushbuttons, limit switches, pressure switches, etc. (Figure 15-8). Discrete devices can have only two states—they are either ON or OFF. In a PLC system, discrete devices are connected to terminals on an input card. The input module then communicates the status of the discrete inputs to the processor.

Devices such as transducers and thumbwheel switches may also be connected to PLC input cards. However, these inputs are not considered discrete since their input signal may represent many states (not only ON or OFF). These devices require the

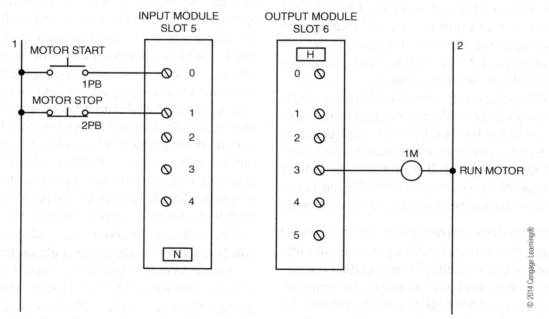

Figure 15-6 PLC I/O wiring for motor control circuit.

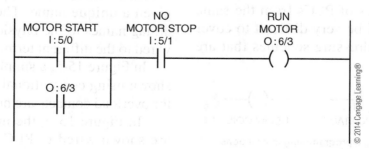

Figure 15-7 PLC logic for motor control circuit.

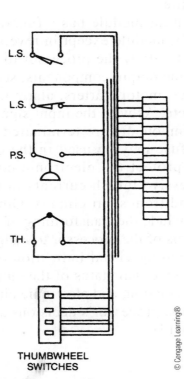

Figure 15-8 Discrete inputs.

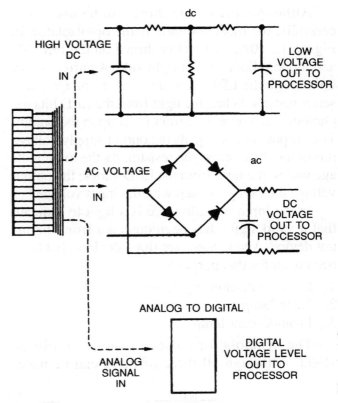

Figure 15-9 Typical voltage interface modules.

use of special inputs cards such as analog input or BCD input cards to interface with the PLC.

The signal from the discrete input must be prepared for access to the processor. Field-supplied voltages from the input may cover a range of 24–240 VAC/VDC. The input signal must be converted to a low-level DC logic voltage for use in the processor. This voltage will vary with different manufacturers, but it is in the area of 5–12 V. When the input voltage is AC, a bridge rectifier converts the AC to DC, and resistors drop the voltage to the required level. If the input is DC, only resistors are used to drop the voltage level. When the signal is analog, such as from temperature, speed, pressure, and so on, the incoming analog signal is converted to a binary value that can be represented in the PLC as a number through the use of an analog to digital converter (Figure 15-9).

Because of "noise" or large voltage spikes in the input signal, it is necessary to isolate this incoming signal before it enters the processor. Three devices are used:

1. Optoisolator (Figure 15-10A)
2. Reed relay (Figure 15-10B)
3. Transformer (Figure 15-10C)

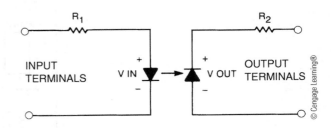

Figure 15-10A Optoisolator.

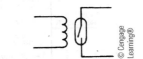

Figure 15-10B Reed switch.

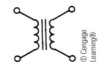

Figure 15-10C Transformer.

Although any of the three can be used successfully, the most common is the optoisolator. In Figure 15-10A, an LED is shown on the left side and a photodiode on the right side. A current is set up through the LED from the source voltage and a series resistor. When the light from the LED hits the photodiode, a reverse current is set up in the output. The output voltage equals the output supply voltage minus the drop across the resistor. As the input voltage varies, the light from the LED varies; the output voltage thus keeps in step with the input voltage.

In the input module there is a light to indicate that a valid signal has been detected from the input device. Again, there are three devices that have been used for this purpose:

1. LED (light-emitting diode)
2. Neon lamp
3. Incandescent lamp

The symbols for these are shown in Figure 15-11. Although all three of these can be used, the most common is the LED owing to its long operating life.

The output module has a function similar to the input module except in reverse order. It functions to allow the processor to communicate with the output components, such as solenoid valves, motor starters, alarm horns, LED displays, etc. As with the input signals, all the output components are terminated on output cards. A fuse may be used in the output circuitry to protect the electronic components from excessive inrush currents in the output devices and from short circuits. Other considerations include the transforming of low-power logic signals of the processor to a higher power level, providing visual indication of both the fuse and functional status of the output points, electrical isolation, and electronic circuitry that indicates the processor's decisions and actions (Figure 15-12).

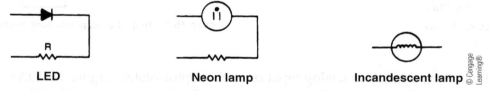

LED **Neon lamp** **Incandescent lamp**

© Cengage Learning®

Figure 15-11 Symbols for PLC indicator lights.

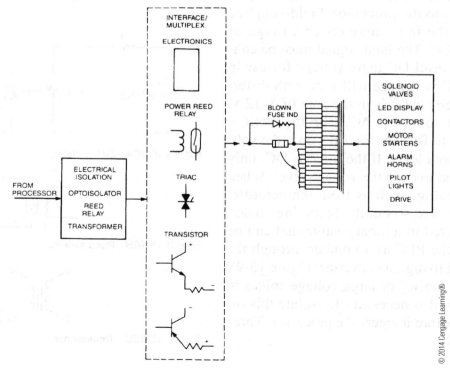

Figure 15-12 Output in block form.

15.5 Processor

The processor module is considered the brain of the PLC. This unit consists of a state-of-the-art microprocessor and memory chips. There are also circuits necessary to store and retrieve information from the memory and communication circuits required for the processor to interface with the programmer, printer, and other peripheral devices.

The processor module is designed to support a fixed number of discrete, analog, or specialty I/O ports. Also, processors determine variations in the available program instructions, fixed mathematical functions such as proportional-integral-derivative (PID) control, and communication link with other processors.

The processor serves to monitor the status (ON/OFF) of the input and output devices. It also scans and solves the logic of the user program. Scanning, then, is the process of examining inputs—solving logic—and updating outputs. The PLC will perform this scanning sequence over and over as long as the PLC is running. The time required for the PLC to complete one scan (read inputs, solve logic, and update outputs) is called the scan time. A typical value for scan time is in the order of 20 msec. This time will vary based on the speed of the processor, the size of the logic program, and the types of instructions used in the program. The scan time may also vary in every scan of the processor depending on the logic that must be solved during the scan. The PLC will usually not continue to solve a rung of logic once it has been determined that there is not a path available for the PLC to solve the logic as TRUE.

15.6 Memory

The memory section of the processor serves two important functions:

1. It remembers information that the processor may need to make decisions. This part of the memory is sometimes referred to as *storage* or *data table*. It is in this area that the ON/OFF status of all the discrete inputs and outputs is stored. Numeric values of timers and counters may also be stored in this memory.

2. It remembers the instruction given by the user, telling the PLC what to do. This section of the memory may be referred to as the *user program memory* and contains the ladder logic program. It is generally many times larger than the storage or data table. Instructions are placed in the memory via a programming device. For small-end PLCs, a handheld device or a small screen attached to the processor can be used for programming. Usually, however, a PC with the appropriate software is used to program the logic, to upload programs from the PLC, and to download programs to the PLC.

Figure 15-13 shows the location of the data table and the program storage in the memory.

There are two general classes of memory:

1. Volatile
2. Nonvolatile

The following describes the various types of memory within these two classes. It is important to learn how their characteristics affect the manner in which programmed instructions are retained or altered.

Volatile memory will lose its programmed contents if operating power is removed or lost. It is therefore necessary to have battery backup power at all times. One type of volatile memory is RAM, which means read/write, solid-state *random access memory*. Information stored in RAM can be retrieved or "read." "Write" indicates that the user can program or write information into the memory. "Random access" refers to the ability of any location in the memory to be accessed or used.

There are several types of RAM memory; two of them are:

1. Metal-oxide semiconductor (MOS)
2. Complimentary metal-oxide semiconductor (CMOS)

Nonvolatile memory retains its information when power is lost. It thus does not require a battery backup. A common type of nonvolatile memory is ROM (*read-only memory*).

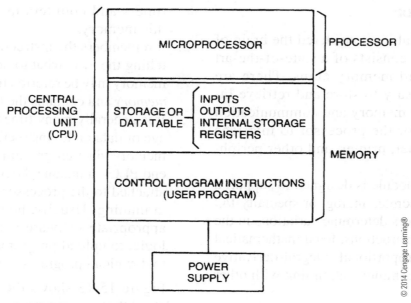

Figure 15-13 Location of units within CPU (block form).

As with volatile memory there are several types of ROM:

- PROM
- EPROM
- EAROM
- EEPROM

PROM means *programmable read-only memory*. It is a special type of ROM and is generally programmed by the manufacturer. It has the disadvantage of requiring special programming equipment and once programmed cannot be erased or altered.

EPROM means *erasable programmable read-only memory*. A program can be completely erased by the use of an ultraviolet light source. After the program chip is completely erased, program changes can be made.

EAROM means *electrically alterable read-only memory*. An erasing voltage applied to the proper pin of an EAROM chip will completely erase the program.

EEPROM means *electrically erasable programmable read-only memory*. Although it is a nonvolatile memory, it offers the same programming flexibility as RAM. The program can easily be changed with the use of a PC with EEPROM software or an EEPROM programming unit or manual programming unit.

PLCs may have different memory sizes. The actual size will depend on the user's application. Sizes are generally expressed in K (1024) values (2K, 4K, 16K, etc.). These numbers represent the number of words available. The range may be from 256 words for small PLCs to 32M for large PLCs. A word is usually 8, 16, or 32 bits (Figure 15-14).

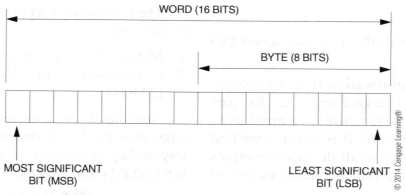

Figure 15-14 Basic information unit–16-bit word.

15.7 Power Supplies

The power supply on the PLC is used to provide power to the CPU and all of the cards installed in the system. Analog cards and specialty cards will usually require more current for proper operation than the standard input and output cards. Therefore, it is important to make sure the power supply is adequately sized to properly run the system. The power supply typically provides power to the cards through the backplane on the chassis. When cards are plugged into the backplane, the various voltage levels are available for use. The power supply does not supply power required to operate the field devices. Therefore, the input/output devices themselves must be powered by the control system. I/O cards will draw power from the power supply for their proper operation but not to power the I/O. Remote input/output systems also require the use of additional power supplies.

A satisfactory power supply will take the incoming line power and rectify and filter. In some power supplies the high-voltage DC will be switched at a high (ON/OFF) rate to the filter section. The filter will remove any switching transients that may be on the DC. Many manufacturers will include additional electronics to monitor the incoming line voltage. If the incoming power source has excessive noise or voltage fluctuations then it may be advisable to use an isolation transformer to protect the PLC from these disturbances.

In the discussion on memory we noted that some systems require a backup battery power source to enable the processor memory to maintain a program in the event the power source is lost. The batteries may be of several types, including lithium, alkaline, or nickel-cadmium. The lithium battery is used by many manufacturers because of its excellent operating characteristics. It has a

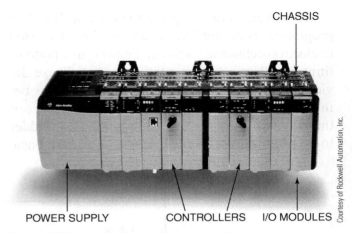

Figure 15-15 **A PLC System complete with processor, power supply, and I/O cards** (used with permission from Rockwell Automation Inc.).

high power density for its size and a long shelf life. Figure 15-15 shows a PLC system complete with processor, power supply, and I/O cards.

15.8 Programming

In Chapter 12 several circuits were shown using relay ladder-type schematic diagrams. When the PLC was introduced, the manufacturers developed a language of commands, instructions, and operations that were presented in a form similar to the relay ladder-type diagram. This programming language is called ladder logic. Although other programming languages are available (structured text, instruction list, function block diagram, and sequential function chart), ladder logic is the most prevalent in industry.

A simple electromechanical relay control circuit is shown in Figure 15-16. A verbal description for the circuit as drawn would in effect say: If the push-button is pressed, then the green light (1LT) will be energized. If the limit switch remains actuated, then the red light (2LT) remains energized.

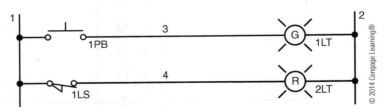

Figure 15-16 Simple relay control circuit.

However, in drawing the logic for a PLC, the programmers do not have symbols like those used in electromechanical circuit diagrams (push-buttons, limit switches, pressure switches, etc.). These devices will be shown in the control drawings as being physically wired to the I/O cards. However, these symbols are not available in the PLC ladder logic program. Refer to Figure 15-4 for the basic programming elements. The relay circuit from Figure 15-16 may look like Figure 15-17 in a PLC logic program.

If we were now to take this circuit and show it in a PLC with the inputs, outputs, and bit addresses, it would appear as shown in Figure 15-18.

Remember how the PLC scan is executed. The PLC reads the status of the inputs, solves the

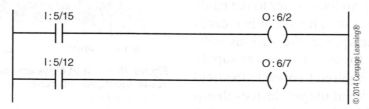

Figure 15-17 PLC logic for relay circuit in Figure 15-16.

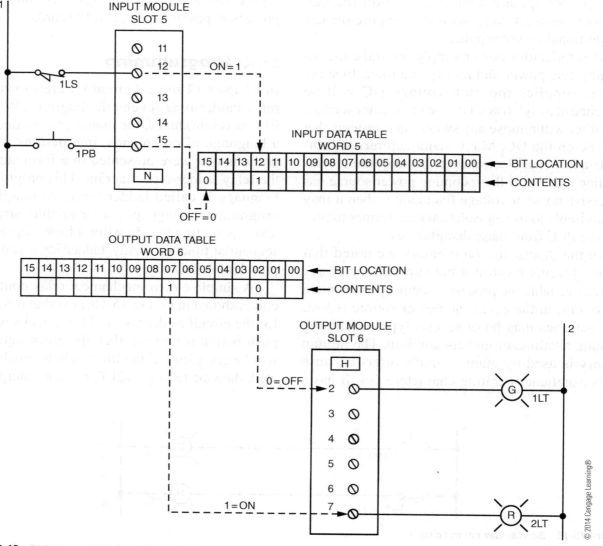

Figure 15-18 PLC data table information for circuit in Figure 15-17.

logic, and then updates the outputs. This process continues as long as the processor is running. For the example provided, the PLC will read the inputs. Suppose the push-button (I:5/15) is not being pressed and the limit switch (I:5/12) is being actuated. The PLC will place a 0 in the bit location for the push-button (I:5/15) since the PLC input is OFF. Similarly, the PLC will place a 1 in the bit location for the limit switch (I:5/12) since the PLC input is ON. After the inputs have been read, the PLC will solve the logic. Since the bit address in the input table for I:5/15 contains a 0, the NO contact will solve FALSE. Therefore, coil O:6/2 will be OFF and the PLC will place a 0 in this bit location (O:6/2). Similarly, since the bit address in the input table for I:5/12 contains a 1, the NO contact will solve TRUE. Logic continuity exists to coil O:6/7. Therefore, coil O:6/7 will solve TRUE and the PLC will place a 1 in this bit location (O:6/7).

Finally, the PLC will update the outputs. The PLC will look at the status of the bits in the output table and update the outputs accordingly. Since the bit in location O:6/2 contains a 0, the green light will be OFF. However, the bit location O:6/7 contains a 1. Therefore, this output will be ON and the red light will be illuminated.

Now suppose that the push-button is pressed. When the PLC updates the input status, a 1 will now be placed in the input table for bit location I:5/15. As the PLC solves the logic, the NO contact will now provide continuity to coil O:6/2 and the PLC will place a 1 in the bit location of the output table corresponding to this coil address (O:6/2). Finally, when the PLC updates the outputs a 1 will be in bit location O:6/2 and the green light will be energized.

A program statement consists of a condition and an action; for example, "If the limit switch is closed, then energize the red light." In general, then, it can be said that each action is represented by a specific instruction or set of instructions. The instruction tells the processor to do something with the information stored in the data or user table.

Another relay circuit diagram is presented in Figure 15-19. The main function of any ladder-type diagram is to present the conditions that are available to control outputs based on inputs. The conditions existing in the circuit shown in Figure 15-19 when no actions have been initiated are as follows:

- The NC STOP push-button is not pressed.
- The NO START push-button is not pressed.
- The NC limit switch is not actuated.
- Relay coil 1CR is not energized.
- Relay contact 1CR-1 is open.

The sequence of operations required to energize the relay coil 1CR are as follows:

1. Press the START push-button.
2. Relay coil 1CR is energized.
   a. Relay contact 1CR-1 closes, sealing around the START push-button.
3. Relay coil 1CR remains energized until either the limit switch 1LS is actuated or the STOP push-button (2PB) is pressed.

From this sequence it can be stated that for an output to be energized or activated, there must be a complete path for electrical energy from wires 1 to 2 through any path in the ladder rung. This can be expressed as logic continuity.

If the circuit shown in Figure 15-19 were programmed for a PLC, it would appear as shown in Figure 15-20. The sketch combines the inputs, outputs, logic, physical wiring of the I/O devices, and data tables in the memory.

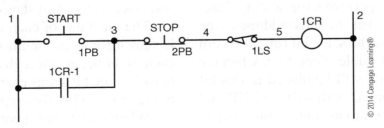

Figure 15-19 **Relay circuit for controlling 1CR.**

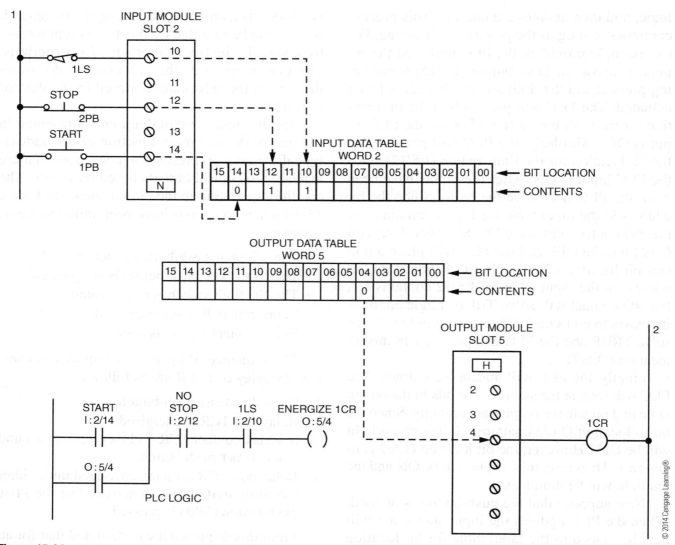

Figure 15-20 PLC I/O drawing, logic, and data table information for circuit in Figure 15-19.

Each contact and coil is given an address that references it to the data table. This address can be either a connected input or output or an internally stored bit location.

In the relay circuit, contact 1CR-1 was used to seal around the START push-button. In PLC logic, the NO contact O:5/4 performs the sealing function. NO contact O:5/4 has the same address as the output; that is, they use the same bit (bit 4 of word O:5) in the output data table. Therefore, when the output is energized, a 1 will be placed in this bit location. The NO contact will solve TRUE and continue to seal in the circuit until continuity is lost to the relay coil.

The PLC checks the status of all discrete input devices for OFF or ON (0 or 1). It then scans the user program and solves the logic. The next step is to set the discrete output devices ON if conditions indicate that there is continuity to the coil. The process continues: examine inputs, solve logic, and determine output status.

The bit status of each input and output device is shown for start conditions in Figure 15-20. As each event occurs through a complete cycle of the machine, the bit status changes to reflect the status of the input and/or the output device.

When no actions have been initiated, the input devices and the resulting PLC scan provide the following information from reading the inputs:

- The NC STOP push-button is not pressed (I:2/12 contains a 1 since the PLC input is ON when the NC STOP push-button is not pressed).
- The NO START push-button is not pressed (I:2/14 contains a 0 since the PLC input is OFF when the NO START push-button is not pressed).
- The NC limit switch is not actuated (I:2/10 contains a 1 since the PLC input is ON when the NC limit switch is not actuated).

When solving the logic, the PLC will not show continuity to coil O:5/4. This is because there is a 0 in locations I:2/14 and O:5/4 so there is no path of continuity to the coil. Therefore, the coil is OFF, a 0 is placed in bit location O:5/4 of the output data table. When the PLC updates the outputs, the relay 1CR (mapped to address O:5/4) will be OFF (or de-energized).

From this circuit, it can be shown that pressing the START push-button will place a 1 in the bit location I:2/14 of the input data table. A path of continuity will now exist to the coil since all NO contacts in logic have a 1 stored in their bit locations. Coil O:5/4 will energize, placing a 1 in this bit location of the output data table. The relay 1CR will now energize as soon as the PLC updates the outputs during this portion of the scan.

In addition, since bit location O:5/4 contains a 1, the NO seal contact will now solve TRUE allowing the operator to release the START push-button while still maintaining continuity to the coil and maintaining the output (relay 1CR) in the ON (or energized) state.

The output (relay 1CR) will remain ON until the limit switch is actuated or the STOP push-button is pressed. Remember that either one of these conditions will load a 0 in the corresponding bit of the input data table. A zero in either of these bit locations will open the path for continuity to coil O:5/4.

We can now look at an industrial application involving a plastic injection molding machine with cores in the mold.

A piston and cylinder assembly is shown in Figure 15-21. Fluid power is supplied to the cylinder through a two-position, single-solenoid, spring-return operating valve. Piston and cylinder assemblies #2 and #3 (associated with the cores) can be operated with fluid power, pneumatic power, or a mechanical device. Their motion must be detected by limit switches.

Piston #1 must be in position 1 for start conditions. The piston is to move to the right as shown. At the position that #1 piston engages and actuates limit switch 1LS, if either #2 or #3 piston has moved to a position actuating limit switches 3LS or 4LS, then #1 piston stops and returns to

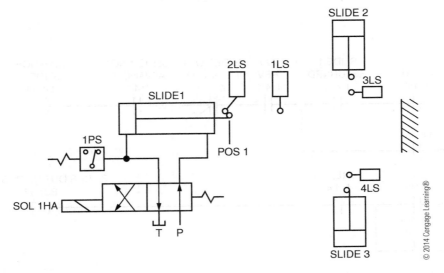

Figure 15-21 **Hydraulic circuits for three slides.**

position 1. If limit switches 3LS or 4LS are not actuated when #1 piston actuates 1LS, the piston continues its travel and builds pressure to a pre-set level on a work piece. The piston then returns to position 1. The REVERSE push-button causes the piston to return whenever the push-button is pressed.

Figure 15-22 shows the electromechanical control circuit for the assembly shown in Figure 15-21. Figure 15-23 shows the same circuit arranged for a PLC. Figure 15-24 shows the connections of the "real world" devices as they would be connected into the PLC's I/O cards.

15.9 Examine On/Examine Off

One manufacturer of PLCs uses the EXAMINE ON/ EXAMINE OFF terminology to define the basic instructions for the PLC. The EXAMINE ON instruction will pass power when the reference address is ON (or has a 1 in the bit address location). The EXAMINE ON instruction is identical to the NO contact described in the text. In a like manner, the EXAMINE OFF instruction will pass power when the reference address is OFF (or has a 0 in the bit address location). The EXAMINE OFF instruction is identical to the NC contact described in the text.

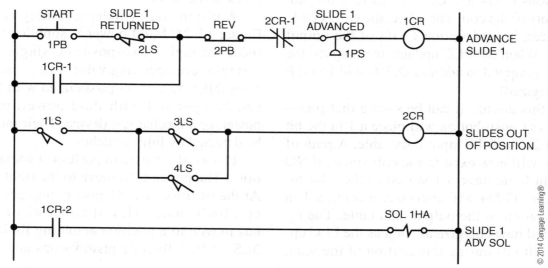

Figure 15-22 Relay control circuit for hydraulic circuit shown in Figure 15-21.

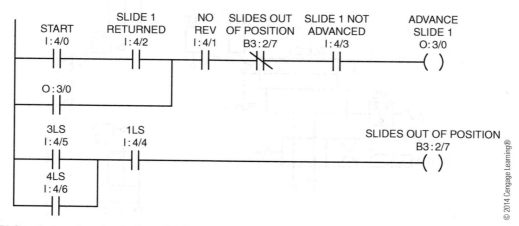

Figure 15-23 PLC logic for relay circuit shown in Figure 15-22.

© 2014 Cengage Learning®

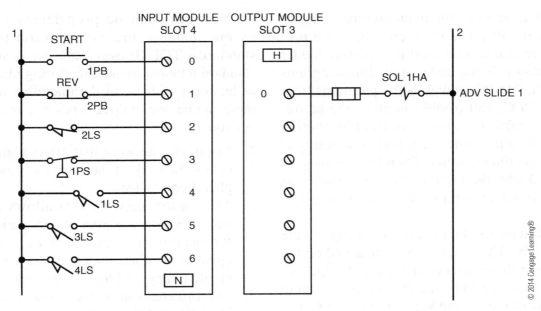

Figure 15-24 **PLC I/O drawing for relay circuit shown in Figure 15-22.**

15.10 **Peripheral and Support Devices**

By definition, *peripheral* devices are all devices outside the processor, including all hardware. However, by convention peripherals and support devices have come to include only more sophisticated devices such as PCs, printers, programming devices, modems, etc.

Since the number and design of such devices is large and many devices are complex, only brief descriptions of a few are provided in this text. Every year many changes are made in methods, and new devices are introduced. As a result, it would be a waste of time to describe all of the currently available devices, as many of them will quickly become obsolete.

15.10.1 Printers

The documenting of a given program for troubleshooting or as a record is important. The printer must be compatible with the processor, which may include such items as the size of the paper and number of columns to be printed.

An "intelligent printer" has an internal microprocessor to control the various options that the unit may contain.

If the printer is to operate correctly, the communication link between the programming terminal and the printer must be properly installed. A shielded cable is used with the shield grounded at *one end only.*

15.10.2 Modems

The term *modem* is derived from MOdulator-DEModulator. It is used to connect the processor via telephone lines to a computer, printer, or another PLC.

The modem accepts digital data from either a processor or programming terminal. It is then converted into a frequency-modulated signal, which is then sent over a transmission cable and converted back to a digital signal by another modem.

Telephone modems are used to troubleshoot a remote PLC-based control system. The major difference from the transmission cable used for remote field devices is that the transmission is over commercial telephone lines. Lower transmission frequencies are generally used (300–3300 Hz). This range is the optimum bandwidth of a telephone system.

15.10.3 Programming Devices

The programming device may be a handheld programming unit, a small display with push buttons that mounts to the front of the PLC (smaller PLCs), or a PC. The use of the PC as a programming terminal is by far the most prevalent in the industry, especially for larger applications.

Proper software from the manufacturer must be loaded on the PC. The PC then becomes a portable device that can be used for creating ladder logic, editing existing code, uploading programs from the PLC to PC, downloading programs from the PC to PLC, and troubleshooting the equipment. The software is ideal for troubleshooting problems on a machine since logic continuity is displayed on the terminal. Therefore, it may be determined why the logic is not being solved as it should (or which inputs are not being actuated properly).

The PC may also be used to save a copy of the PLC program. The program may be saved on the hard drive of the computer or on any storage medium that is supported by the PC. It is important to save a copy of the current version of the PLC program that is running on the machine. Therefore, if the PLC program is lost, the program may be reloaded into the PLC, restoring the system to a running condition in a matter of minutes.

15.11 Data Communications

The data highway is a communication link that enables PLCs to communicate with other PLCs, various computers throughout a plant, process sensors, and operator consoles. Within a plant, a data highway network consists of twisted-pair cables, coaxial cables, and fiber-optic cables. They carry the digital data between various PLCs and other intelligent microcomputer-based devices within packets of information.

Within a given plant these networks generally extend a maximum of approximately 15,000 feet (cable length). Networks allow information such as control settings and status values to be shared between a wide range of devices. Through a sophisticated system of digital signal transmission and reception, binary data can be organized in a structured format called a *protocol*. A common type of protocol is transmission control protocol and Internet protocol (commonly referred to as TCP/IP). This data format is then electronically moved (or switched) over fiber-optic cable or wire cable. A common system of switching devices and cables is called *Ethernet*.

Networks can be proprietary (supported by one manufacturer) or conform to industry standards. TCP/IP over Ethernet is an industry standard adopted by manufacturing plants as well as business offices. In the manufacturing plant, there can be one to three types of communication networks:

1. The device network that offers high speed access to plant floor data from a broad range of plant floor devices.
2. The control network that allows intelligent automation devices to share information required for supervisory control, work-cell coordination, operator interface, remote device configuration, programming, and troubleshooting.
3. The information network that gives higher-level computing systems access to plant floor data.

Components of Allen-Bradley's Data Highway Plus are shown in Figure 15-25. It is an industrial local area network designed for factory floor applications and allows connection of up to 64 devices, which may include:

- Programmable logic controllers (PLCs)
- Human machine interfaces (HMI)
- Personal computers
- Larger host computers
- Numerical and process controllers

The result is an automation system of computer-controlled devices that can help increase

Figure 15-25 **Data Highway-Plus communication interface.** *(Courtesy of Rockwell Automation, Inc.)*

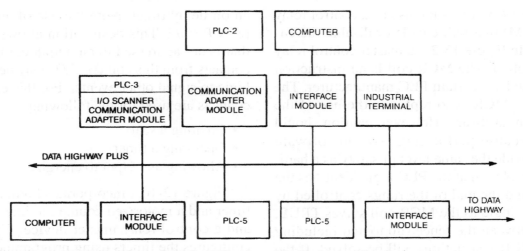

Figure 15-26 **Typical Data Highway-Plus configuration.** *(Courtesy of Rockwell Automation, Inc.)*

output, improve quality, reduce costs, and speed products to market. A typical Data Highway-Plus configuration is shown in Figure 15-26.

15.12 Converting from Relay Logic to PLC

The programming language adopted by PLC manufacturers uses basic ladder logic symbols. For more complex symbols, such as timers, a generic rectangle is used in this text. The programming instructions and the operation of these devices vary drastically between PLC manufacturers. Therefore, a simplified representation will be used here for discussion purposes.

To understand the operation of a control circuit implemented with a PLC, this text shows the conversion of the relay logic circuits in Figures 12-6A, 12-7A, 12-8A, 12-9A, 12-10A, 12-11A, and 12-13A to typical PLC circuits.

In the conversion of a hard-wired, relay-logic circuit to a PLC circuit, one of the critical principles to understand is the state of the input devices. Whether the input switches are NO or NC in hardware will affect the instructions that are used in logic to perform the operation. For example, an NC STOP push-button in a motor control circuit is necessary to interrupt the flow of current to the motor controller. However, in a PLC circuit the push-button could be hard-wired either open or closed.

STOP devices should always be hard-wired using NC contacts to the PLC. This is important so that the circuit will fail to a condition of being safe— this is termed *fail-safe*. For example, if an NC contact from a STOP push-button is wired to an input card, an NO contact must be programmed in the logic for the motor starter in order for the circuit to function properly. If the wire to the PLC input card was lost, the input will turn OFF and the motor circuit will de-energize. The circuit, therefore, fails to a condition of being de-energized (OFF). If an NO contact from the STOP push-button were wired to the input card, an NC contact would need to be used in logic for the circuit to function properly. In this case, if the wire connected to the input terminal were disconnected, the input could not be energized and the output could not be turned OFF. The circuit could not be stopped by pressing the STOP push-button since it is no longer connected to the input card. This circuit fails to an unsafe condition since there is no means available to control the circuit once it has been started. Therefore, NC contacts should always be used for STOP devices.

In industrial applications where the PLC will replace the relay logic circuit, the operator controls and process switches will generally remain as they are wired. Therefore, the PLC logic will need to be written so the process functions correctly.

The following descriptions focus on where the PLC can be applied in completing the circuit requirements.

Figure 12-6A shows the use of a control relay master (CRM) that will enable or disable the entire circuit. In Figure 15-27, a master control relay MCR is applied. The MCR coil is an instruction used in logic by a certain PLC manufacturer. The intent of the MCR is to replicate the use of the CRM coil in hardware. However, caution should be taken because performing tasks in software may not provide the same level of safety as a hardware circuit. Note that the PLC logic identifies the beginning and the end of the rungs controlled by MCR. If the logic to the MCR coil solves TRUE, the logic between the fence (up to and including the END MCR instruction) will be solved. If the logic to the MCR coil solves FALSE, then the logic between the MCR fence will not be solved. All non-retentive outputs will be turned off.

Figure 12-7A shows the use of a motorized timer. With separate clutch and motor circuits, motorized timers can be configured in many ways. For PLCs, timers are either on-delay or off-delay.

Therefore, in Figure 15-28 an on-delay timer, TON, was used to simulate the action of the motorized timer. When using a programmed timer, the programmer must place a value in the preset, time base, and accumulated value. The time base is the increments the timer will use to time, for example, .01 sec, .1 sec, or 1 sec. The preset is the number of counts (in increments of the time base) for the timer to time out. The actual preset time may be determined by multiplying the preset by the time base. For example, in Figure 15-28, the timer will time for 5 seconds before it times out. The accumulated value is the current timing count stored in the timer. This value will continually increase as the timer times and will stop when the preset value is reached. An initial value of zero is usually entered in the accumulated value.

Figure 12-8A is a relay only, switching logic circuit. Therefore, Figure 15-29 will reflect the same operation given the condition of the input devices and the contacts used in the logic program.

Figure 12-9A, similar to Figure 12-7A, incorporates a motorized timer. Figure 15-30 employs

an on-delay timer. Note the use of an "internal" relay B3:5/2. This designation allows for a relay that is not addressed to an output card. However, contacts from this coil (B3:5/2) may be used in the logic to signal other events. For this example, the contacts are used for the following:

- Creating a seal
- Operating a timer
- Allowing an output to energize

Figure 12-10A incorporates both, a motorized timer and a motorized counter, a time delay relay, and a control relay master device. Figure 15-31 configures the timers using programmed on-delay timers. The counter is also programmable. Like a timer, a counter will require a preset value (count).

Figure 12-11A utilizes a safety switching logic technique of the START switches wired such that both switches will need to be actuated to extend the cylinder. Also the switches must be released before the next cycle can start. In Figure 15-32 the straightforward application of an on-delay timer and appropriate logic contact configuration will cause the PLC to operate exactly the same as the relay logic circuit.

Figure 12-13A incorporates standard motor control logic. Figure 15-33 shows a technique for controlling the motor circuit external of the PLC circuit. The entire motor control circuit is shown external to the PLC. The fact that the motors are ON has no impact on the PLC circuit controlling the valves. There may be a safety issued involved, especially if the motor must be controlled when the PLC should be disabled for any reason.

Authors' Note

This chapter on programmable logic controllers is a general introduction into a multifaceted subject. Therefore we recommend the following textbook for more detailed coverage of the concepts, programming, and implementation of PLCs: *Introduction to Programmable Logic Controllers*, by Gary Dunning (Thomson/Delmar Learning, 2000).

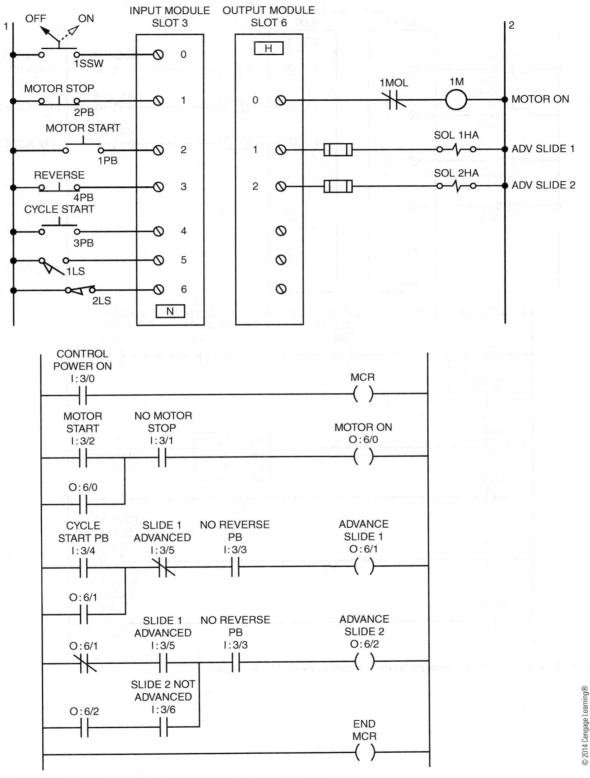

Figure 15-27 **PLC logic and I/O drawing for relay circuit shown in Figure 12-6A.**

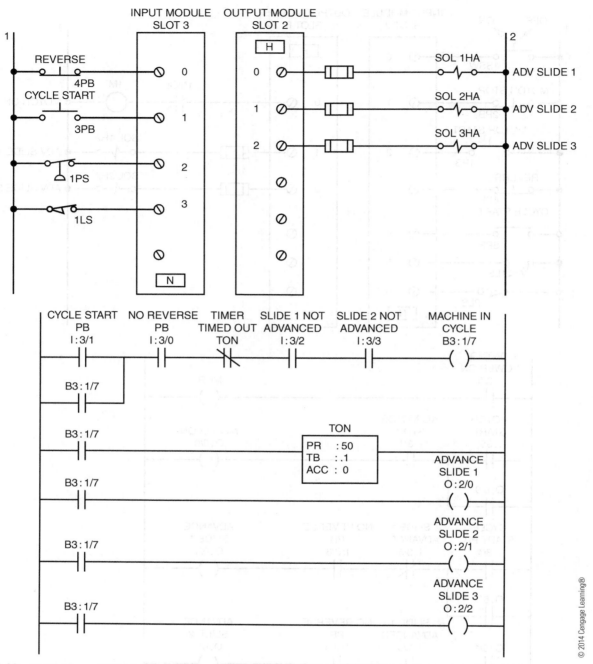

Figure 15-28 PLC logic and I/O drawing for relay circuit shown in Figure 12-7A.

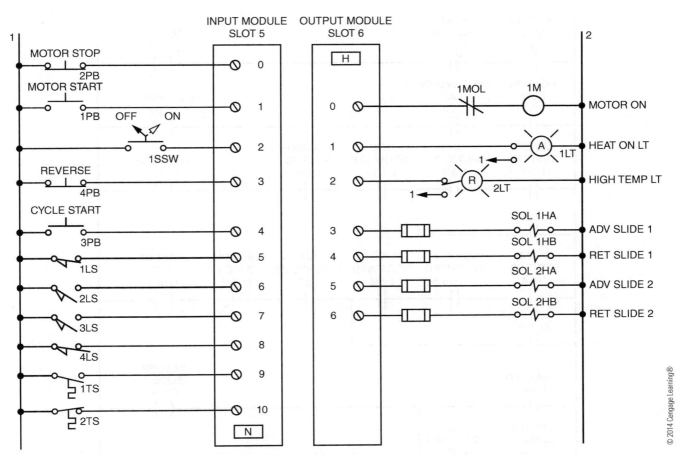

Figure 15-29A PLC I/O drawing for relay circuit shown in Figure 12-8A.

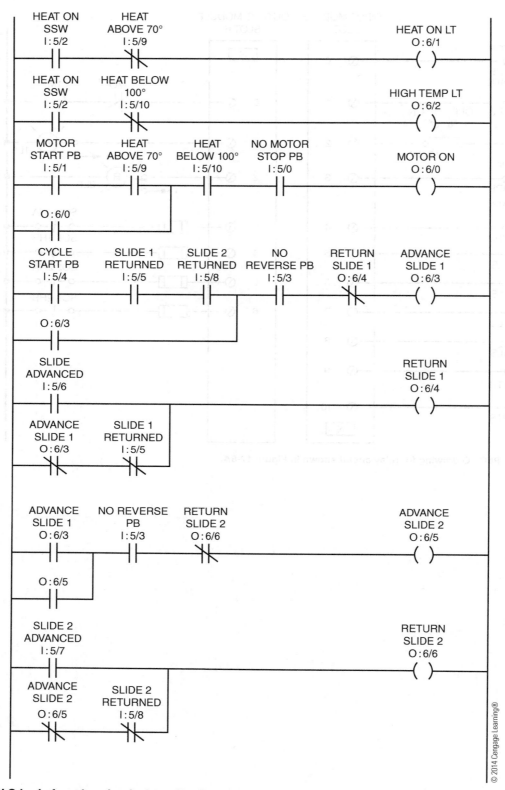

Figure 15-29B PLC logic for relay circuit shown in Figure 12-8A.

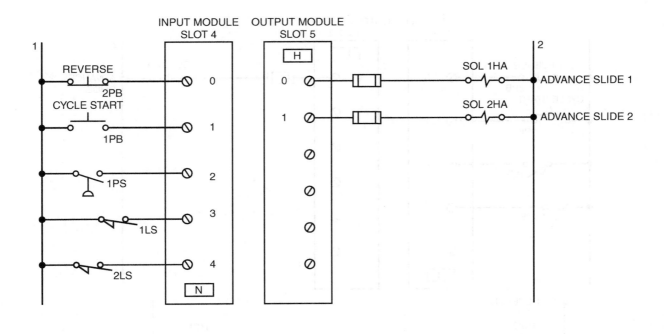

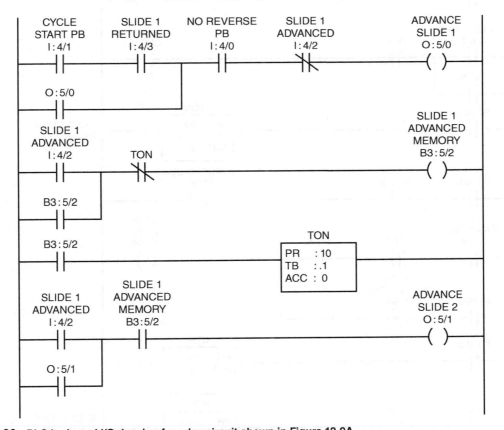

Figure 15-30 PLC logic and I/O drawing for relay circuit shown in Figure 12-9A.

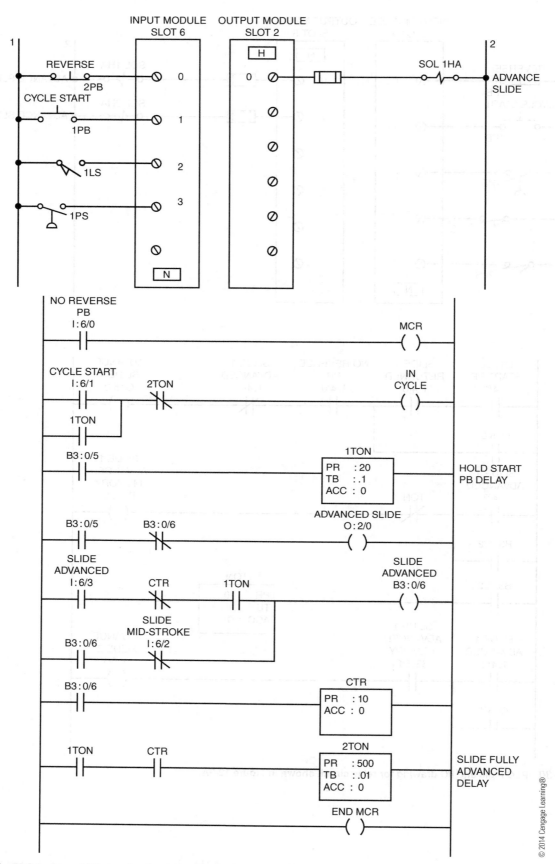

Figure 15-31 **PLC logic and I/O drawing for relay circuit shown in Figure 12-10A.**

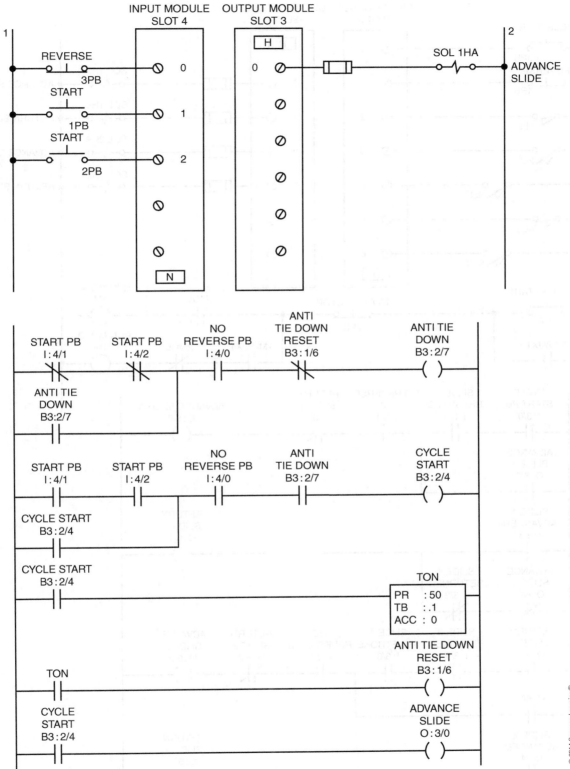

Figure 15-32 PLC logic and I/O drawing for relay circuit shown in Figure 12-11A.

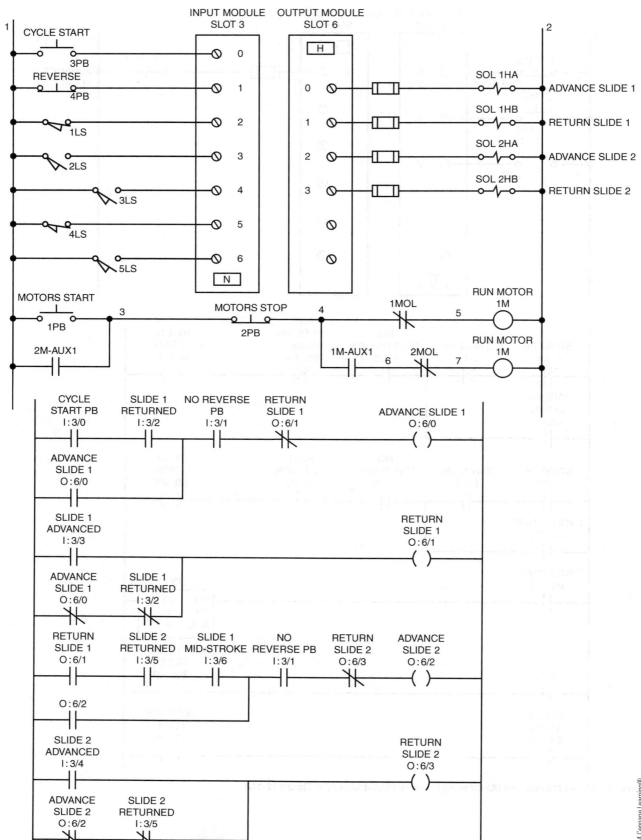

Figure 15-33 PLC logic and I/O drawing for relay circuit shown in Figure 12-13A.

Recommended Web Links

Students are encouraged to view the following Web sites as a supplement to the concepts presented in this textbook. Review and analyze the array of products that are available for electrical control applications. Many of these sites offer technical information that can help in converting the principles to practical applications. To view catalogs, your PC may require Adobe Acrobat Reader software.

Programmable Controllers

1. Allen-Bradley
   www.rockwellautomation.com
   Review: Products/Product Catalogs/Automation Systems/Programmable Controllers

2. Cutler-Hammer
   www.eaton.com
   Review: Products & Services/Cutler-Hammer/Automation & Control/PLCs and HMI-PLCs

3. Idec
   www.idec.com
   Review: Automation/PLCs

4. Mitsubishi
   www.meau.com
   Review: Products/Programmable Logic Controllers

5. Toshiba
   www.tic.toshiba. au
   Review: Industrial Product Systems and Services/Controls/Programmable Logic Controllers

6. GE Fanuc
   www.gefanuc.com
   Review: Products/Control Platforms

7. Omron
   www.omron.com
   Review: Products & Services/Industrial Automation/Automation Systems

8. Modicon
   www.schneiderelectric.com
   Review: Products & Services/Automation and Control/PAC, PLC & other controllers

9. Siemens
   www.siemens.com
   Review: Product Groups/Automation

Achievement Review

1. Name and describe the operation and function of four electrical control devices that provide information or input to a control system.

2. Name and describe the operation and function of three electrical control devices that provide decision making or logical sequencing of a control system.

3. Name and describe three electrical control devices that provide work or output from a control system.

4. Given the control circuits developed and explained in prior chapters, using the PLC symbols, convert the following circuits to PLC control: (a) assume all NO inputs and (b) assume that all input devices must remain as originally designed.

1. Figure 7-6	Pg 90
2. Figure 7-7	Pg 91
3. Figure 7-8	Pg 92
4. Figure 8-11	Pg 123, 124
5. Figure 8-12	Pg 125, 126
6. Figure 9-23	Pg 142
7. Figure 9-24	Pg 143
8. Figure 10-6A	Pg 149
9. Figure 10-6B	Pg 150
10. Figure 10-6C	Pg 150
11. Figure 10-12	Pg 154
12. Figure 10-13	Pg 155, 156
13. Figure 11-3	Pg 159

5. List four advantages of a PLC over hard-wired relay logic circuits.

6. Explain the function of the processor in a PLC. What are the differences in the processor of a small limited function PLC verses a processor for a larger multifunction PLC?

7. Explain how the PLC I/O functions. What are three types of I/O modules?

8. List three different types of devices that can electrically "condition" the input signals to a PLC.

9. Explain how the size of the PLC memory will determine the number of functional I/O points.

10. Name two types of computer memory used in PLCs.

11. Using Figure 15-18 as a guide, diagram the input, outputs, and bit addresses of the following circuits in the nonactive state (assume that all input devices must remain as originally designed):
   1. Figure 7-6
   2. Figure 7-7
   3. Figure 7-8
   4. Figure 8-11
   5. Figure 8-12

12. Why is the data communication highway important to large manufacturing processes?

13. For the circuit shown in Figure 15-28, how will the logic change if an NO contact is used for 1LS?

14. For the circuit shown in Figure 15-29, how will the logic change if the MOTOR STOP push-button is changed to an NO contact? For each design, explain what would happen to the motor if the wire connected to input terminal I:5/0 was lost or disconnected. Which design is fail-safe?

15. For the circuit shown in Figure 15-30, what is the time delay for the timer shown?

16. For the circuit shown in Figure 15-31, what happens when the REVERSE push-button is pressed?

17. For the circuit shown in Figure 15-32, what will happen if 1PB is tied down with a screwdriver (actuated all of the time) once the cycle is initiated?

18. For the circuit shown in Figure 15-33, modify the PLC logic and I/O cards to include the control of the motors in the PLC.

16

CHAPTER

Industrial Data Communications

OBJECTIVES

After studying this chapter, you should be able to:

- Understand the concept of a distributed data factory.

- Describe the architecture, media, and protocol of a local area network.

- Explain the difference between synchronous and asynchronous data transmission.

- Describe interference problems encountered by networks in industrial applications.

- Understand the components of an industrial data communication system.

- Explain the difference between an open network system and a proprietary network.

- Give three reasons why Ethernet is a preferred protocol.

- Name the four layers of an industrial network model.

- Identify five differences between device-level and control-level networks.

16.1 Overview

Industry relies on production processes and machines that produce a great deal of data. These data are used to create information such as product quality and production results. Some of the factors that necessitate the conversion of industries into the so-called distributed data factories include:

- Production systems are becoming integrated; in other words, a complete manufacturing operation may be composed of several individual processes.
- Processes are dependent on other processes for information to complete their objectives.
- Processes are controlled by any number of individual controllers (PLCs).
- Sensors and actuators are becoming smarter. They produce computer digital signals that can communicate directly with a controller.

- Data are used in both production processes and in business data processing.
- The amount of data created in the factory, and electronically processed into control and management information, has broadened the term *control technology* to *information technology* (IT).

The distributed data factory becomes a complex set of computer-controlled processes. Like building architects, engineers develop the unique design of the interrelated control systems to create an architecture perspective to information technology. This architecture is a road map illustrating how the individual sensors, actuators, displays, and controllers are combined to form a distributed data factory.

Figure 16-1 illustrates a simple production organization that represents a distributed data

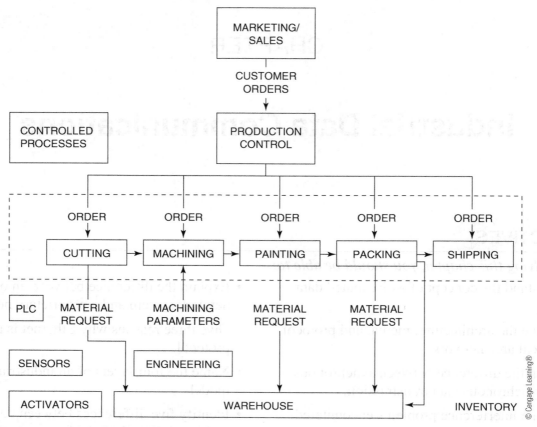

Figure 16-1 **Distributed data factory architecture.**

factory. This facility has many processes, each of which has a need to communicate with other processes for instructions and coordination. This architecture shows the flow of data between the processes.

We are familiar with the control functions of a PLC. This chapter covers the communication tasks the PLC can perform with sensors, actuators, other PLCs, and business computers. Effective communication between machines and other equipment is essential for coordinating the functions of automation systems such as computer-aided manufacturing (CAM), computer-integrated manufacturing (CIM), and control area networking (CAN).

16.2 Industrial Information Technology Architecture

Industrial data communications is an important part of the industrial IT architecture. *Data*

communications is defined as the transmission of data converted into a digital coded format (1s and 0s) to and from one place to another over an interconnected system. When applied to industrial processes, *industrial data communications* can be defined as the movement of data to and from any place in the factory. The term *factory* must include the surrounding facilities such as management and business offices, warehouses, and material supply facilities. Also, any process found in construction, transportation, and hospitals can be applied to the data factories concept discussed in this chapter.

For industrial data communications, the architecture of the communication system becomes very important to the success of the manufacturing processes and business objectives by providing data when and where it is needed. Therefore, the concept of digital pulses electronically moving through wires from an originating device to a destination device is an important principle of data communications.

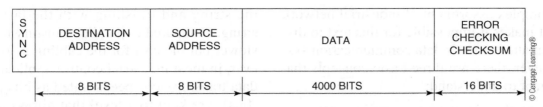

S Y N C	DESTINATION ADDRESS	SOURCE ADDRESS	DATA	ERROR CHECKING CHECKSUM
	8 BITS	8 BITS	4000 BITS	16 BITS

Figure 16-2 **Information packet.**

An industrial IT architecture must consider the following issues:

- *Connection*—How are we going to connect process and control devices together so that data will move when and where needed?
- *Transmission*—As digital data are electrical signals, what is the best electronics and media (wire or cables) we can use to move data?
- *Access*—As we have many devices trying to communicate to each other, which device will have priority? How will the devices be recognized?

16.3 Data Communication Network Concepts

Data communication networks involve the movement of digital data through transmission cables or wires. The data originate from some source such as a sensor, move through transmission lines such as twisted-pair cable, they are switched or routed to appropriate network segments, and terminate into a controller. The data path would reverse if the data originate at a controller and terminate into an actuator.

Figure 16-2 illustrates the arrangement of a packet of information that is "stamped" with a digital code, which represents source and designation identifiers corresponding to a network protocol. At the end of the packet is an error checking code that follows the data through the network. This code is checked at every point in the network to ensure that data remain the same.

With a set of identifier codes, the movement of the data is switched to and from the process equipment throughout the factory. Figure 16-3 illustrates how a network switch will interpret the data string for the destination and the error checking code. The overall control of the data movement is done by network controllers, such as a switch, that follow the rules of the protocol and, acting on the designation identifiers, switch to specific locations through programmed software at the PLC or process computer level.

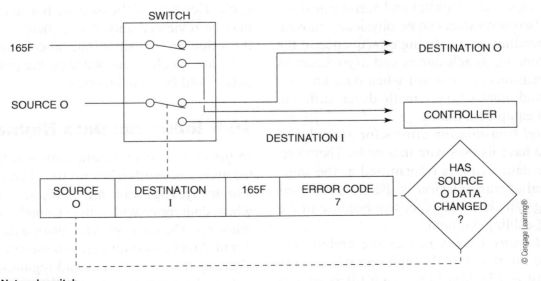

Figure 16-3 **Network switch.**

The complex assortment of industrial network equipment makes it impossible for this text to discuss each manufacturer's data communication system. However, there are three basic concepts that are important to understand:

1. For a control device to be connected to a network, it must be network compatible. That is, the device must have the physical connection such as the correct connector and it must have the electronics to create and interpret strings of digital data.
2. The string of digital data is structured in a manner that places the device identifier code, the destination code, an error checking code, and the operating value for the device all in the same string.
3. The process of determining the way in which the string fits together, how the communication is started and ended, whose turn it is to send or receive data, and how messages are confirmed is controlled by a *protocol*. The protocol is the traffic cop of the network. Many different protocols exist for industrial data communication applications. Four common protocols are Ethernet, Modbus, ControlNet, and DeviceNet.

16.4 Data Transmission

Since data from some industrial devices may be more important than other devices in the process, the amount of time required for the data to move becomes important. Parallel and serial transmissions are two ways data can be physically moved. Also, depending on the timing or sequence of the data movement, synchronous and asynchronous communications can control when data are sent and received. Each of these methods has different speed and equipment issues.

Parallel transmission allows for every bit of the data to have its own wire in a cable. Therefore, the entire data string is transmitted at the same time. Parallel transmission is the fastest means of moving data, but it is expensive because of the amount of cabling required.

Serial transmission moves the coded data string one bit at a time through one wire. If the data string is 128 bits long, each bit is moved through the wire beginning with the first bit in the string and finishing with the last bit in the string. Compared to parallel transmission, serial is slower but the cost for the cabling is lower. However, in most industrial control applications serial data transmission speeds (rated as bits per second [bps]) are kept to a level that allows controllers to monitor process activities and make appropriate adjustments more frequently. Common serial data transmission standards are RS-232, RS-422, and RS-485. Each of these standards identifies the maximum data rates using specific cables, connectors, and terminators.

Within the different channels for data to move, the issue of coordinating the flow of data is important. *Synchronous data transmission* is the movement of data that is controlled by a clocking device that determines exactly when data should move. This type of transmission ensures that the source and destination devices are looking for the data all at the same time.

Asynchronous data transmission allows for the source data to start and stop when ready. If the destination device is not ready to receive the data, the source keeps sending requests to the destination device until the destination acknowledges the source device. Only after the acknowledgement has taken place does the data move to the destination. The request-acknowledgement process is called *handshaking*.

In process control applications, data such as I/O status, which has time-critical implications, are scheduled to be reviewed during the PLC scan cycle. However, if the data are not so critical, they may be reviewed only where there is a change in the value. In this situation, time-critical applications are synchronous whereas the not-so-critical data would be asynchronous.

16.5 Industrial Data Highway

Industrial data communication implemented through a network takes on the characteristics of a highway. Highways are configured in channels where data move within the channels (serial transmission). The channels can move data in two different directions (bidirectional transmission). The highway has certain rules and regulations for priority and accuracy (protocol). When conditions are right, the flow of data may increase (bit rate).

The type of highway is also interesting to analyze. Secondary channels where data are originating or channels where data are being displayed are called *network segments*. This channel is to get the data to and from its intended source and destination. Primary channels where a large amount of data is switched and controlled are called *network buses* (also called *backbone*).

For industrial data networks, the *device level* is the secondary channel. Data traffic is slower and the data are programmed as to where they are going. The device level is where I/O devices are connected.

The data are moved to the *control level* where controllers analyze the state of the devices and compare this data to the programmed control logic, and corrections to the I/O devices are initiated. This level of the network is faster than the device level because the controller is evaluating the data strings of many I/O devices.

Periodically, the controllers send information to computers that evaluate the production data with other data such as warehouse inventory levels (once we produce it, where are we going to store it?), customer orders (how many of these items should we produce?), and product transportation needs (when are we going to pick up the product and move it to the warehouse or ship it to the customer?). This level of the network is called the *information level*.

16.6 Network Topologies

A *topology* is the pattern of interconnections among nodes. How devices are connected to a network can influence a network's cost and performance. Industrial networks must be rugged and heavy-duty because of environmental conditions such as heat, moisture, and dust.

Networks have three types of topologies: bus, star, and ring. Figure 16-4 provides an example of the connectivity for each of the three topologies.

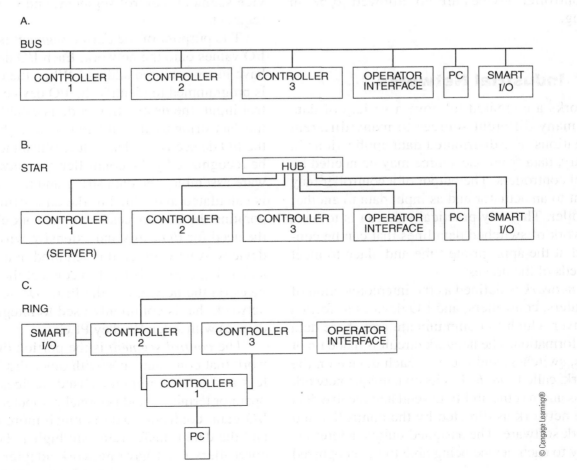

Figure 16-4   Network topologies.

Besides the connection arrangement, there are significant differences in the topologies.

Industrial networks are designed to fulfill the object of data distribution. If the data are distributed locally within the production process, then the bus topology shown in Figure 16-4A is widely used. Most industrial networks are well-suited for bus topology.

Some proprietary networks may use a different topology such as a star. Star has the advantage of higher data rates because of its direct wiring between the I/O devices and the controller. The controller serves as the network controller. Figure 16-4B shows all the network devices connected to a central hub. Controller 1 serves as the network controller.

In a ring topology, Figure 16-4C, the goal is to achieve the maximum data transfer rate between the devices. In this concept the controllers are designated to be on the ring because of their processing requirements and complex data. Generally, noncontroller devices are not allowed to be on the ring.

16.7 Industrial Networks

Networks are created to move a variety of data from many different sources to many different destinations. For distributed data applications in industry, data from one source may be needed by several controllers. The output of a controller may be sent to an actuator and as input data to another controller. The only practical solution is to have a network of switched data lines that can be controlled at the appropriate time and place to meet the needs of the process.

A *network* is defined as the interconnection of computers, controllers, and I/O devices to form a path over which to communicate and share data and information. The network circuit is made up of cables, switches, and routers. Each device on the network, called a *node*, has its own unique network address and has the ability to send and receive data on the network as directed by the controller and network software. The assigned unique address is the key to each device being able to be recognized

as the source or destination of the data. For many factory automation open-system architectures, the address of the device is automatically assigned as units are connected to the network.

The ability for data to know where they are to go once they are on the network is determined by the switching method used by the network circuit. Generally, the data are formed into data strings that contain the source address, the destination address, the data value in binary coded form, and an error checking code. All of this binary data is formed into a packet that moves together through the network. The network circuit devices recognize the source and destination addresses and make the correct circuit routes to ensure that the data are moved to the right location.

Figure 16-5 illustrates network architecture for an industrial application. To review this diagram we should look at the two major parts of the network and then focus on the details. First, there are three segments, or parts, to the network: device segment, control segment, and information segment.

The purpose of the *device segment* is to move I/O values onto the network. Each I/O device will have its own source address. When the controller is programmed to identify the I/O device as a control input, the destination code is established for the data string used by the I/O device. The state of the I/O device is the data value. When the I/O is to be recognized by the controller, the network software assembles the data string and the error code is calculated and added to the data string. Incorporating the I/O devices into a network eliminates the need for I/O cards and extensive wiring to the devices. A network card is required in a PLC slot and a single cable is used to connect the field devices on the network to the PLC. DeviceNet is a network that is commonly used to integrate field devices with Allen Bradley PLCs.

The *control segment* is the portion of the network that communicates with other PLC controllers and other computer-based devices such as operator terminals and personal computers. Unlike I/O data, controller data is much more complex and the data transfer rates are higher. Therefore, controllers must have network adapter cards to

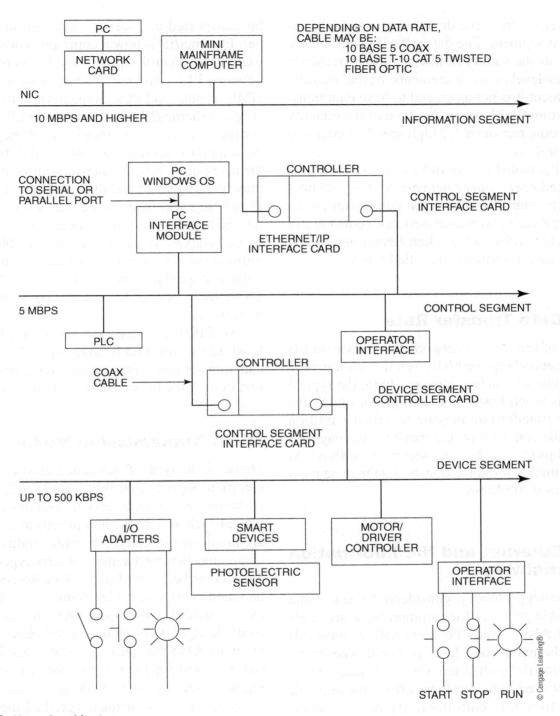

Figure 16-5   **Network architecture.**

match the appropriate network connection. ControlNet is a common network used in the control segment to communicate between Allen Bradley PLCs and other devices.

The *information segment* is considered to be the interface to the computers that are not process controllers. These computers can be various computers located in the distributed data factory as shown in Figure 16-1.

Second, each segment has a defined data transmission rate. This rate is determined by the type of cable used to connect the devices, the network protocol, and the specifications of the network interface card in the controller. In Figure 16-5, the data rate increases from the device segment to the information segment. The data string becomes more complex as the variety of information increases in the higher-level network segments. At the information segment it is not as crucial to have data transferred promptly. However, because of the quantity of data being transferred, a high speed transfer rate must be utilized.

Finally, industrial networks use a network concept called *peer-to-peer network*. In this architecture, each controller is an equal participant on the network; there is no master network controller. In business networks such as client/server networks a master network controller is called a *server*.

16.8 Data Transfer Rate

The unit of measure for network data transfer rate is bits per second (bps or b/s). Even though the rated transfer rate of a network is very high, the typical bit rate is much lower. For example, at the device level the transfer rate may be less than 1 million bps. At the control level the transfer rate may be 5 million bps (5 megabits per second or 5Mbps). At the information level, the transfer rate may be rated at 10Mbps to 100Mbps.

16.9 Ethernet and the Information Highway

The Ethernet protocol is considered to be a stable and reliable protocol for common local area networks (LANs). Many business offices have adopted Ethernet as their LAN protocol. Therefore, when industrial controllers send information to a computer used in the business office, the network segment from the controller to the business computer uses Ethernet. Ethernet uses a bus or star topology.

For companies having several factories and the need to control data among all factories, the network becomes a wide area network (WAN). Transmission control protocol/Internet protocol (TCP/IP) is widely used. This protocol is the standard for the Internet. Data utilizing this protocol can be transported to a server connected to the Internet. Ethernet/IP is now a common protocol that is used at the control segment level to communicate between PLCs and other devices such as drives, HMIs, robots, and PCs. Some devices may require special Ethernet/IP adapters to be installed in order for the devices to communicate on the network. An adapter is an interface device that transforms the data on the Ethernet network into a format that may be recognized and understood by the device. Data being sent from the device is also converted into an appropriately formatted Ethernet message by the adapter. Many devices are capable of communicating on an Ethernet network and do not require a special network adapter interface card. These devices may be directly connected to the network.

An Ethernet network at the control segment level allows for data transfer at 100 Mbps. This data transfer rate is comparable to the communication rates at the information segment level.

16.10 Transmission Media

Media is the type of substance used to carry the electrical signals that represent the digital data. Although telecommunications systems use exotic medial such as satellite and microwave, industrial applications typically use cable. Industrial data communication media may use three types of transmission media: twisted-pair (TP) wires, coaxial cables, and fiber-optic cables. Some applications may use unshielded twisted-pair wires. A common data media is eight wires (four twisted pairs) terminating in an RJ45 data jack. In some cases fiber-optic cable is used for high transmission rate network segments such as the network backbone.

As examples of media usage, the Ethernet standard can be implemented in five different ways:

10Base5	Standard "thick" coaxial cable
10Base2	"Thin" coaxial cable
10BaseT	Unshielded twisted pair (UTP)
100BaseT	Unshielded twisted pair (UTP)
10BaseFL or FOIRL	Fiber optics

16.11 Troubleshooting Networks

Transmission errors are a common problem with industrial data communications. The failed messages must be sent again, increasing the amount of data to be sent and slowing down the system. When exposed to faulty connectors or external electrical interference, network controllers, like all sensitive computer circuits, can lose data and create transmission errors. Voltage drops, power surges, and power outages are the most common problems. Fiber-optic cable is not affected by electromagnetic interference. Also, with the advances in microwave technology, more industrial applications are using wireless technology.

Cable routes should be planned to avoid fluorescent light fittings and power cables (exceptions can be made in the case of fiber-optic cable). They should not be run in the same conduit as power, or in the same cable tray of a trunking system. Crossing power cables is allowed but it must be at right angles, and some form of bridge should be used. Another problem encountered in networks is the occasional failure of the data to reach its destination. Intermittent problems are usually caused by worn cables, connectors, or electrical noise. The connectors should be verified to ensure a good connection is being made between the network and each device. The ground and cable shield connections should be verified to ensure that noise is not presenting a problem to the system.

Complete failure of data transfer is typically caused by an open (or deteriorated) cable or connector. Additionally, a network component failure can cause the complete failure of the network. Adapter modules and switches should be checked for proper operation.

16.12 Open Systems versus Proprietary Systems

The electronics, which control data networks, are a mix of computer systems, some of which follow industry standards (called *open systems*) whereas others are *proprietary* (unique) *systems* designed by a specific company. Proprietary systems are constructed of data equipment following the same proprietary software and hardware. The benefit of a factory data communication network architecture using proprietary network equipment is that all of the computer systems within the factory can be connected to the same network and the transmission of information is controlled through specific software. However, proprietary systems are only available from a specific manufacturer.

The benefit of open-system architecture is that the data communication equipment can be purchased from several different manufacturers. However, incompatibility between equipment is a problem that technologist must solve and technicians must maintain.

16.13 Network Layers

When you carefully examine how data move from a series of bits coming from a sensor to a numerical display on an operator interface panel, you begin to realize that data are constantly being pushed, tested, and transformed as they move through a network. The concept of layering of information breaks down the complex process of transporting data into pieces that can be managed by single network devices.

Here is an example of network layering:

Layer 7	Application	Defining the meaning of data and determining where data originates and terminates
Layer 6	Presentation	Building the data string (handled by controller software)
Layer 5	Session	Opening and closing communication paths (handled by controller software)
Layer 4	Transport	Error checking (handled by controller software)
Layer 3	Network	Determining data paths in the network (handled by controller software)
Layer 2	Data Link	Managing data transmission, source, destination, and checksum (handled by controller software)
Layer 1	Physical	Defining voltage levels and signal connections (handled by interface cards, repeaters, and hubs)
Layer 0	Transmission	Defining physical data transport such as the type of media to use

From this example you begin to see that industrial data communication occurs at four critical points: Layer 7, where data originates; Layers 6, 5, 4, 3, and 2, where the controller software adjusts the data string; Layer 1, where the network electronics control the voltage and signal levels; and Layer 0, where the data are actually conducted through wires. Therefore, for the technician, troubleshooting a network is reduced to four areas: (1) the sensor or actuator; (2) the controller program and operating system; (3) the network interface cards and switching electronics; and (4) the network wires and connectors.

16.14 Typical Network Systems

Figure 16-6 describes three network systems produced by Allen-Bradley.

	DeviceNet Network	ControlNet Network	EtherNet/IP Network
Function	Connects low-level devices directly to plant-floor controllers without interfacing them through I/O modules	Supports transmission of time-critical data between PLC processors and I/O devices	Plant management system tie-in (i.e., material handling); configuration, data collection, and control on a single high-speed network; time-critical applications with no established schedule
Typical devices Networked	Sensors, motor starters, drivers, PCs, push-buttons, low-end human–machine interface (HMI), bar code readers, PLC processors, and others	PLC processors, I/O chassis, human–machine interface (HMI), PCs, drives, robots	Mainframe computers, PLC processors, robots, human–machine interface (HMI), I/O and I/O adapters
Data transmission	Small packets; data sent as needed	Medium-size packets; data transmissions are deterministic and repeatable	Large packets; data sent regularly
Number of nodes (max)	64	99	No limit
Data transfer rate	500, 250, or 125kpbs	5Mbps	10Mbps to 100Mbps
Architecture	Open system	Open system	Open system
Example Functions	Control, configure, and collect data; networking sensors and actuators to a PLC or a PC to reduce field wiring and increase diagnostics	Control, configure, and collect data; PLC processor controlling remote I/O chassis, peer-to-peer messaging with other controllers using redundant media connections for time-critical applications	Control, configure, and collect data; using a single PC for data acquisition from many PLC processors, *or* using a single PC to program up/download multiple PLC processors for non-time-critical messaging between controllers

Figure 16-6 Types of Allen-Bradley network systems.

Recommended Web Links

Students are encouraged to view the following Web sites as a supplement to the concepts presented in this textbook. Review and analyze the array of products that are available for electrical control applications. Many of these sites offer technical information that can help in converting the principles presented to practical applications. To view catalogs, your PC may require Adobe Acrobat Reader software.

1. Allen-Bradley
   www.rockwellautomation.com
   Review: Products/Literature Library/Networks and Communication

2. Modicon
   www.modicon.com
   Review: Products/Automation and Control/ Bus, Networks & Communication

Achievement Review

1. Describe how a distributed data factory is different from a factory that does not use computers to control its processes.

2. What is the purpose of taking an architecture approach to documenting a production process?

3. What are the three issues that are addressed by the industrial IT architecture?

4. What is the central goal of the industrial data communication architecture?

5. Describe at least three principles of data communications.

6. What is the purpose of an error checking code in the packet of information flowing through a network?

7. What are the three basic concepts all manufacturers require of their data communication systems?

8. Describe the difference between parallel and serial data transmission wiring methods.

9. Describe the difference between synchronous and asynchronous data.

10. Provide a general description of an industrial data highway.

11. Contrast and compare the three network topologies.

12. Give three reasons why Ethernet is a preferred network protocol.

13. Name the four layers of an industrial network model. Also, list typical devices responsible for the operation of each layer.

17

CHAPTER

Quality Control

OBJECTIVES

After studying this chapter, you should be able to:

- Define quality.

- Explain why quality is important.

- Describe the role of control systems in quality control.

- Design a simple alarm circuit.

- Understand the importance of control parameters and standards.

- Describe the purpose of a normal distribution chart.

- Describe the purpose of a frequency distribution chart.

- Explain the purpose of a data acquisition system.

- Provide several reasons why quality control systems can be, or become inaccurate.

17.1 Defining Quality and Quality Control

The issue of product quality and cost effectiveness is critical to surviving in business. Competitions among companies throughout the world for limited markets have driven the need to produce products of the highest quality and at the lowest price.

Quality is a term that can mean many things to different people. A consumer may define quality as long lasting, maintenance-free, good appearance, and low cost compared to the benefit received. However, for the manufacturer the concern is how to meet the expectations of their customers, within a price that is acceptable by the customer, keeping in mind that the manufacturer's customer may be another manufacturer or an assembler.

As readers of this book, the question you are looking to have answered is: How are electrical controls used to maintain the quality of a product?

The answer is: through the use of controls, operated by qualified personnel.

The objective of quality control systems is to utilize technology in a manner that will allow process operators to maintain the standards defined as quality by the company. Most companies strive for "zero defects." However, the nature of the company's manufacturing process (production rate, tolerance standards, level of technology being utilized, and so on) will determine whether "zero defects" is a justifiable goal.

17.2 Electrical and Electronic Circuits Used in Quality Control

Circuits are used to monitor the real functional tolerances required for engineering specifications. The data obtained from the monitoring process are evaluated with statistical analysis. The result of the analysis will inform quality control personnel of the production of defective parts beyond an acceptable limit. Variations in product tolerances can be attributed to: (1) changing conditions within the fabricating equipment, (2) quality of the raw materials being used, and (3) operator error.

Today, all manufacturing companies employ one of two types of sample monitoring processes: (1) where only a few parts out of the production of many are evaluated or (2) continuous quality monitoring process where every part is evaluated. As process control techniques utilizing computer technology improves, the goal of achieving continuous (100%) quality monitoring can be economically achieved.

It is the goal of every manufacturing process to produce good parts that are within the product specifications. Rejected parts can be costly to a manufacturing facility. The company will spend additional operational costs to rework any bad parts that can be salvaged. They will also incur additional costs for materials for the rejected parts that must be scrapped.

Although rejected parts cost facilities money, the costs associated with allowing a bad part to be shipped to a customer is even more detrimental. The release of defective parts to the consumer could result in a bad image for the manufacturer, could lead to expensive warranty costs, or could even potentially cause injury to the customer. Therefore, the proper design, implementation, and maintenance of the quality control system are crucial to industry.

There are numerous methods for checking the quality of a part. In an automated system, transducers may be used to measure specific characteristics of the part against test limits. For this type of system, some of the items to consider are:

- How repeatable is the test (will the system respond with the same result every time the part is tested)?

- Is the test algorithm valid (does the test provide accurate results that eliminate or reduce the impact from other variables)?
- How is the transducer calibrated to ensure it is reading properly and what is the frequency of calibration?
- How is the electronic circuitry checked to ensure the reading has not drifted and is still accurate?
- Can a part (with known test values) be run through the system periodically to verify proper functionality?
- Are the readings from the transducer stable?
- Does the test accurately represent the conditions for the part once it is in the field?

Another method of automating the quality check of a part is to have a third party system analyze the part and send the results to the control system. For this type of system, there are two issues to consider—the integrity of the signal and whether the system is fail-safe. The use of a normally closed push-button (as a STOP device) is one example of a fail-safe design. This same concept applies to a quality signal. If the BAD signal is sent from the test system, the corresponding logic may be shown in Figure 17-1A.

The potential problem with this circuit is the condition where the testing circuit has stopped functioning properly. In this case, the BAD signal would never be sent causing all of the parts to be called GOOD regardless of their actual status. Therefore, a fail-safe system would require the GOOD signal to be energized from the test system to consider the part acceptable. This logic may look like Figure 17-1B.

To verify the integrity of the signal, one additional step is necessary. At some point during every

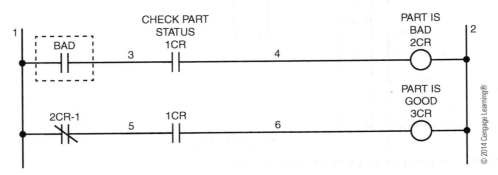

Figure 17-1A   **Quality logic using "Bad Part" signal.**

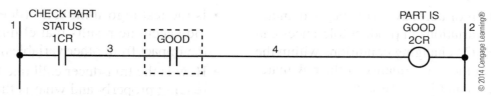

Figure 17-1B Quality logic using "Good Part" signal.

cycle of the machine, the GOOD signal should be verified that it has gone OFF. In summary, in every cycle the GOOD signal should go OFF, then transition to ON if the part being tested is GOOD, and then go OFF again. This method of checking the integrity of the signal and using a fail-safe design will ensure that all components of the quality system are functioning properly during every cycle. These concepts may be applied to all switches on a machine to make sure that all of the components are working properly.

17.3 Quality Achieved Through Machine and Process Monitoring

In processes in which there is a reliance on an operator to control the actuators that influence the quality of products, alarm circuits are provided. These circuits use process sensors and visual or audible alarm devices.

Figures 17-2A and 17-2B illustrate the switching of pilot lights at designated temperature points

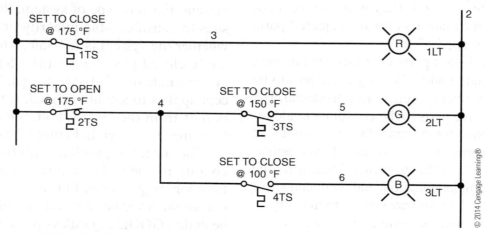

Figure 17-2A Temperature indicating circuit.

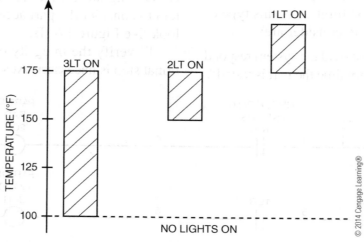

Figure 17-2B Pilot light sequence for circuit shown in Figure 17-2A.

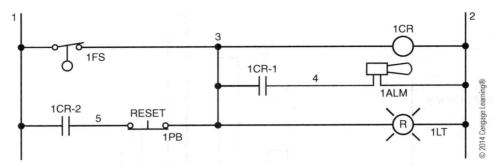

Figure 17-3 Fluid level alarm circuit.

to provide an operator with a visual representation of the process's current temperature. This temperature indicator circuit relies on the operator to be fully trained on the significance of the lights and about any actions needed when certain lights are on or off.

Figure 17-3 is an example of alarm circuit that can provide both a visual and an audible indication of the status of the float switch. Alarm horns are important in situations where the operator can be some distance from the process actuators.

17.4 Process Tolerance (Standards)

Tolerance is the measurement that defines the range of values acceptable in meeting the specifications of quality. In machining processes, tolerances can be measured one part at a time or continuously. Each method requires a different application of sensors and indicator circuits.

For example, tolerances can be measured using depth gauges and calipers with digital display.

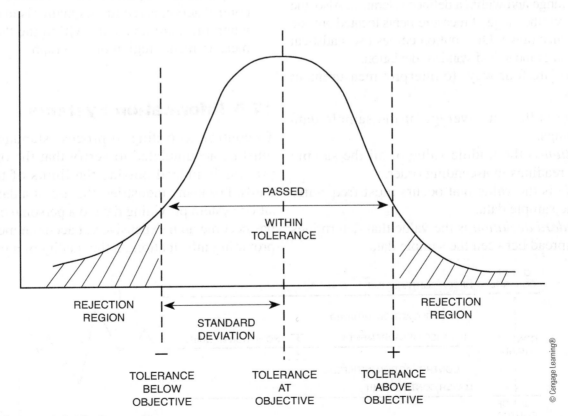

Figure 17-4 Normal distribution graph.

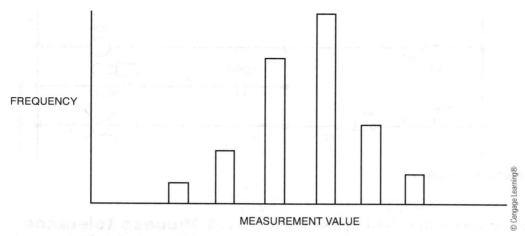

Figure 17-5 **Frequency distribution bar graph.**

These gauges can use linear displacement transducers or linear potentiometers as measurement sensors.

To classify the data obtained by measurement devices, people who are responsible for maintaining or assuring quality need to present the data in a format that will easily identify problems and show how well or how badly the process is operating. One method of graphing the measurements is with a *normal distribution*. Figure 17-4 illustrates how the measurements are within a "pass" range and within a defined tolerance. Also, the curve shows the range of measurements located outside the tolerance range. Distribution curves use statistical calculation of mean and standard deviation.

There are four ways to interpret measurement readings:

1. *Mean* is the true average of the sample data readings.
2. *Median* is the middle value of all the sample data readings in ascending order.
3. *Mode* is the value that occurs most frequently in the sample data.
4. *Standard deviation* is the value that determines the spread between the sample data.

Other graphic techniques are:

- *Frequency distribution.* Figure 17-5 illustrates a bar graph that shows the measured values and the frequency (number of times this value was measured).
- *Control chart.* These charts are applied to continuous measurement tolerances. Figure 17-6 shows this very important graph that identifies the points in the process measurements where a control action is required. Quality is maintained when measurements are within the "specified measurement" region of the graph.

17.5 Information Systems

To control according to process standards, data must be accumulated to verify that the operating process is indeed outside the limits of the standards. For many industries, the use of a data acquisition system providing data to a personal computer has become an inexpensive yet accurate method for providing information on the quality of a process.

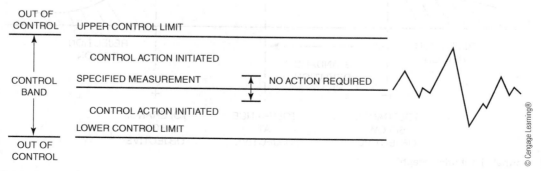

Figure 17-6 **Control chart with limits.**

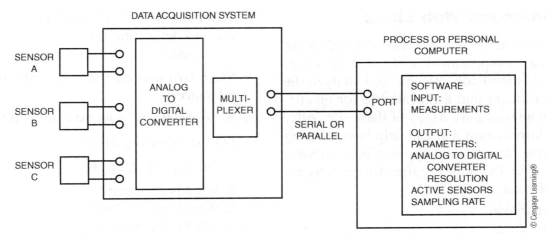

Figure 17-7 **General block diagram of process measurement and control system.**

17.5.1 Data Acquisition Systems

A data acquisition system is generally a self-contained circuit that allows for the direct connection of process sensors. The output is a multiplexed digital signal. Figure 17-7 illustrates the general block diagram of a data acquisition system. Many manufacturers have developed circuits that plug directly into a PC card slot.

The manufacturer's specifications usually define the type of sensors that can be connected to the input channels, the sampling rate of that channel, and the bit resolution of the analog to digital converter. The acquisition system has the ability to receive the measurement parameters from an operator using a personal computer. Generally, manufacturers of acquisition systems will develop PC software to allow the control of the acquisition system.

17.5.2 Personal Computer Software

As the personal computer receives the data from the acquisition system in sequence with the PC's clock, the manufacturer's software will store, print, or display the data. The software is generally adequate for basic quality analysis. However, for advanced statistical analysis, the data files will need to be analyzed by other software such as statistical analysis software or an electronic spreadsheet. Graphics software can be used to display the data in many graphic forms, including 3-D in multiple colors.

17.6 Maintaining Quality

Maintaining quality is the ability to correct any process or machine errors quickly. Therefore, quality is related to the type of control being used. The application of closed-loop control (see Chapter 6) to manufacturing systems is used extensively to keep parts within quality tolerances because tolerances ensure better fit with other parts resulting in less maintenance and longer product life.

Sensors are the eyes of the control system. Without working sensors, the ability of a closed-loop system to correct errors is lost. Therefore, good maintenance must be practiced. The following is a list of commonly found problems that cause quality control systems to become inaccurate:

- Sensors not calibrated properly
- Worn-out parts (i.e., leaking pressure lines)
- Sensors not properly bonded to the process
- Loss of power or intermittent power
- Voltage differential between ground points (faulty grounds)
- Incorrect closed-loop parameters
- Operators not being observant—reliance on open-loop control
- General lack of maintenance

Recommended Web Links

Students are encouraged to view the following Web sites as a supplement to the concepts presented in this textbook. Review and analyze the array of products that are available for electrical control applications. Many of these sites offer technical information that can help in converting the principles to practical applications. To view catalogs, your PC may require Adobe Acrobat Reader software.

1. Mettler Toledo
   www.mt.com
   Review: Web site

2. InfinityQS
   www.infinityqs.com
   Review: Software/Data Collection

Achievement Review

1. Interview a working member of your family, a neighbor, or a supervisor, and develop reasons why quality is important to their company.

2. With three or four other students in your group, provide a list of criteria that define the quality of a product. You can use common products such as stereos, automobiles, computers, and so on.

3. Develop a list of four sensors that can be used to monitor the quality of a product. Each sensor should relate to the criteria that defines the quality.

4. Design an alarm circuit that will sound a horn and activate an alarm light when the fluid level in a tank is too low.

5. Explain the role of the analog to digital converter in a data acquisition system.

6. Develop a list of five reasons why quality control systems can be, or become, inaccurate. From these reasons, develop a way to correct the problem.

7. What is the objective of quality control circuits?

8. What mathematical method of analysis is used to determine the level of quality being processed?

9. State two causes for variations in product tolerances.

10. Is "zero defects" a realistic goal? Why?

11. For the following data set:

    4, 28, 3, 15, 9, 7, 28, 12

    a. What is the mean?
    b. What is the median?
    c. What is the mode?

12. For the following data set:

    33, 17, 14, 28, 33, 16, 5, 33, 18, 17

    a. What is the mean?
    b. What is the median?
    c. What is the mode?

13. The frequency distribution below shows a graph for the overall length of 57 parts. A part is considered good if the overall length is between 2 and 4 cm. If a part is too long, it may be reworked. If a part is too short, it must be scrapped.

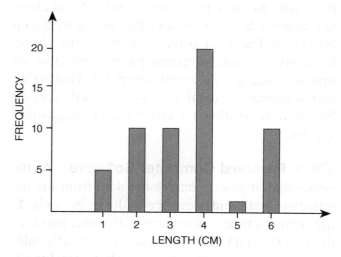

For this data set:
   a. How many good parts were produced?
   b. How many rejected parts were produced?
   c. Of the rejected parts, how many may be reworked?

14. For the circuit shown in Figure 17-3, modify the circuit so the alarm will begin to sound 30 seconds after the red light is illuminated, if the fault still exists. The alarm will remain sounding until the RESET push-button is pressed.

15. For the circuit shown in Figure 17-3, modify the circuit so the alarm will begin to sound immediately when the fault exists. The alarm will sound for 30 seconds while the fault still exists. After a 30 second delay, the horn will automatically silence but the red light will stay illuminated until the RESET push-button is pressed.

18 CHAPTER

Safety

OBJECTIVES

After studying this chapter, you should be able to:

- Explain the problems associated with unsafe worker areas.

- Describe three ways workers can be protected from moving machinery utilizing electrical controls.

- Name two organizations that publish electrical safety literature.

- Describe two advantages for utilizing a programmable controller to monitor industrial safety.

- Diagram a motor control circuit with multiple stop switches.

- Explain why machine safety and protection should be a concern of a company.

- Describe three types of sensors used in machine safety applications.

- Identify the sequence of steps within a control program to recover from a machine malfunction.

18.1 Worker Safety

Worker safety is a major concern for businesses, particularly those engaged in the production of industrial machines and manufactured products. Besides personal injury to workers, the direct and indirect costs of accidents on the job can be a large financial loss for a company. Therefore, it is very important for companies to incorporate monitoring and shielding devices within their machines that will guard workers from mechanisms that could cause bodily harm. Approximately one-third of all general industry citations handed out by the Occupational Safety and Health Administration (OSHA) point to the lack of appropriate machine guarding.

Safety for the operator and the machine are important considerations during the design phase of a piece of equipment. The best method for implementing a safe system is to design out any pinch points or hazards that exist on the machine before it is built. However, there are situations where the hazards cannot be eliminated. An example of this is when the tooling must touch the part to perform the required process.

It is imperative to protect personnel from any hazards that may potentially exist on a piece of equipment. Today, it is common practice to perform a risk assessment for a machine during the design phase of the equipment. During this assessment every motion on the machine is evaluated and every potential hazard is looked at in the event of any possible failure that may occur. The assessment team will also determine every person who may be exposed to a potential hazard, such as operators, coordinators, skilled-trades, or a person walking in the area. Each hazard will be evaluated based on two factors—the degree of injury if exposed to the hazard and the frequency of exposure for every person identified by the team. A rating is then given to each motion that determines the level to which the safety circuit must be designed for that motion. The circuit may be a simple STOP circuit as shown in Figure 18-1A or as complex as the safety circuit shown in Figure 18-1B.

Figure 18-1B uses a light curtain and a dual-channel safety relay with appropriately rated safety devices to control the circuit.

Safety circuits may be designed in many different ways and using various types of components. Safety devices are specially designed and built to comply with a certain safety rating. Therefore, it is crucial when replacing any components in a safety circuit to select appropriate devices that comply with the safety rating of the circuit.

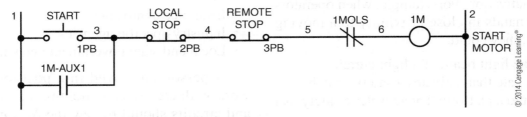

Figure 18-1A Motor control circuit with STOP push-buttons to de-energize circuit.

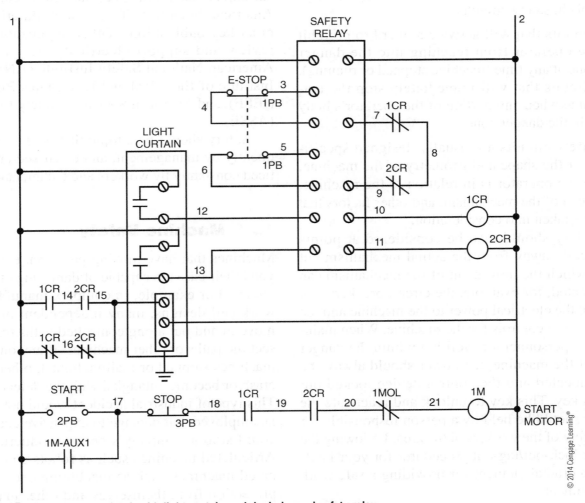

Figure 18-1B Motor control circuit with light curtain and dual-channel safety relay.

Devices used for safety circuits include:

- Light curtains
- Safety relays
- Safety plugs
- Pull cords
- Safety mats
- Safety switches
- E-stop push-buttons with mushroom head operators

The type of switched device will depend on the manner of activation. For example, when operators place their hands in close proximity to a moving mechanism, they could:

- Break the light beam of a light curtain
- Move a gate that activates a safety switch
- Move onto a platform that activates a safety mat

There are two methods for utilizing electrical controls in safety circuits:

1. Circuits that will activate a guard to prohibit the operator from reaching into the danger zone at any time (machine stopped or running).
2. Circuits that will immediately stop the machine when any portion of the operator's body is in the danger zone.

Safety circuits are usually designed specifically for the shape and geometry of the machine, where the operator is in relation to the machine, the speed of the mechanism, and other factors that must be taken into consideration.

Safety should also be considered at points physically away from the actual mechanism but from which the movement of the mechanism can be affected, for example, the circuit breakers that control the electrical power to the machine and the push-button controls for the machine. When maintenance personnel are working within the danger area of the machine, the power should always be disconnected and the control device locked out with a key. This key to unlock and reactivate the power should be held by a person responsible for the safety of the workers in the area. Following the proper lock-out/tag-out procedures for your facility is a crucial element for providing a safe work environment.

Under the law, employers are obligated to provide their employees with a safe place to work and rules for them to perform their work safely. The federal government enacted the Occupational Safety and Health Act, which created OSHA as an agency within the Department of Commerce. This group is responsible for monitoring the safety of workers and prosecuting any company violating the policies of the Agency. Besides OSHA, many other agencies are involved in worker safety, including the following:

- Insurance companies
- Industry-specific agencies
- Local and state government agencies

A person employed in a position requiring work with electrical switching, control devices, and circuits should review the *National Electrical Safety Code*®, the National Electrical Code®, American Society of Testing and Materials, and other key publications. Other organizations that review and support electrical safety are: The American National Safety Institute (ANSI), The Institute of Electrical and Electronics Engineers (IEEE), and American Society of Safety Engineers (ASSE).

Safety should be a topic that is reviewed frequently by management, and discussed and practiced constantly by workers and technicians.

18.2 Machine Safety

Machines that have moving parts that extend beyond their base are capable of damaging other machines. For example, in complex manufacturing work cell designs, many independent machines move in and out along controlled paths that intersect the paths of other machines. If any one of the machines stopped or malfunctioned, others could crash or become entangled with the defective one. This type of industrial accident would not involve an employee, but damage to expensive equipment could stop a company's entire production line. Articulated machines such as robots and coordinated machines such as machining centers need protection from themselves and other machines, for example, in machining cells, robot arms frequently pass through the line of travel of adjacent machines.

Machine safety in a coordinated and programmed manufacturing system involves the use of detectors and programmed safety sequences. Examples of common detectors are:

- Zero speed switches on motors for detecting stopped motors
- Overtorque switches for detecting jammed parts
- Proximity switches to detect overtravel
- Flow switches to detect loss of lubricating oil
- Pressure switches to detect jammed hydraulic actuators
- Temperature switches to detect overheating

Safety sequences are difficult to implement because each machine is designed to work independently. Therefore, when they are placed in a coordinated work process with other machines, a master controller is needed. This master controller is not only responsible for the correct sequencing of each machine in the process pattern, but also the recognition of a problem, the determination of the severity of the problem, and the correct action to be taken to ensure machine safety. The corrective action is in the form of a programmed sequence of moves given to each machine in the process. The sequence would involve:

1. Stop all machines.
2. Evaluate the current positions of each machine.
3. Begin to retract each machine to a home position.
4. Notify the operator of the detected problem.
5. Identify the defective machine and the source of the problem.
6. Wait for a restart command from the operator.

18.3 Diagnostic Systems

Diagnostic systems are used for the detection of faults with a manufacturing process and appropriately notifying the machine operator. In this system, the operator has the responsibility of identifying the problem and making any corrective actions. For example, a diagnostic system could detect a pump malfunction through the detection of loss of pressure in a tank.

More advanced diagnostic systems will store the number of times the fault event has taken place and what were the conditions when the event occurred. From this type of information, maintenance personnel can better predict when fault events may take place and take corrective action before any damage to the process can occur or poor quality material is produced.

18.4 Machine Safety Circuit

As an illustration of machine safety, Figure 18-2 shows a circuit designed to monitor the pressure across hydraulic fluid filters to detect and warn operators of clogged filters, automatically switch to a clean filter, and shut down the electrical motor driving the hydraulic pump. The circuit operates as follows.

Valves A and B are solenoid valves with a position switch. Pressure switches PS1 and PS2 are differential pressure switches that activate when the pressure across the filters increases because of a clogged filter. When the pump motor is started, a secondary contact from the START PUMP push-button will energize coil 1TR, which seals through the instantaneous NO contact 1TR-1 and NC contact 1CR-1. Then the NOTO contact 1TR closes immediately and stays closed to energize valve A.

When a filter gets dirty, the differential pressure across the switch goes up and associated relay contacts deactivate one solenoid valve while activating the other. The operator is alarmed. By taking the filter off-line, the opposite filter carries the load of the fluid.

Should both filters become clogged, valve A and valve B close, thereby alarming the operator and shutting down the pump motor. Additional contacts of 3CR and 4CR will ensure complete shutdown.

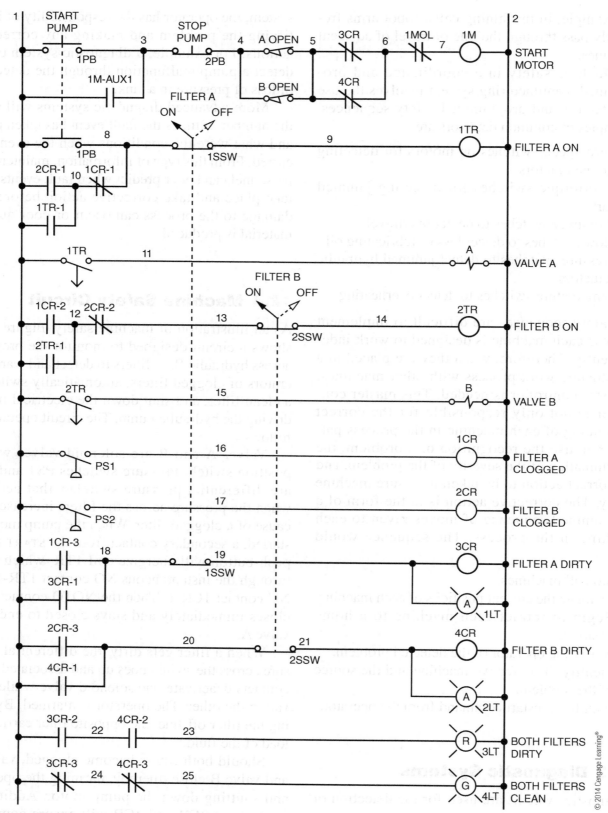

Figure 18-2 Hydraulic fluid filter monitoring circuit.

18.5 Programmable Controllers in Safety

Programmable controllers have been used to monitor safety. As an intelligent device acting as a "Safety Engineer," the controller can be programmed to monitor machine as well as human safety. However, the controller must have the appropriate detection devices strategically located to detect a safety violation or hazardous condition. The benefits of the safety controller are:

- Immediate response to a change in conditions
- A stored "preplanned action" to the response
- Notification to appropriate personnel
- Maintenance of an on-going log of safety violations
- Evaluation of conditions prior to a problem to identify the actual fault that created the problem

Figure 18-3 illustrates the principle of disabling the power to a programmable controller's input, output, and processor modules. The power can be disabled by the operator or by a PLC fault. Alarm horns are effective in gaining an operator's attention.

However, the outcome of the risk assessment will dictate how safety circuits must be implemented. Depending on the hazards and the frequency of exposure, the safety circuit may be required to be implemented in a hardware circuit only. In this case, the PLC logic may not be used in the safety circuit.

18.6 Other Safety Conditions

Many industries employ processes in operating environments conducive to explosion and fire, and that are very wet. Therefore, electrical sensors and actuators must be utilized that have been designed and certified to operate within these environments. Only devices meeting or exceeding environmental specifications should be used; for example:

- Intrinsically safe devices in explosive environments
- Electrically safe devices used underwater
- Ground fault sensors
- Lightning arrestors

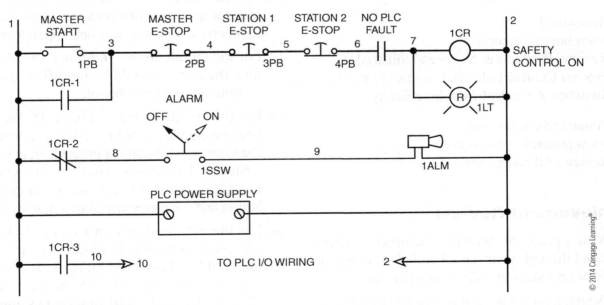

Figure 18-3 PLC control circuit.

© 2014 Cengage Learning®

Recommended Web Links

Students are encouraged to view the following Web sites as a supplement to the concepts presented in this textbook. Review and analyze the array of products that are available for electrical control applications. Many of these sites offer technical information that can help in converting the principles to practical applications. To view catalogs, your PC may require Adobe Acrobat Reader software.

1. Allen Bradley
   www.rockwellautomation.com
   Review: Products/Product Catalogs/Safety Products

2. Modicon
   www.schneiderelectric.com
   Review: Products and Services/Automation and Control/PAC, PLC & Other Controllers/Safety Controllers and Modules

3. Pilz
   www.pilz.com
   Review: Products/Sensor Technology and Control Technology

4. Honeywell
   www.honeywell.com
   Review: Products & Services/Industrial Process Control/Industrial Control/Sensors, Switches & Controls/Machine Safety

5. Pinnacle Systems, Inc.
   www.pinnaclesystems.com
   Review: All categories

Achievement Review

1. With a group of three or four other students, travel through your school and observe any areas where student safety can be at risk.

2. Design a motor control safety circuit with two operator stop switches, a bearing overheating thermal switch, and a zero speed switch.

3. Go to your local public library and obtain a copy of the *National Electrical Code*®. Describe four points where electrical safety is emphasized.

4. Should or should not programmable controllers provide all safety detection and alarm?

5. Make an appointment with a manager of a local manufacturing company and ask him/her why safety is important to the company and what is done to ensure safety.

6. Describe three types of sensors used in machine safety applications and provide examples of how they could be used.

7. Go to your school's machine shop and prepare a "safety manual" for people who will be using the machines.

8. What is the advantage of storing the number of times a fault event has taken place and what the conditions were when the event occurred?

9. Provide three advantages for a "safety controller."

10. Besides OSHA, what other agencies are involved in matters related to worker safety?

11. For the circuit shown in Figure 18-1A, show how the STOP push-buttons would be modified to incorporate mushroom head operators.

12. For the circuit shown in Figure 18-1A, show how the circuit would be modified to add an additional STOP push-button.

13. For the circuit shown in Figure 18-1A, show how the circuit would be modified so that when the START push-button is pressed, an alarm will sound for 10 seconds. After the 10 second delay, the operator may release the START push-button and the motor will start automatically.

14. For the circuit shown in Figure 18-3, show how the circuit could be modified to add a light that will be illuminated along with the alarm. The alarm will sound for 15 seconds and then silence. The light will remain ON until the operator acknowledges the situation by pressing a RESET push-button.

19
CHAPTER

Troubleshooting

OBJECTIVES

After studying this chapter, you should be able to:

- List five areas that should be considered when starting a troubleshooting job.
- Describe two methods that can be used in troubleshooting a job.
- Explain a procedure for checking fuses.
- Discuss why a loose connection in an electrical power circuit can be a major problem.
- Explain why contacts on electrical operating equipment should not be filed.
- List several problems resulting from low line voltage.
- Discuss the merits of good housekeeping.

- Show how an open in a common line can result in a faulty circuit operation.
- Explain one method for checking an electrical control circuit.
- List four steps of the procedure to locate problems resulting from momentary faults.
- Explain the relationship among voltage, current, and resistance or impedance expressed in Ohm's law.
- List several problems that can occur in an electric motor.
- Explain some of the problems in a control circuit that can be checked with a meter.

19.1 Safety First

Testing, care, and maintenance of electrical equipment should be performed by trained and experienced electricians who are thoroughly knowledgeable about electrical systems. High voltages and currents are present that may cause injury or extensive equipment damage.

All possible precautions that must be taken cannot be anticipated in testing all the different equipment. Always be certain that all power has been turned off, locked out, and tagged in any situation in which you must actually come in contact with the circuit or equipment. Be sure the equipment cannot be turned on by anyone but you.

Use only well-designed and well-maintained equipment to test, repair, and maintain electrical systems and equipment. Use appropriate safety equipment, such as safety glasses, insulating gloves, flash suits, hard hats, insulating mats, and so on, when working on electrical circuits.

Make sure that multimeters used for working on power circuits contain adequate protection on all inputs, including fuse protection on *all* current measurement input jacks.

19.2 Analyzing the Problem

Effective troubleshooting starts with an analysis of the problem. Too frequently, troubleshooting is

289

approached in a hit-or-miss fashion. This approach generally creates more expense and wastes time.

To analyze any problem, it helps to break it down into types of sections to limit the size of the job. In troubleshooting, the following causal areas might be considered:

- Electrical, electronic
- Mechanical
- Fluid power
- Pneumatic
- Personnel

In many cases, the problem may be a combination of two or more of these areas.

The following three examples illustrate how breaking down a problem into types of causes can simplify the troubleshooting procedure.

1. From advance information supplied by a machine operator or from early examination by an electrician, what first appears to be an electrical problem may turn out to be a mechanical one.

2. The failure of a limit switch to function properly may be caused by problems in the electrical contacts or the mechanical operator. The result of a cycle failing to complete is the same, however, regardless of which of these two items caused the problem.

3. As long as people operate machines, problems will arise that do not respond to the usual form of troubleshooting. Such problems may be intentional or unintentional. They may stem from misunderstanding, lack of cooperation, or lack of knowledge of the machine.

Whatever the cause of a problem, it will be recognized quickly by the troubleshooter who takes the approach outlined here. The problem should be handled carefully and diplomatically so that the machine can be returned quickly to its intended job.

Problems can be further separated into physical location or type of operation. For example, in a large machine in which the cycle of operation moves from one section of the machine to another, the trouble may be localized in one section. Localizing the area of the problem may immediately eliminate 75% of the total machine as a possible

trouble source. A practical application here could be trouble developing in loading or unloading equipment on a press. If the proper clearance signal has been given to the loader or unloader from the press control, trouble developing after this cycle starts can generally be localized in the loading or unloading control.

Success in troubleshooting is the ability to segregate the problem area from other unrelated circuitry.

Before getting into the mechanics of troubleshooting, let us examine some of the problem spots.

19.3 Major Trouble Spots

It would be impractical, if not impossible, to list all potential trouble spots. However, the following areas contribute to a large percentage of troubles.

19.3.1 Fuses Checking fuses is generally a good place to start when a problem has occurred. Too often this is overlooked. The details for checking fuses in a complete power circuit is covered in Section 19.6.

The replacement of an open (defective) fuse can be an important safety factor. As discussed in Chapter 2, there are three different types of fuses. Within each type there are different voltage and current ratings. Too often just any type of fuse is used as a replacement. Unless changes have been made in the machine circuit and components, the replacement fuse should be exactly the same type, voltage, and current rating as the fuse removed.

The policing of fuses can be a problem. However, the replacement policy given here must be rigidly maintained if safety to personnel and the machine is to be realized.

One extremely important case involves a machine connected to a power source that has a high short-circuit current available. In such cases, it may be advisable to use current-limiting fuses with high interrupting capacity.

19.3.2 Loose Connections There may be hundreds of connections on a machine. Each of these spots may be a source of trouble. Many advancements have been made in terminal block and component connectors to improve this condition. The use of stranded conductors in place of solid conductors has, in general, improved the connection problem.

The problem starts when the machine is built and continues throughout the life of the machine. It may be of greater importance in power circuits, as the current handled is of greater magnitude. A loose connection in a power circuit can generate local heat. This heat spreads to other parts of the same component, other components, or conductors. An example of where direct trouble can arise is thermally sensitive elements. These can be overload relays or thermally operated circuit breakers.

For the correction of loose connections, the best advice is to follow a good program of preventive maintenance in which connections are periodically checked and tightened.

19.3.3 Faulty Contacts
Potential problems with faulty contacts exist in such components as motor starters, contactors, relays, push-buttons, and switches.

A problem that appears quite often and one of the most difficult to locate is the NC contact. Observation indicates that the contact is closed but does not reveal if it is conducting current.

Any contact that has had an overload through it should be checked for welding.

Such conditions as weak contact pressure, dirt, or an oxide film on a contact will prevent it from conducting. Many times contacts can be cleaned by drawing a piece of rough paper between the contacts. **Caution:** Use only a fine abrasive to clean contacts. Do not file contacts. Most contacts have a silver plate over the copper. If this plating is destroyed by filing, the contact will have a short life. If contacts are worn or pitted so badly that a fine abrasive will not clean them, it is better to change the contacts.

Another problem that may occur with a double-pole, double-break contact is cross-firing, that is, one contact of the double break travels across to the opposite contact, but the other remains in its original position. If both the NO and NC contacts are being used in the circuit, a malfunction of control may occur.

19.3.4 Incorrect Wire Markers
The problem of incorrect wire markers usually appears on the builder's assembly floor or in reassembly in the user's plant. The error can be difficult to locate, as a cable may have many conductors running some distance to various parts of the machine.

One common problem is the transposition of numbers. For example, a conductor may have a 69 marked on one end and a 96 on the other end. Another problem that may occur is in connecting conductors into a terminal block. With a long block and many conductors, it is a common error to connect a conductor either one block above or below the proper position.

19.3.5 Combination Problems
Reference has been made to combination problems, but their importance should be emphasized. The following are typical types of combination problems:

- Electrical-mechanical
- Electrical-pressure (fluid power or pneumatic)
- Electrical-temperature

The greatest problem is that the observed or reported trouble is not always indicative of which aspect of the combination is at fault. It may be both.

It is usually faster to check the electrical circuit first. However, both systems must be checked as both may contribute to the problem.

As an example, very few solenoid coils burn out due to a defect in the coil. Probably over 90% of all solenoid trouble on valves develops from a faulty mechanical or pressure condition that prevents the solenoid plunger from seating properly, thus drawing excessive current. The result is an overload or a burned-out solenoid coil.

19.3.6 Low Voltage
If no immediate indication of trouble is apparent, one of the first checks to make is the line and control voltage. Due to inadequate power supply or conductor size, low voltage can be a problem.

The problem generally shows up more on starting or energizing a component, such as a motor starter or solenoid. However, it can cause trouble at other spots in the cycle.

A common practice in small shops is the addition of more machines without properly checking the power supply (line transformers) or the line conductors. The source and line become so heavily loaded that when they are called on for a normal temporary machine overload, the voltage drops off rapidly. This drop may result in magnetic devices such as starters and relays dropping off the line (opening their contacts) through undervoltage or overload protective devices.

Heat is one result of low voltage that may not be noticed immediately in the functioning of a machine. As the voltage drops, the current to a given load increases, producing heat in the coils of the components (motor starters, relays, solenoids), which not only shortens the life of the components but may cause malfunctioning. For example, where there are moving metal parts with close tolerances, heat can cause these parts to expand to a point of sticking. In cases in which electrical heating is used, the heat is reduced by the square of the voltage. For example, if the voltage is dropped to one-half of the heating element's rated voltage, the heat output will be reduced to one-fourth.

19.3.7 Grounds

19.3.7.1 Typical Locations There are many locations on a machine where a grounded condition can occur. However, the following are a few spots in which grounds occur most often.

- *Connection points in solenoid valves, limit switches, and pressure switches.* Due to the design of many components, the space allowed for conductor entrance and connection is limited. As a result, a part of a bare conductor may be against the side of an uninsulated component case. Where bolt and nut connections are made, the insulating tape may not be wrapped securely, or it may be of such quality that age destroys its insulating properties. This problem may occur in the field (user's plant), where due to the urgency for a quick change, sufficient care is not taken when handling the wiring and connections on a replaced unit.
- *Pulling conductors.* In pulling conductors through conduit where there are several bends and 90-degree fittings, or into pull boxes and cabinets, the conductor insulation may be

scraped or cut. If care is not taken to eliminate sharp edges or burrs on freshly cut or machined parts, cuts and abrasions occur. Avoidance of scrapes and cuts is one good reason for the use of the insulation required for machine tool wiring.

- *Loose strands.* The use of stranded wire has greatly reduced many problems in machine wiring. However, care must be taken when placing a stranded conductor into a connector. All strands must be used. One or two strands unconnected can touch the case or a normally grounded conductor, creating an unwanted ground. Even if the ground condition does not appear, the current-carrying capacity of the conductor is reduced.

19.3.7.2 Means of Detection For the best operation of an electrical system, some means of detecting the presence of grounds should be available. There are two methods shown here. Each has merits.

Using the grounded method shown in Figure 19-1, the circuit is de-energized by the opening of the fuse. This condition means that the machine will be down until the ground is located and removed. In the small shop this inconvenience is usually not too serious. The ungrounded method shown in Figure 19-2 may be used in large production shops. This method has the advantage that production can continue with one ground until there is time to locate the ground and remove it.

In the first method, all coils are tied solidly to a common line, and this line is grounded. The opposite side of the control power source is protected by a fuse or circuit breaker.

If a ground condition should appear in the circuit shown in Figure 19-1 on wire number 8 between contact 1CR-2 and solenoid 1PA, the load (coil) is bypassed because there is a direct path to ground on both sides (relay 1 CR energized—1CR-2 closed). This direct path to ground puts a direct short on the control power source, opening the protective device and removing the control power from the circuit.

When the ground is located and removed, the fuse can be replaced or the circuit breaker reset.

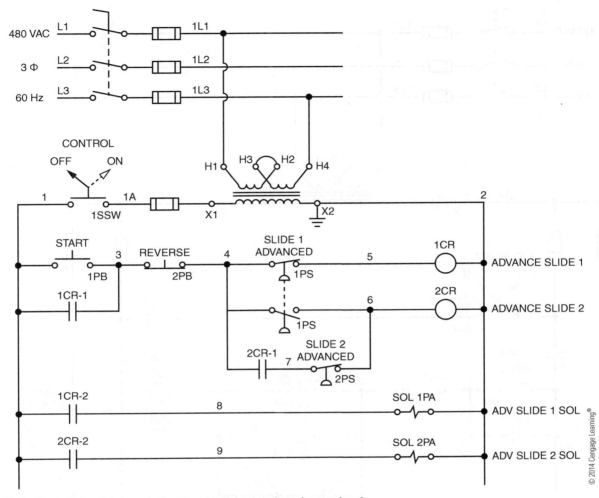

Figure 19-1 Control circuit (grounded) with potential ground at wire number 8.

Thus, the control circuit is again ready for operation. Note that the ground condition must be removed first.

In Figure 19-2, the common side is not grounded. A set of two ground detector lights (standard 120-V indicating lights) are connected in series across the power source. A solid ground is then placed between the two lights.

With the system showing no grounds, both lights will glow at half brilliance, since two 120-V bulbs are connected in series across 120 VAC (60 V on each).

If a ground appears in the system, one of the lights will go out. The other will go to full brilliance. The determining factor of how these two lights perform depends on which side of the load the ground appears. For example, if a ground

should appear on wire number 16 between the fuse and solenoid 1PA, pilot light 1LT will go out when contact 3CR-1 closes. Pilot light 2LT will glow at full brilliance.

An advantage of this method is that in some cases a ground can be present but not cause any immediate trouble. A circuit that keeps ground faults from disabling a machine gives maintenance time to check for the ground without immediately taking the machine out of production.

The important point is that some system should be used to detect the presence of a ground.

19.3.7.3 Wiring and Grounding   Incorrect wiring and grounding can result in major problems for companies with computers and computer controlled equipment. These problems often show up

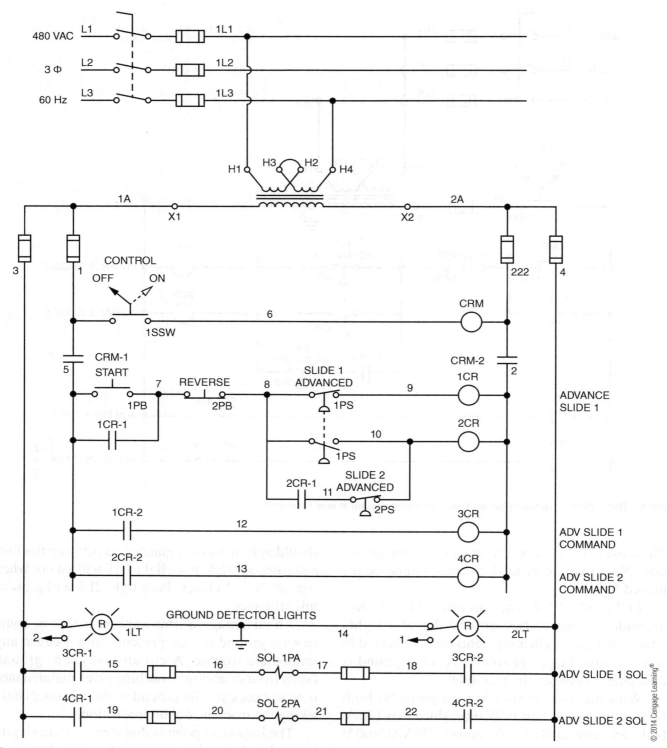

Figure 19-2 Control circuit (ungrounded) with ground detector lights.

as unexpected shutdowns, burned-out equipment, scrambled data, and equipment lockups.

The Electric Power Research Institute estimates that 80% of all power quality problems are the result of poor grounding, wiring, or load placement on the user's circuits. By making a few simple checks with your digital multimeter (DMM) you may be able to eliminate some common power quality problems caused by wiring and grounding.

It is recommended that all circuits that serve critical loads should be on dedicated circuits and neutrals. Check wiring for correct bonding, tied neutrals and grounds, overloaded neutrals, and shared neutrals to critical loads and to ensure that the neutral is not used for equipment grounding.

A visual check of wire condition may be all that is necessary to eliminate many problems associated with poor equipment performance.

19.3.8 Poor Housekeeping
Poor housekeeping leads to more work for the troubleshooter. There is an overall economy in having a clean machine and a well-organized and well-executed preventive maintenance program.

Dust, dirt, and grease should be removed periodically from electrical parts. Their presence causes mechanical failure and forms paths between points of different potential, causing a short circuit.

Moving mechanical parts should be checked, particularly in large motor starters. Such items as loose pins and bolts and wearing parts are sources of trouble.

Overheated parts generally indicate trouble. Without proper instruments, it is difficult to determine the temperature of a part or how high a temperature can be sustained. Certainly any signs of smoke or baking of insulation are cause for immediate concern.

Manufacturers of components have done considerable design work to prevent dust, dirt, and fluids from entering.

When it is necessary to remove a cover or open a door for troubleshooting, immediately replace it after the trouble is corrected.

Many users have gone to great lengths to develop and rigidly enforce a good electrical maintenance program. Records are kept of each reported trouble and the work that was done to correct the problem. These records are compiled periodically and should be available to the skilled-trades department. Such records not only lead to faster troubleshooting in the future, but also give the production supervisor an indication of why the output in a given department may be down.

19.3.9 Trouble Patterns
As troubleshooting work progresses over a period of time with a particular machine or group of machines, a pattern of issues may develop. For example, it may be necessary to increase the production rate with a particular machine. Certain areas of this machine may not have been originally designed to handle an increased rate of operation. Unless the machine is redesigned and rebuilt, a pattern of trouble may develop.

Another example is a machine that is relocated to another section of the user's plant where the environment may be different. A change in the atmosphere and a presence of dust, dirt, or metal chips may create a pattern of trouble for a machine if it is not designed to operate under these conditions.

19.3.10 Opens in a Common Line
Many circuits have two or more common connection points for multiple connections. For example, from the circuit diagram shown in Figure 19-3, wire number 1 has two points of connection. Wire number 2 has four points of connection. Wire number 4 has six points of connection. As it is not good wiring practice to place more than two conductors under any one terminal, connections are generally jumpered on the panel components or brought back to the terminal block. It is not unusual for a condition to arise in which a jumper is omitted. For example, suppose the jumper connecting the common wire for relay 4CR is omitted (Figure 19-3). The schematic circuit is shown as it would appear with this jumper missing to illustrate the effect on the condition of the circuit and the resulting faulty operation.

In this rather simple circuit, it does not appear that this error would be committed. However, there are cases in which there may be many of these

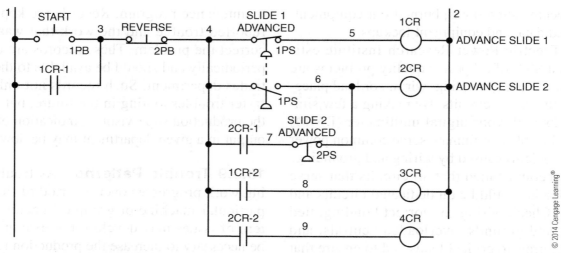

Figure 19-3 Control circuit with open common.

common connections. In larger and more complicated circuit diagrams, the probability of jumpers being missed is much greater.

19.3.11 Wiring the Wrong Contact
Many components such as relays, limit switches, pushbuttons, and temperature switches have NO and NC contacts available for use.

A wiring error may be made, particularly when only one of the two available contacts is used in the circuit. The error consists of wiring the wrong side of the contact; that is, the NO contact may be put into the circuit where the NC should be used. The reverse of this situation may also be true.

Unless the person who does the wiring is completely familiar with the component and double-checks the work, this error will occur frequently.

The routine checks discussed in the following section will reveal this error quickly.

19.3.12 Momentary Faults
The momentary fault is one of the most difficult problems to troubleshoot. With this type of fault, the machine or control can be under close observation for hours with no failures. However, the fault may occur at any time, and if the observer's attention is even briefly diverted, any direct evidence that might have been seen is lost.

There is no direct solution to this problem. The best approach is a well-organized analysis. The following steps might be helpful:

1. Attempt to localize within the total cycle. If the fault always occurs at the same place in the cycle, generally only the control associated with that part of the cycle is involved. If the fault occurs at random spots through the cycle, then the spots to examine are those that are common through the entire cycle. Examples are a drum or selector switch used to isolate an entire section of control.

2. Examine for loose connections, particularly in the area of the fault. Attention should be paid to areas where mechanical action may have damaged conductors, pulled them loose from connectors, cut, or broken them. The complete break of a stranded conductor within the insulation is rare.

3. Localize attention to components. Many times casual observation will not disclose the trouble. In these cases the complete replacement of the component(s) in question is the quickest and best solution. Here, the plug-in components have a distinct advantage in returning a machine to operating condition in a minimum of time.

4. Examine the circuit for unusual conditions. This type of trouble rarely occurs in the user's plant. However, there are cases in which, either through an oversight on the part of the circuit designer or by a change of operating conditions on the machine, a circuit change is indicated as a solution to the problem.

19.4 Equipment for Troubleshooting

Some of the most important tools for a trouble-shooter include:

- A working knowledge of Ohm's law.
- Meters that can be used to measure/indicate the important units involved in Ohm's law.
- Meters that can be used to measure/indicate temperature.
- Safe and applicable test probes.

Figure 19-4 outlines the relationship among important areas of electrical work: voltage, current, resistance or impedance, and power.

Control circuits, relay or PLC, should have visual indicators to identify logic and actuator status. They may be LEDs, incandescent lamps, or neon lamps. Their operation supplies important information to the troubleshooter. However, there is still important information needed when a system or machine fails to operate properly. This information may include the voltage, current, and resistance/impedance in a given circuit.

For example, an open circuit will show infinite resistance, and a closed circuit will show zero resistance. Figures 19-5, 19-6, and 19-7 show some of the instruments available for troubleshooting. Figures 19-8 and 19-9 show examples of well-designed test probes.

19.4.1 Multimeters

Many tools are available for troubleshooting problems in electrical circuits. For measuring analog signals that vary over time, an oscilloscope is a useful tool. Probes may be connected to the circuit and a real-time graph of

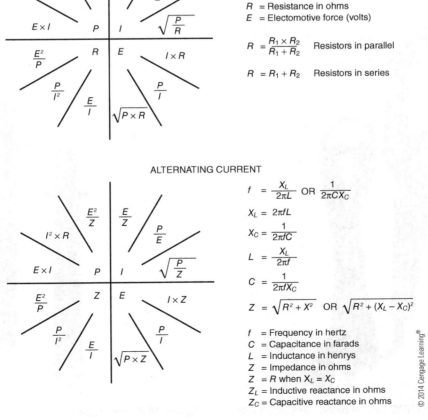

DIRECT CURRENT

P = Power in watts
I = Current in amperes
R = Resistance in ohms
E = Electomotive force (volts)

$R = \dfrac{R_1 \times R_2}{R_1 + R_2}$ Resistors in parallel

$R = R_1 + R_2$ Resistors in series

ALTERNATING CURRENT

$f = \dfrac{X_L}{2\pi L}$ OR $\dfrac{1}{2\pi C X_C}$

$X_L = 2\pi f L$

$X_C = \dfrac{1}{2\pi f C}$

$L = \dfrac{X_L}{2\pi f}$

$C = \dfrac{1}{2\pi f X_C}$

$Z = \sqrt{R^2 + X^2}$ OR $\sqrt{R^2 + (X_L - X_C)^2}$

f = Frequency in hertz
C = Capacitance in farads
L = Inductance in henrys
Z = Impedance in ohms
$Z = R$ when $X_L = X_C$
Z_L = Inductive reactance in ohms
Z_C = Capacitive reactance in ohms

© 2014 Cengage Learning®

Figure 19-4 Ohm's law.

Figure 19-5 Multimeter. *(Reprinted with permission from the Fluke Corp.)*

Figure 19-7 Digital thermometer. *(Reprinted with permission from the Fluke Corp.)*

Figure 19-6 Clamp-on AC/DC current probe. *(Reprinted with permission from the Fluke Corp.)*

Figure 19-8 Industrial test probes. *(Reprinted with permission from the Fluke Corp.)*

Figure 19-9 **Banana-plug test probes.** *(Reprinted with permission from the Fluke Corp.)*

the signal may be displayed on the screen. Some oscilloscopes also offer a storage capability so the signal may be captured over a length of time.

A general-purpose tool that is frequently used to troubleshoot electrical circuits is a digital multimeter (DMM). Most DMMs are capable of displaying voltage, current, and resistance measurements. The user must select the appropriate measurement function by turning a rotary dial switch to the proper position. It is also important to select appropriately between an AC and DC signal. When measuring an AC signal, the DMM will display the root mean square (rms) value of the signal. The rms value is the effective or equivalent DC value of the AC signal. Most meters will provide an accurate reading only if the AC signal is a pure sine wave. If the signal is nonsinusoidal, a *true-rms* meter will need to be used to obtain an accurate reading. In addition, the specifications for the meter should be consulted to ensure it is capable of accurately reading the frequency of the AC signal being measured.

19.4.2 Voltage Measurements
A DMM may be used to measure voltage in an electrical circuit. The meter must be set to the appropriate voltage function and to the correct range. The probes should be placed at the two points where the voltage must be measured across the circuit. The display will indicate the potential difference between these two points in the circuit (Figure 19-10).

Since voltage measurements must be taken while the circuit is live, it is important to give attention to the proper positioning of the test probes. It is crucial to avoid slipping off the test points and to keep your hands clear of high voltage and moving machinery.

It is also critical to ensure that proper voltage is supplied to all electrical equipment on the circuit. Additional loads (new machinery, new outlets, and so on) that have been added to existing circuits may cause excessive voltage drop and energy losses. The *National Electric Code*® provides the allowable voltage drop for a branch circuit (Figure 19-11).

To conserve energy and reduce equipment costs, motors are typically matched closely to load requirements and have very little reserve power. Therefore, with motors operating close to full load, the chance of motor burnout and energy waste due to unbalanced voltage is increased. Voltage imbalance results in unbalanced currents in the stator windings and can be caused by single-phase loads that have been tapped onto three-phase supply circuits (Figure 19-12).

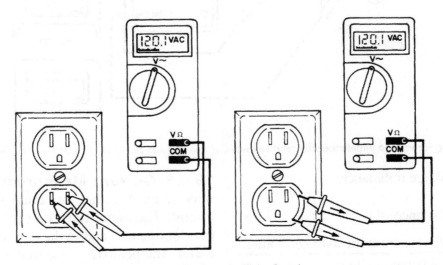

Figure 19-10 **Measuring voltage.** *(Reprinted with permission from the Fluke Corp.)*

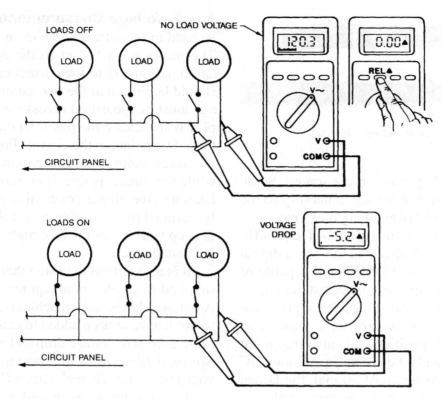

Figure 19-11 Measuring voltage drop in branch circuits. *(Reprinted with permission from the Fluke Corp.)*

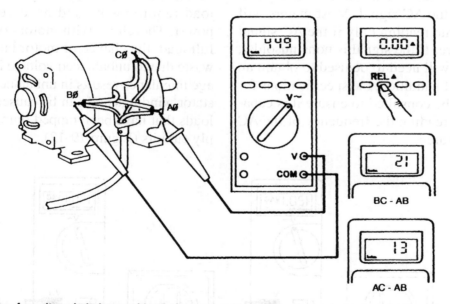

Figure 19-12 Checking for voltage imbalance with relative mode. *(Reprinted with permission from the Fluke Corp.)*

To calculate voltage imbalance:

Percent voltage unbalance =

$$\frac{\text{Maximum deviation from the average voltage}}{\text{Average voltage}}$$

19.4.3 Current Measurements

A DMM may be used to measure current in an electrical circuit. The meter must be set to the appropriate current function and to the correct range. For most meters the positive probe may need to be repositioned to the appropriate location on the meter. The electrical circuit must be turned off and the circuit must be broken. The probes should be placed in series with the current path. For a DC circuit the positive probe should be placed in the position where the potential difference is the most positive. The circuit may now be energized. The display will indicate the current flowing in the circuit. A current probe similar to the clamp-on unit shown in Figure 19-6 is frequently used in industrial applications. The current probe is placed around the wire where the current flow is to be measured. The advantage to this type of probe is that the circuit does not need to be broken to wire the meter in series with the circuit.

For motors, current tests should be made to ensure that the continuous load rating noted on the motor's nameplate is not exceeded and that all three-phase currents are balanced. If the measured load current exceeds the nameplate rating or the current is unbalanced, the life of the motor will be reduced because of high operating temperature. Unbalanced current may be caused by voltage imbalance between phases, a shorted motor winding, or a high resistance connection.

19.4.4 Resistance Measurements
A DMM may be used to measure resistance in an electrical circuit. The meter must be set to the appropriate resistance function and to the correct range. Power to the electrical circuit must be turned off. Otherwise, damage to the meter may occur. The probes should be placed at the two points where the resistance must be measured across the device. The device should be disconnected on one end to remove the resistance from the remainder of the circuit. The display will indicate the resistance across the device.

19.5 Motors

Properly troubleshooting a motor will depend on the type and rating characteristics of the motor. Human senses, such as sight, sound, feel, and smell should also be incorporated into the troubleshooting techniques.

19.5.1 Single-Phase Motors
There are several types of single-phase motors that require a centrifugal switch to open the start circuit after the motor is up to speed. If the motor hums but will not start, try rotating the shaft by hand. If it now starts, then the centrifugal switch is not operating properly or the start winding is open. Check the winding with an ohmmeter.

19.5.2 Polyphase Motors
If a polyphase motor does not start, the applied voltage may be too low (less than 10% rated voltage). Low voltage causes low starting torque. It is also possible that one phase is open. This condition is called *single phasing*. Check each phase with an ohmmeter. This condition may be recognized by excessive noise level and rapid heat buildup.

If the motor is overloaded, heat buildup may result. Motors will generally operate for short periods on overload but not continuously. Check each phase with a tong ammeter and compare with the nameplate rating.

19.5.3 Overheating of Bearings
Bearings can overheat due to lack of oil, dirty oil, or oil not reaching the shaft. There is also the problem of misalignment of shaft and bearing.

19.5.4 Noise
Improper balance can cause excessive vibration. Be sure that the motor is properly mounted and coupled to the load. In some cases an uneven air gap due to worn bearings may be the problem, or dirt in the air gap may be the problem. Compressed air can be used to blow the dirt away.

19.5.5 Speed
If a rotor of a squirrel-cage induction motor has an open circuit or high resistance, the speed will drop. If the speed in a universal motor is high, the problem may be a shorted field coil.

19.5.6 Component Coils (Relays, Contactors, and Full-Voltage Starters)
Figure 19-13 shows a flow chart for use in troubleshooting component coils. In many cases, shortcuts can be made and success realized. However, whatever system is used, a systematic approach should be followed. Too many times a "hunt-and-try" method only leads to wasted time and possible damage to equipment.

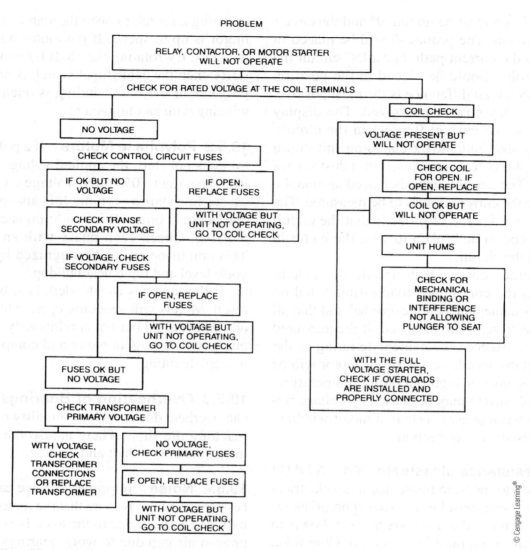

Figure 19-13 **Use of flow chart in troubleshooting.**

19.6 Troubleshooting a Complete Control Circuit

Figure 19-14 shows a power and control circuit in which two 480-VAC, three-phase, 60-Hz motors with magnetic full-voltage starters are used. The control voltage is taken from the secondary of an isolated secondary transformer and is 120 volts.

Depending on the desired machine speed, one or two motors can be used. Heating elements are used in the machine operating fluid to bring the fluid temperature up to 70°F as required by the machine. The machine must be in a given position for start condition and the operation reverses by means of a pressure.

The approach to effective troubleshooting, as noted earlier, is to first segregate the section where the trouble is observed to be occurring. In the circuit shown in Figure 19-14, there are three sections indicated: motors control, heat control, and cycle control. In a particular problem you have observed that the motors and heat control are okay. The problem is that solenoid 1HA is not energizing at the proper time. You should then be concerned only with the cycle-control section.

However, let us first take a look at this entire circuit. Always keep in mind the possible causes of trouble as outlined earlier in the chapter.

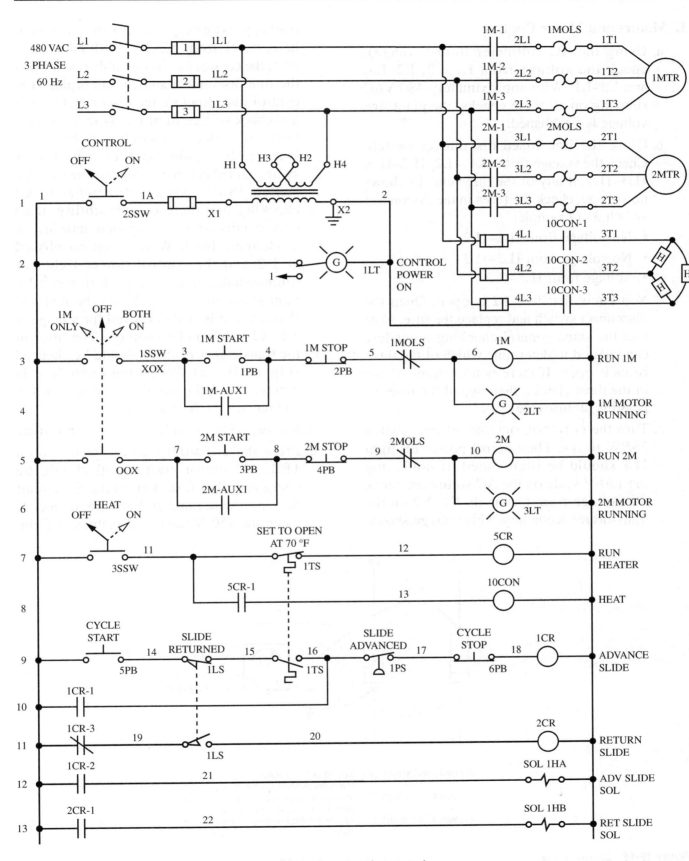

Figure 19-14 Control circuit for motor control, heat control, and motion control.

1. Motors and Motor Control

a. Using the AC voltmeter (600-V range), check the voltage from L1–L2, L2–L3, and L3–L1. With approximately 480 VAC showing on all three checks, the plant line voltage is established.

b. Close the line fused disconnect switch. Check the voltage from 1L1–1L2, 1L2–1L3, 1L3–1L1. If any of the three checks shows no voltage, check the fuses in the disconnect switch. For example:

- No voltage from 1L1–1L2
- No voltage from 1L2–1L3
- Voltage from 1L1–1L3

You know that fuse #2 is open. Open the disconnect switch and replace the fuse. Note that the same general checking procedure can be used to determine which of the three fuses is open. If there is no voltage at any of the three checks, then two of the fuses or possibly all fuses are open.

c. Turn the CONTROL OFF/ON selector switch 2SSW to ON. The control power on light 1LT should be illuminated. If not, using the 150-V scale on the AC voltmeter, check the voltage from terminals X1–X2 on the transformer secondary. This voltage should read approximately 120. If there is voltage here, check from X2–1A. With no voltage here, the control fuse is probably open. Open the line disconnect switch and replace the control fuse. Reclose the disconnect switch and check voltage from 1-2. With no voltage here, the selector switch may be wired incorrectly. Again, open the disconnect switch and check the wiring on the selector switch 2SSW. Always remember in all checking that there is a possibility of an open conductor at a component terminal or at a terminal block. With voltage established at 1-2 and the control power light not illuminated, depress the push-to-test pilot light to determine if the bulb is burned out. When you originally checked for voltage at X1–X2 you found no voltage. The fuses in the transformer primary should be checked (Figure 19-15). When the open fuse is located, open the fused disconnect switch and replace the fuse or fuses.

d. Rotate 1SSW to the left (1M only) position.

e. Press the 1M MOTOR START push-button 1PB. 1M motor starter coil should be energized. If not, start checking the circuit line from wire number 2 to wire number 1, using the 150-V range AC voltmeter. From

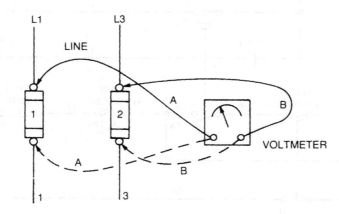

Put lead A on L1; B on L2 — Should read line voltage.
If not, the problem is in the supply.
Put lead A on L1; B on 3 — With line voltage reading, the #2 fuse is good.
With no reading, #2 fuse is open.
Put lead A on 1; B on L2 — With line voltage reading, #1 fuse is good.
With no reading, the #1 fuse is open.

Figure 19-15 **Checking fuses.**

this point on, there can be two methods used.

- *Series checking.* With this method, one probe is placed on wire number 2 and the other progressively moves from wire number 6 to wire number 5 to wire number 4, to wire number 3, and finally to wire number 1. At the point where voltage is indicated, the open circuit is in the first component to the right. For example, with the MOTOR START push-button held operated, there is voltage at 4 but not at 5; the open, then, is in the 1M STOP push-button 2PB.

- *Half-split checking.* With this method, one probe is placed on wire number 2 and the other directly to wire number 4. Thus, in the series checking indicated above, only one additional move with the second probe is required to locate the problem. Although in this particular case it does not seem to have an advantage, there are cases in which there are many components in series. In these cases, considerable time can be saved with the half-split method.

  f. Rotate 1SSW to the right (BOTH ON) position. Press the 1M MOTOR START push-button and the 2M MOTOR START push-button 3PB. 1M motor starter coil and 2M motor starter coil should be energized. With one or both of the motor starter coils not energized, follow the same general procedure in checking as indicated for 1M only.

Apart from the possibility of an open motor starter coil, about the only other problem that may come up in the motor starter circuits is the possibility that one of the overload relay contacts in series with the motor starter coil is defective and open or a jumper is missing on the starter.

2. Heat control

Rotate the HEAT OFF/ON selector switch 3SSW to the ON position. With the temperature of the machine operating fluid above 70°F, neither relay coil 5CR nor contactor coil 10CON will be energized.

With the temperature below 70°F, temperature switch NC contact will be closed, allowing relay coil 5CR to be energized. Relay contact 5CR-1 closes, energizing contactor coil 10CON. Contactor contacts 10CON-1, 10CON-2, and 10CON-3 close, energizing the heating elements.

If the contactor 10CON is not energized:

a. Check to see if relay coil 5CR is energized.

b. Check the voltage from 2–12. If voltage is present, the coil is open or there is an open conductor.

c. No voltage at 2–12 but voltage at 2–11. Check the temperature switch. It may be defective or incorrectly wired.

d. No voltage at 2–11, the HEAT OFF/ON switch may be incorrectly wired or there may be an open conductor.

3. Cycle Control

The same general pattern is followed in checking the cycle control as was followed with the motor and heat control. Press the CYCLE START push-button 5PB.

a. Control relay 1CR does not energize.

Using the series checking method explained, place one voltmeter probe on wire number 2. Then, using the other probe, proceed consecutively through points 18, 17, 16, 15, 14, and 1. At the point that voltage is detected, the open circuit is in the component immediately to the right of this point. Remember that the CYCLE START push-button 5PB must be held operated during these checks.

An example of a problem would be voltage was noted at point 14 but not at 15. There can be three possibilities:

a. Operating dog not holding the switch 1LS in the actuated condition.

b. Limit switch 1LS is incorrectly wired.

c. Open conductor.

Using the half-split method, you would probably start with the second probe at point 16. In this case it would have required only two moves to locate the problem, thus reducing the time to troubleshoot.

The problem of open conductors has been mentioned several times. Remember that push-buttons, selector switches, and pilot lights are generally located on an operator's panel. Relays, contactors, and motor starters are located in a control cabinet. Limit switches, pressure switches, and temperature switches can be mounted remotely on the machine. These placements result in several terminal block connections. Be sure these connections are kept tight.

19.7 Troubleshooting the Programmable Logic Controller

It is difficult to offer a complete and comprehensive approach to troubleshooting the PLC since each PLC manufacturer has taken a slightly different approach to the design of the unit. These differences mean that a diagnostic chart that would apply to one unit would not necessarily apply to another.

There is one area that will generally apply to all units: the input and output devices. This hardware at least will be the same in most cases. It will consist of components supplying input information for motion, pressure, and temperature, and components controlling all outputs, such as solenoids, relays, motor starters, visual indicators, and alarm systems. The area between the inputs and outputs is the responsibility of the central processing unit. This area is seldom approached by the troubleshooter, apart from replacement of units within the processor unless, of course, the problem is with the PLC logic program.

In most cases the PLC manufacturer will supply visual indication of all incoming and outgoing signals. However, there are several areas in which the use of a meter to check volts as well as an ohmmeter to check continuity is useful. Starting with the power supply and continuing through all the hardware, information can be obtained even from the absence of visual indication.

Position indicators that are mechanically attached to a machine or process line may be out of alignment with the operator. Temperature sensors, such as thermocouples, may not be seated properly or may be open. Pressure sensors may be defective and not responding correctly to a given pressure. All these components can be checked for correct

voltage with a meter. Continuity can be checked with an ohmmeter. Problems with output components, such as actuator solenoids, motors, and indicating lamps can be resolved by checking the output signal voltage and the continuity of a suspected circuit.

One manufacturer of PCs has provided diagnostic monitoring I/O modules. They provide the troubleshooter with information to quickly identify the possible source of a problem. In practice, the module actually "learns" the operation of a process. When the operator is satisfied that the process is correct, the monitoring module is used to observe the PC and the system hardware being controlled.

19.8 Electronic Troubleshooting Hints

The electrician does not necessarily need to be an electronics technician to locate and repair many of the problems associated with solid-state equipment. The following hints will help you locate common problems associated with solid-state equipment.

1. Actual test voltages and other measurements should be recorded when the equipment is in good working order. They can then be compared to measurements during troubleshooting. Having records to refer to will speed up troubleshooting.

2. Look for obvious problems by visual inspection (physical damage, overheated terminations, broken wires, corrosion, etc.).

3. Measure supply voltage and current. Is it within the limits of the equipment?

4. Check all circuit breakers and fuses. If circuit breakers or fuses are open, they should not be closed until the fault has been cleared. Otherwise, if the circuit is closed on a fault, operator injury or equipment damage could result.

5. High temperature is the leading cause of failures of electronic equipment. Equipment should be placed in a well-ventilated area so that heat can escape.

6. Transient AC voltage peaks or voltage surges in control circuits may last only milliseconds but can cause problems in the control circuit. A DMM with peak-hold function can be used to determine whether a transient AC peak problem exists.

7. Do not change control settings unless you are familiar with the equipment. If the settings and adjustments are incorrect, resetting them can be a time-consuming process. If you cannot fix the equipment or you are unsure of proper and safe troubleshooting procedure, call the manufacturer.

Recommended Web Links

Students are encouraged to view the following Web sites as a supplement to the concepts presented in this textbook. Review and analyze the array of products that are available for electrical control applications. Many of these sites offer technical information that can help in converting the principles to practical applications. To view catalogs, your PC may require Adobe Acrobat Reader software.

1. Fluke Corporation
   http://www.fluke.com
   Review: Solution Centers/Calibration

2. BK Precision
   http://www.bkprecision.com/
   Review: Test & Measurement Instruments

3. Tektronix
   http://www.tektronix.com
   Review: Products

4. Test Equipment Depot
   http://www.testequipmentdepot.com
   Review: Test Equipment Catalog

Troubleshooting Allen-Bradley Products

1. Allen-Bradley
   www.rockwellautomation.com
   Enter the word "troubleshooting" in the Search box. Troubleshooting guides are provided for individual products.

Achievement Review

1. Explain two different methods of troubleshooting an electrical circuit.

2. Draw a sketch that explains how to check fuses.

3. What types of contacts may cause trouble and yet appear to be in good condition?

4. Why should you never file a contact?

5. What is a *combination problem* as applied to machine control?

6. Explain two methods of detecting grounds in an electrical circuit.

7. Explain how an open in a common line can cause trouble.

8. List a few locations where you may be able to find grounds that have been created in wiring a machine.

9. What are some of the electrical problems with machines that could have been avoided by good housekeeping practices?

10. What condition may lead to a low-voltage problem in an industrial shop?

11. What is the value of a good preventive maintenance program?

12. In the circuit shown in Figure 19-14, the electrician has incorrectly wired the NO and NC limit switch contacts. NO contact is wired to 19–20 and the NC contact is wired to 14–15. With this change, can a cycle be started? If not, why not?

13. In the circuit shown in Figure 19-14, the NC contact on relay 1CR is found in which circuit line?
    a. 10
    b. 11
    c. 12

14. In the circuit shown in Figure 19-14, with the selector switch 1SSW set to OFF, which motors can be started?
    a. M1
    b. M2
    c. Neither

15. In the circuit shown in Figure 19-14, the following condition exists:

    • 120 VAC is measured between wires 5 and 2
    • Pilot light 2LT is not illuminated

    What could be the potential problems?

16. In the circuit shown in Figure 19-14, the following condition exists:

    • The cylinder will not return
    • Relay coil 1CR is OFF
    • Relay coil 2CR is energized

    What could be the potential problems?

17. In the circuit shown in Figure 19-14, the following condition exists:

    • The cylinder will not return
    • Relay coil 1CR is energized
    • 120 VAC is measured between wires 20 and 2

    What could be the potential problems?

18. Using Ohm's law, calculate the current in line A–B of the following diagram:

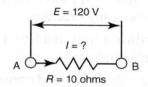

19. Using an ohmmeter, how can you tell if a circuit is open or closed?

20. What is the purpose of the centrifugal switch in some single-phase motors?

21. What can cause noise to develop in a motor?

22. What is recommended for all circuits that serve critical loads?

23. Readings are taken across 3 loads as indicated:

    $V_1 = 10$
    $V_2 = 8$
    $V_3 = 18$

    Calculate the percent voltage unbalance.

20 CHAPTER

Designing Control Systems for Easy Maintenance

OBJECTIVES

After studying this chapter, you should be able to:

- List several important points in the assembly of a machine using electrical control.

- Show how to cross-reference relay contacts and switches with circuit line numbers.

- List several items that can be of help in making better drawings for electrical control.

- Draw a typical control panel showing where components such as relays, motor starters, and disconnecting means should be placed.

- List the advantages and disadvantages of returning all circuit connections to the panel.

- Discuss several concerns with the use of mechanical limit switches.

- Explain the advantages of having all electrical components accessible.

- Name several places where indicating lights can help the troubleshooter.

20.1 Design Considerations

In its final form, a design generally involves some compromise. Considerations such as safety, cost, manufacturing, assembly, and operating conditions enter into the final machine. All of these considerations, and others, show that careful design planning is a prerequisite to ease of maintenance.

Maintenance problems are considered in the design through analyzing items such as:

- The economics of building a competitive product
- The environment in which the machine is to be used
- The problem of installing electrical components after other mechanical and fluid power work has been completed

- The possible presence or development of vibration, shock, or other mechanical motion that may affect the electrical components

For the best chance of success in dealing with maintenance problems, the designer should:

- Present adequate wiring diagrams, layouts, and instructions
- Display evidence of experience in the electrical control field
- Avoid sloppy workmanship

In the assembly of a machine, several points are important:

- Circuit wiring should be exactly as called for on the electrical prints. In case of an error on the prints, corrections should be made before the machine is checked out.

- All conductors and terminals should be properly marked with numbers corresponding to the wiring diagram.
- A sufficient number of terminal block checkpoints should be provided.
- All terminal connections should be double-checked to be sure the conductors are tight in the connectors.
- Care should be taken in pulling conductors through conduit and fittings so that no conductor insulation is cut or scraped.
- Conduit fill, as set up in electrical standards, should not be exceeded.
- In long or difficult-access conduits, three or four spare conductors should be pulled.

In the field installation, a greatly improved overall job can be obtained by close cooperation between the user and the builder. The user must know:

- Total power requirements with the amount of possible low-power factor loads (induction motors)
- Foundation and assembly with relative location of separate control enclosure, if used
- Location of main power connection
- Location of terminal disconnect points on the machine (where machine is disassembled for shipment)

The builder must know:

- User's power voltage, phase, and frequency
- Any special conditions or limitations on power supply
- Unusual ambient temperature condition
- Unusual atmospheric conditions
- User's specification of manufacturer or components (if any)
- User's specification on motor starters, if not full voltage
- A complete and accurate sequence of functions the machine is to follow, which can be difficult to obtain because of:

  a. Inexperience, particularly in the case of a new process
  b. General lack of knowledge
  c. Problems of communication, generally best solved by a cooperative effort on the part of the user and the builder

In all cases, a well-studied, careful analysis and procedure should be followed in selecting, applying, and installing the electrical control. This approach always pays off in terms of reduced maintenance.

In providing some practical help as a guide to designing for easy maintenance, there are two major areas to examine:

1. Diagrams and layouts
2. Locating, assembly, and installing components

20.2 Diagrams and Layouts

There are aids to making circuit diagrams and layouts more useful and to cutting maintenance and troubleshooting time. The numbering of lines on the drawing and cross-referencing the relays and their contacts are two such aids. As shown in Figure 20-1, a normally open (NO) relay contact is cross-referenced by providing the line number where the contact is used in the drawings. A normally closed (NC) relay contact is cross-referenced in the same manner with the addition of a bar drawn under the line number where the contact is used. Switching devices (limit switches, push-buttons, pressure switches, etc.) utilize a similar cross-referencing scheme that shows the line number where additional contacts from the same device are used in the drawings. The difference in these methods is that switching devices use a rectangle, or similar symbol, and do not differentiate between an NO or an NC contact. Note that the line numbers are enclosed in a geometric figure to prevent mistaking them for wire numbers.

All conductors should be properly numbered. A numbering scheme should be carried throughout the entire electrical system. The wire number must change when going through a device such as a limit switch, push-button, relay contact, and so on. The incoming and outgoing conductors as well as the terminal blocks should be labeled with the proper electrical wire numbers. If at all possible, connections to all electrical components should be taken back to one common checkpoint (such as a sequentially ordered row of terminals).

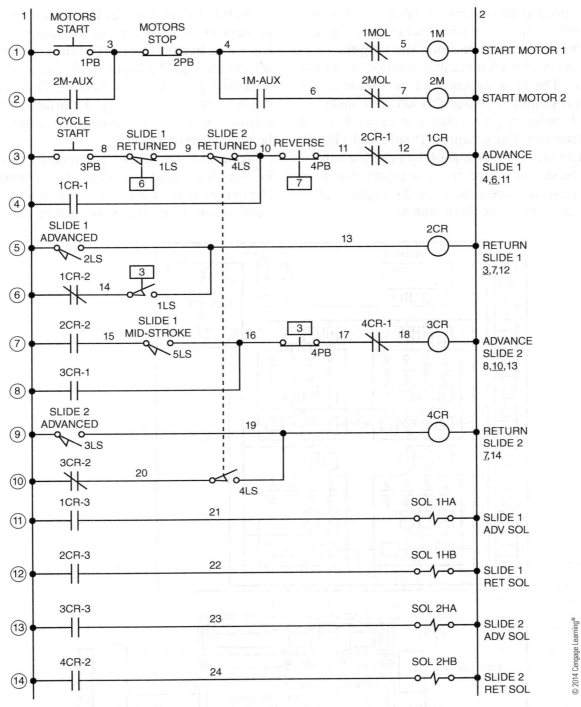

Figure 20-1   **Control circuit shown in Figure 12-13A including line numbers and cross-referencing.**

All electrical elements on a machine should be correctly identified with the same marking as shown on the wiring diagram. For example, if a given solenoid is marked "SOL 1HA" on the drawing, the actual solenoid on the machine should be labeled with the same marking, "SOL 1HA." It is important that these labels are made of a material that will withstand the environment. Also, the labels should be mounted to the machine base instead of being attached to the device itself. If the labels were attached to the devices, the label would be discarded when a defective component was replaced.

A drawing should be made showing the relative location of each electrical component on the machine. The drawing need not be a scale drawing, but it should be reasonably accurate in showing the location of parts relative to each other and in relative size. For example, if solenoid 1HA is located on the left-hand end of the machine base, it should be shown on the machine layout drawings in this position (see Figure 20-2). Figures 20-2 through 20-4 illustrate these points.

Notice from the panel build drawings that all the devices are wired using a consistent scheme. The hot or live wire enters the device from the left or from the top. The load wire then exits the device from the right side or the bottom. This consistent approach to wiring is very important so that any person working on the machine is accustomed to the wiring method used.

A panel layout drawing should be included in the electrical control drawings (see Figure 20-4). The panel layout should show the mounting location for all components inside the cabinet.

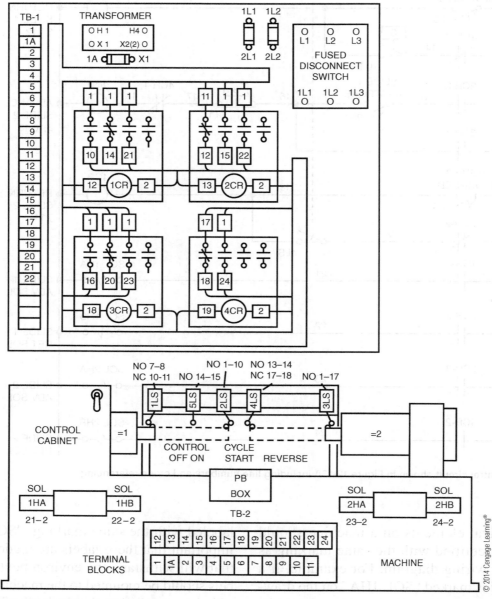

Figure 20-2 **Machine and panel build drawings.**

A device number identifying each component should be shown in the drawings and a label with the device number should be attached to the sub-plate near the component. A cross-referencing scheme should also be provided that links each component to an item on the electrical parts list for the machine.

Push-button layouts should also be provided for all operator interface devices (see Figure 20-3). The layout should include the nameplate information, device number, and cross-referencing to the parts list.

It is important to list all of the components used on the machine on the electrical parts list. Each component should be cross-indexed in some manner with the components as they appear in the circuit diagram or layout.

The component should be described so that the user can obtain a replacement if necessary. An example is shown in Figure 20-4. Only basic information is provided here to describe the components and to cross-reference them to the circuit. In an actual parts list for the machine, the manufacturer along with the part number must be listed for each device. In addition, if any modifications need to be made to the component, the changes should be indicated on the control drawings or on the parts list.

Within any organization, maintenance or troubleshooting can be assisted by standardization of circuit numbers and component numbers of designation. For example, in a motor starter circuit, the numbers 1, 2, 3, and 4 can always be used in the same location. After some experience, the maintenance worker or troubleshooter can remember that the coil is always 4,1; the seal and START button is 3, 4; and the STOP button is always 2,3.

It is understandable that a given arrangement cannot always be used completely. However, with a group of machines that are similar in design, this pattern of standard circuit numbers can be followed to a large extent.

It follows that components can carry a similar standardization. For example, solenoid 4HA may be a clamp forward solenoid. A limit switch 4LS may be a clamp forward stop.

In a group of similar machines, it helps the maintenance personnel if the same designations for specific electrical components can be maintained on all the machines. The worker can then become acquainted with the functions of specific components, such as solenoids and limit switches, regardless of the machine involved.

Another help is to advance circuit numbers for similar components in the tens place. For example, if more than one motor starter is used, the #1 motor starter would utilize wire numbers 1, 2, 3, and 4. The #2 motor starter would use wire numbers 21, 22, 23, and 24. The #3 motor starter would use wires 31, 32, 33, and 34. The numbers should mean more than just numbers. For example, with the proper numbers in the ones and tens places, the troubleshooter can immediately spot the circuit numbers of 32 and 33 as the STOP button on the #3 starter.

In a control circuit diagram having more than one solenoid, each solenoid should be assigned a reasonable description of its function. This description will help maintenance or service personnel to quickly locate a possible source of trouble without referring to the fluid power circuit. The description will aid in quickly locating the solenoid on the machine. This suggestion can also apply to the relays, in which a specific relay controls the initial energizing of a solenoid. For example, in Figure 20-1, a circuit is shown with several solenoids, not only indicating their function but also picturing their relative location.

An item that is almost a must with complex circuits is some form of sequence of operations. This information may be written out in detail, listing component by component each step in the

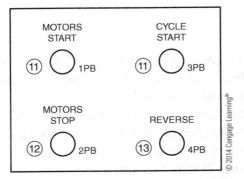

Figure 20-3 Push-button layout showing nameplate information, device numbers, and cross-referencing to parts list.

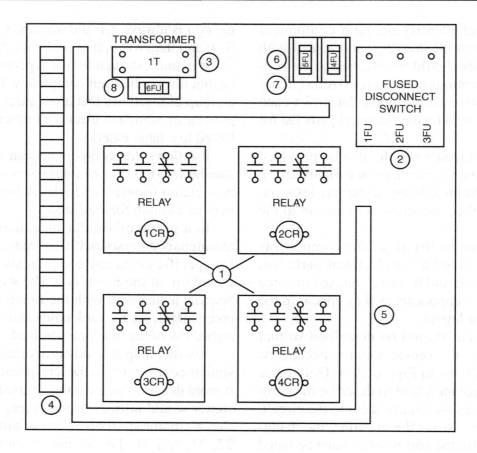

① 600-V MAX, CONTROL RELAY — FOUR-POLE, 120 VAC, 60-Hz COIL, OPEN TYPE

② REMOTE OPERATED FUSED DISCONNECT SWITCH, THREE POLE, 60 A, 600 VAC

③ CONTROL CIRCUIT TRANSFORMER, OPEN TYPE, WITH BUILT-IN FUSE BLOCK; 350 VAC, 240–480 VAC PRIMARY; 120 VAC SECONDARY

④ CHANNEL MOUNTING-TYPE TERMINAL BLOCK, 300 VAC, TUBULAR SCREW TYPE WITH PRESSURE PLATE, NO, 14 WIRE

⑤ 1″ WIDE × 2″ HIGH PANEL WIRING CHANNEL

⑥ TWO-POLE FUSE BLOCK, 30 A, 600-V

⑦ 10-A, 600-V FERRULE-TYPE FUSE

⑧ 15-A, 250-V FERRULE-TYPE FUSE

Figure 20-4 **Panel layout with parts list.**

operation of the circuit. However, a shorter method that gives nearly the same information is the sequence bar chart. (Several examples are shown in Chapter 12.)

The size of the drawing is important, first of all because of the problem of storing them. If every size and shape were allowed, the task of systematic and protective filing of drawings could be burdensome. Page sizes of 8 1/2" × 11" and multiples thereof are generally accepted. The maximum size is usually 34" × 44".

The drawing size can also be a problem for the troubleshooter or maintenance personnel. If the drawing is too large, it is unwieldy to handle at the machine. If it is too small, it is hard to read the schematic.

In the mechanics of drawing the control diagram, there are a few suggestions to "clean up" the drawing and make it easier to follow:

- Evenly space circuit lines. Approximately one-half inch apart is satisfactory in most cases.
- Space component symbols on the line so they are clear and so that space is left for the device identification and the wire numbers. It is also a good idea to leave some additional spare space for future design changes or expansion.
- It is not necessary to carry all circuit lines to their completion. For example, consider the line connection on a push-to-test pilot light. An arrow leading from the line connection on the light and indicating the proper wire number termination of the arrow is sufficient (see Figure 12-12A).
- When the contacts of limit switches or selector switches are separated in a circuit drawing, there may be a problem for the maintenance personnel reading the drawing. There are three things that can be done to help when the circuit is drawn:

a. Arrange the operating switches in the circuit so that the contacts are grouped in close proximity to each other (see Figure 12-13A).

b. If the NO and NC limit switch contacts can be located in the circuit drawing so there are no other lines or contacts between, the two symbols can be joined by a broken line. Do not use the broken line and cross several other circuit lines. This usage only tends to confuse the reader. Figure 20-1 shows an example of the proper use of the broken line connection.

c. A cross-reference can be made between limit switch contact symbols by using an arrow and line number to "tie" the two contacts (Figure 20-5). This arrangement can be used for widely spaced limit switch contact symbols on the same page, or where two contacts of the same limit switch appear on different pages of the circuit diagram. A rectangle may also be used to indicate where additional contacts are used as shown on Figure 20-1.

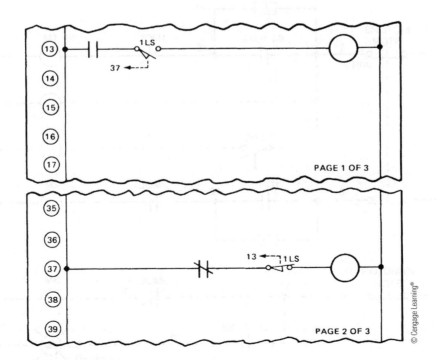

Figure 20-5   Limit switch designation in a control circuit.

Today, electrical drawings are usually generated using a computer software program. The designer should periodically plot a page to ensure the suggestions listed above are complied with and all markings on the drawings are legible.

To increase their utility, many machines are designed with optional features. These may not always be required by the user on the initial job. However, to make it easy to add the option at a later date in the field, provision can be made in the circuit to show its function. The conductors are brought to a terminal block and given wire numbers. The opening in the circuit that is to receive the optional feature is then jumpered. Appropriate notes instruct the user on adding the option. The auxiliary circuit may be shown as a broken line if desired.

Auxiliary circuits are also helpful when the user has equipment to work in conjunction with the machine, for example, a safety device, material feeder, or material removal equipment.

The circuit in Figure 20-6 shows an example of an auxiliary circuit.

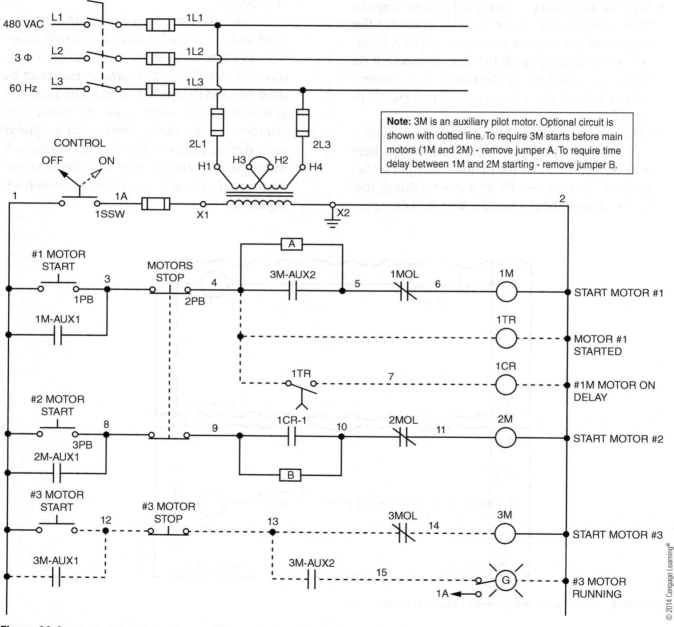

Note: 3M is an auxiliary pilot motor. Optional circuit is shown with dotted line. To require 3M starts before main motors (1M and 2M) - remove jumper A. To require time delay between 1M and 2M starting - remove jumper B.

Figure 20-6   Control circuit showing auxiliary circuits.

20.3 Locating, Assembling, and Installing Components

After the control circuit is designed, one of the first areas to examine is the control panel (Figure 20-7). Although more actual physical work is probably done on the machine, the control cabinet is generally the key to locating trouble.

Maintenance work can be reduced by starting with a systematic and standard arrangement of components. Figure 20-7 shows the layout of a typical control panel with explanatory notes on requirements for good design.

The wiring duct and plastic ties for cabling are universally used. Both save time and expense in the original wiring of the panel and machine. The greatest saving for maintenance is in checking

a circuit. It is time consuming to search out and check a particular conductor that may be at the center of a group of 40 or 50 conductors, all tightly laced. After locating the conductor and changing or correcting the trouble, the cable is usually altered.

Wiring duct is available in various widths and heights. It can be obtained with slots or holes in the sides or open slots to the top. Relay manufacturers use a top strip only, fitted between the relay and supported by center posts. This arrangement further reduces component space requirements.

Where wiring duct is not applicable, plastic ties can be used. They can be installed either by a tool or by hand. If it is necessary to get into the cable, the plastic ties can be quickly cut and new ones applied later in a short time.

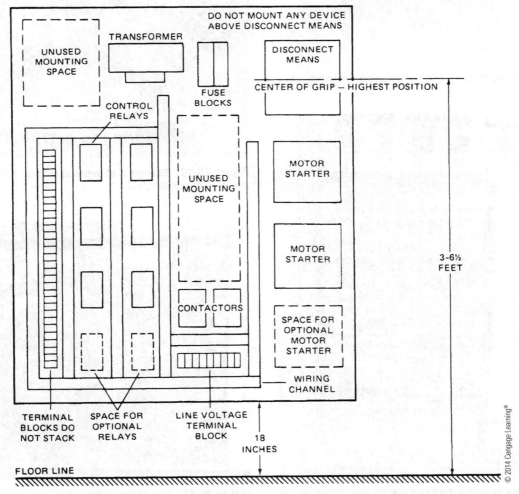

Figure 20-7   Panel layout example. Sub-plate should be removable through enclosure door opening. All line voltage devices (above 120 VAC) should be mounted above or to the right side of control voltage devices. All control devices normally panel mounted should be in one enclosure or a series of multiple door enclosures.

Figure 20-8 shows various types and sizes of wiring duct. Figure 20-9 shows the use of wiring duct and plastic ties.

The use of terminal blocks for electrical control on a machine is important. They can be of tremendous aid in troubleshooting circuits, thus reducing maintenance time. They are valuable when machines must be disassembled for shipment and reassembled in the user's plant. In such an instance, the terminal block on each part containing electrical equipment is a must.

Figure 20-10 shows a variety of terminal blocks available for industrial use. Figure 20-11 shows the use of terminal blocks on a machine.

Terminal blocks provide a central point for troubleshooting by bringing all connections between components back to the control panel. In some cases this design may add expense in the initial wiring of the machine. It also increases the number of conductors used and the size of conduit. Additional terminal blocks and more wiring time are required. However, these disadvantages must be carefully weighed against the maintenance time saved in the user's plant. A much faster job of

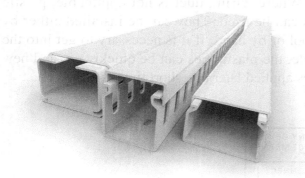

Figure 20-8 **Wiring duct.** (© Peter Sobolev/www.shutterstock. com.)

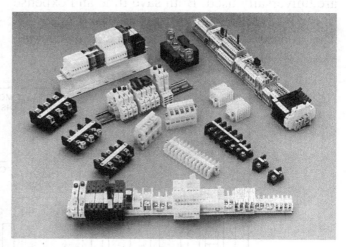

Figure 20-10 **Examples of terminal blocks.** (Courtesy of Rockwell Automation, Inc.)

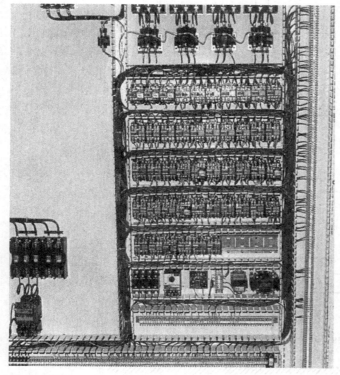

Figure 20-9 **Use of wiring duct and plastic ties.** (Courtesy of HPM Corporation.)

Figure 20-11 **Use of terminal blocks.** (Courtesy of HPM Corporation.)

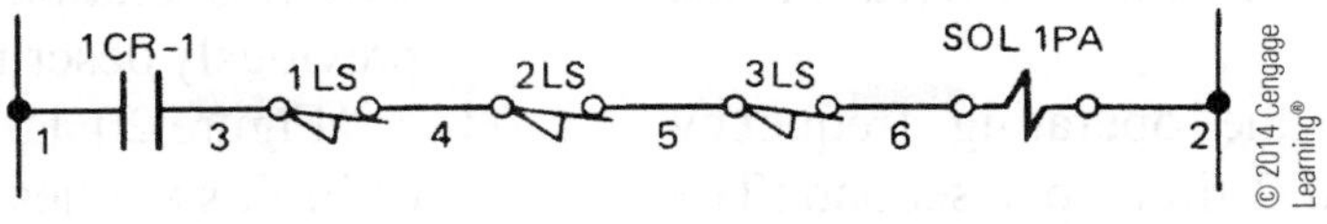

Figure 20-12 Circuit without fuse protection.

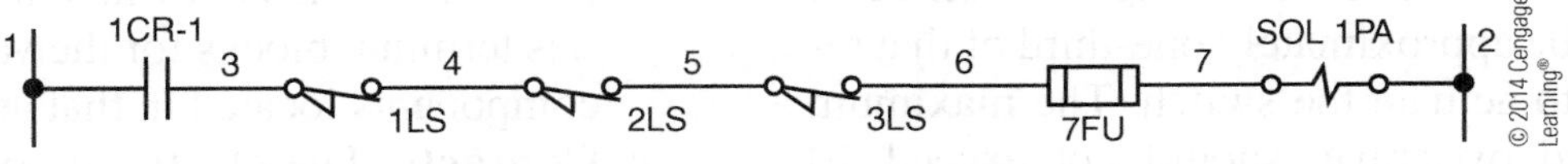

Figure 20-13 Circuit with fuse protection.

troubleshooting can be done if all connections are available in one location.

For example, look at the circuit in Figure 20-12. Limit switches 1LS, 2LS, and 3LS and solenoid 1PA are probably on the machine. Normally, it would be necessary only to bring points 1, 2, and 3 back to the control panel because the source of power and relay 1CR are available here. To provide checkpoints in the control panel for the entire circuit, points 4, 5, and 6 could also be brought back to a terminal strip in the control cabinet.

Many designers find it economical to use either fuses or overload heating elements in the solenoid circuit. They find that it requires less time and is less expensive to replace a fuse or reset an overload than to replace a solenoid coil. In this case the solenoid coil connection would be brought back to the control panel to pick up this protection. An example of this arrangement is similar to that shown in Figure 20-12, except that a protective device is added as shown in Figure 20-13. The fuse selected in this case should have a short time lag to override the momentary inrush current but should open on sustained overload.

20.3.1 Limit Switch Problems   Experience
has shown that a high percentage of limit switch problems results from their application, not from basic design of the unit.

Much of the material in this section is drawn from experience and covers mechanical, vane, and proximity switches.

1. Mechanical Switches

a. Environment. Limit switches are usually located on the machine and are thus subjected to the working area environment. Flying liquids, metal chips and debris, corrosive fumes, excessive temperatures, and shock or vibration are some of the problems.

Generally, a temperature limit of approximately 175°F should not be exceeded. In cases of exposure to higher temperatures, thermally insulated barriers can be of some help.

If heavy shock or vibration is present in the machine, the electrical circuit may be interrupted by the contacts opening. Under these conditions, a low control mass and high contact pressure are important. The switch should be installed so that the direction of the shock forces is in a different plane from the plane in which the contacts operate.

Very few of the housings and/or integral parts of a limit switch will stand up for long in the presence of corrosive fumes. It is helpful to seal the switch at the conduit end, or use a small, light-duty, hermetically sealed switch with a relay.

The problem of flying chips, liquids, or other debris can sometimes be handled by relocation, change of switch operators, or addition of mechanical guards. If the problem is liquids entering the switch through the conduit, change the design of the conduit, or use a multiconductor synthetic rubber cable with a compression seal.

b. Actuation. One of the most common problems is the design of the actuating dog. Such problems as impact forces, direction of travel, speed of the moving part, frequency of repeat operation, and overtravel are always present.

For example, the operating frequency should not exceed 2 times per second. The impact speed should not exceed 400 feet per minute. The minimum designed overtravel should be approximately one-third of that required to actuate the switch. The maximum designed overtravel should not exceed 10 degrees. The angle of the cam engaging the switch actuator should not produce forces that tend to deflect the actuator.

Figure 20-14 shows examples of various cam designs as they are applied to limit switch operation. Solutions to many of the previously described problems are illustrated.

Figure 20-15 shows an actual application of limit switches on a machine. A conductor termination enclosure (referred to as junction box) is also shown. This enclosure carries terminal blocks for the various electrical components located in that area.

c. Contacts. The electrical contacts on limit switches are very similar to contacts on other electrical components. They will not hold up under excessive overloads. Even at normal loads, excessive speed of operation tends to create local heating in the contact,

Excessive impact from improperly designed actuating systems is the leading cause of premature failure of the electromechanical limit switch. At slow speed, impact is rarely troublesome, but as speed increases, impact applied to the switch becomes critical. In today's higher speed machines, therefore, it is important to give proper consideration to correctly designed actuating systems. The following recommendations are designed to assist you in obtaining greater life from limit switches.

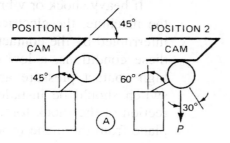

Switch levers should be positioned as nearly parallel with leading edges of cams as possible. Cam A is satisfactory for speeds up to 50 rpm; cam B for speeds up to 200 rpm (nonuniform acceleration of switch lever); cam C for speeds to 400 rpm (uniform or other controlled acceleration).

Cam D is designed with a trailing edge so it can override the switch lever and then return. Cam actuates switch on return also. Lever angle is very important in applications of this type.

The black sector in the roller indicates recommended design limits of angle of pressure *P*. Pressure applied by actuating mechanism to switch operating lever should approximate direction of lever rotation with a variation not to exceed 30° angle of pressure. Changes drastically with rotation of lever. Cam must be designed for proper pressure angles of all positions of lever travel.

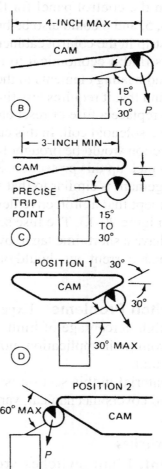

Figure 20-14 **Cam designs for limit switch operation.** *(Courtesy of R. B. Denison Mfg. Co.)*

Figure 20-15 **Industrial application of a limit switch with circuit conductor terminal enclosure.** *(Courtesy of HPM Corporation.)*

which can be damaging. Excessive voltage and high inductive loads will also contribute to contact failure.

A few of the factors that affect sensing and therefore the output switching action are:

- The speed and direction of the material that is to be sensed
- The size, shape, and material that is to be sensed
- Sensitivity adjustment required to properly use the type and amount of energy received by the sensing head

In general, the use of the proximity switch requires greater overall knowledge of electrical work. Because of its relatively complex design as compared to the mechanical or vane switches, more complex maintenance problems can be expected. Maintenance problems generally lie with the amplifier and/or switching unit. Unless there is a complete familiarity with these units, it is better practice to change the complete unit and allow the manufacturer to make the repairs. Plug-in elements make this arrangement relatively simple.

Opposite polarities should not be connected to the contacts of a limit switch unless the switch is designed for such service. Likewise, power from different sources should not be connected to the contacts of a single switch. Normally, if good circuit

design is followed, applying proper polarity does not become a problem.

In many cases the use of double-pole switches can replace two single-pole switches. This arrangement may eliminate a relay that would otherwise be required to interface to the additional circuit.

2. **Vane Switches.** Many of the problems noted for mechanical limit switches can be avoided with the vane limit switch. For example, all integral parts are encapsulated within a cast-aluminum enclosure. This construction prevents the entrance of dust, dirt, coolants, oil, and lubricants into the switch mechanism. **Caution:** Misuse can lead to maintenance problems.

The following are factors to consider when using the vane switch.

a. **Excessive contact load.** The contacts on the vane switch are rated at 0.75 A make and 0.2 A continuous at 115 VAC, 60 Hz. This rating is considerably below the contact rating of the heavy-duty mechanical limit switch. The normal procedure is to use a relay in conjunction with the vane switch to increase its output capacity.

b. **Vane design.** The operation of the vane switch is accomplished by passing an operating vane through the vane slot. The proper size and shape of the vane and the direction and accuracy of the vane entrance into the slot become design problems for each specific application to give maximum operating efficiency.

Figure 20-16 shows various designs of vane switch operators.

3. **Proximity switches.** Most of the information noted for the vane switch can be applied to the proximity switch. In particular, the proximity switch has the advantage of operating in adverse environments.

Under some conditions, the requirements for a separate output unit for switching action may be a disadvantage. One example is supplying space for the amplifier and switching units where many switches are required.

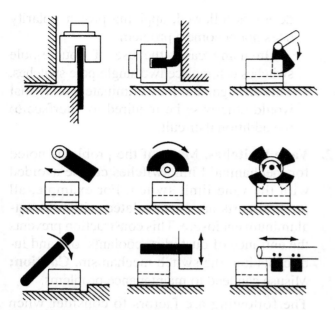

Figure 20-16 Various designs of cams for use with vane limit switches. Arrows indicate direction of cam travel. *(Courtesy of General Electric Company, General Purpose Control, Bloomington, IL.)*

20.3.2 Accessibility of Components

Concerns about accessibility should include the components in the control cabinets, push-button station, and on the machine. Any maintenance problem can be made easier by accessibility.

Problems with accessibility may arise from inadequate planning in the original design state or from lack of cooperation between mechanical and electrical assembly and installation.

For the maintenance worker, the first problem is to locate the trouble. Then the faulty component and/or conductors must be repaired or replaced.

The second problem is time. Downtime is directly proportional to the time it takes to locate the trouble and repair or replace faulty parts. Downtime is a loss both in production and labor. Accessibility of components is important for easy maintenance to minimize downtime.

For the components in the three groups mentioned (the control cabinets, push-button station, and machine), a few suggestions follow:

1. **Control Cabinets**
   a. Keep the components within an easy and safe range of working height: approximately 2 feet to 5½ feet above the floor in the United States. Consult current code requirements for the region where the machine is located.

   b. Design the control panels so they can be removed through the cabinet door opening.

   c. Do not stack terminal blocks on top of one another.

   d. Space components and terminal blocks so that the terminals are accessible for adding or removing conductors.

   e. Use relays with features such as building block assembly for time delay, mechanical latch, and additional standard relay contacts. The changing of NO and NC contacts or the universal contact arrangement is useful.

2. **Push-Button Stations**
   a. Space the individual units so that the wiring to the contact block is accessible.

   b. Where multiple units are used, it may be desirable to hinge the mounting plate to the enclosure so the units are accessible for checking.

3. **The Machine**
   a. Motors should be located so the mounting means are accessible and the motor can be easily removed. The conduit box should be clear so that the cover can be removed.

   b. In the installation of pressure switches, shut-off valves should be provided in the fluid power lines for the removal of the switch unit.

   c. Temperature switches installed in a reservoir for temperature indication or control should be easily removed without draining the reservoir.

   d. Solenoids on valves should be clear of connecting tubes, pipes, conduits, or other mechanical devices. The electrical connections to the solenoids should be accessible for disconnecting. Their connection should be of a bolt-and-nut type and insulated with approved tape. The bolts or screws securing the solenoid to the valve should be accessible for complete removal of the solenoid from the valve.

   e. Planning for conduit and/or wireways and pull boxes within the bed or main structure of a machine always pays off. It clears, from the outside of the machine, a multitude of conduits that might otherwise interfere with component installation. Also, it provides for access to the conduits through pull boxes for easier adding or pulling of conductors within the machine.

20.3.3 Numbering and Identification of Components

The use of the layout drawing is of help in locating the area for components installed on a machine. A further help is to identify these units on the machine with a permanent name plate. The nameplate should carry the identification of the unit exactly as noted on the layout and circuit diagram.

If at all possible, the nameplate should not be attached directly to the component. It should be attached to one side and on a permanent base, because if it is on the component and the component is removed, the nameplate may be lost or may not be changed to the new unit which is installed.

20.3.4 Use of Indicating Lights

It is possible to use many indicating lights on a machine, but this condition is generally the exception. More often, the problem is the absence of indicating lights that give valuable information.

The number of lights is usually in direct proportion to the size and complexity of the control system. For example, a machine with 10 motors may have 10 motors run indicating lights, as compared to one light with one motor.

When a machine goes through multiple sequences, it is valuable to know that each preceding sequence has been completed and the next sequence started.

The machine may have several types of operations available, set up through selector or drum switches. In such cases a light should indicate which type of operation is being used.

Safety and indicating lights are closely tied. For example, with multiple operators, each operator should have a visual indication of the action that the other operators have taken. Other examples of valuable indicating lights are oil temperature, over or under safe operating conditions; safety guards that may have been removed; excessive loading; overtravel; and malfunction of a component in a critical part of a cycle.

For large machinery with many motions, a human-machine interface (HMI) is typically used instead of numerous indicating lights. Each motion, or sequence, may be indicated on the HMI screen to provide the troubleshooter with valuable status indication.

20.3.5 Use of Plug-in Components

Components such as relays, limit switches, solenoids, timers, and pyrometers are available in convenient plug-in designs.

The use of a plug-in component provides many advantages. However, it does not solve all problems. The following points should be considered:

- Does experience indicate a high rate of failure in a particular application? Can a different component give a better life? Can operating conditions be improved, or is replacement the best solution?
- Will downtime for changing the component seriously affect production?
- Does the nature of the component (complexity of circuit) indicate a change rather than repair on the machine?
- Can repair work be accomplished at a bench easier and faster than on a machine?
- Is the machine a prototype or custom design that may require many changes before a final design is reached?
- Is the level of maintenance experience at a given user's shop such that disconnecting and reconnecting circuits on a component may become a problem?
- Are changes being considered by the designer or user for an additional or different type of control after a short time or use?

Recommended Web Links

Students are encouraged to view the following Web sites as a supplement to the concepts presented in this textbook. Review and analyze the array of products that are available for electrical control applications. Many of these sites offer technical information that can help in converting the principles to practical applications. To view catalogs, your PC may require Adobe Acrobat Reader software.

Industrial Control Panels

1. The Panel Shop
   http://www.thepanelshop.com
   Review: Products

2. AEC Corp.
   www.aec-corp.com
   Review: Industrial Control Panels

3. ESL Electrical Systems
   http://www.eslsys.com
   Review: Industrial Control Panels

Enclosures:

1. Hoffman
   www.hoffmanonline.com
   Review: Products

2. Adalet
   http://www.adalet.com
   Review: Products/Explosion-proof enclosures

3. Enclosure Manufacturers
   http://www.enclosuremanufacturers.com
   Review: Enclosures

Terminal Blocks

1. Allen-Bradley
   www.rockwellautomation.com
   Review: Products/Product Catalogs/Industrial Controls/Terminal Blocks and Wiring Systems

Wire

1. Okonite
   http://www.okonite.com
   Review: Product Catalog

2. USA Wire and Cable
   http://www.usawire-cable.com
   Review: Products/600V Single Conductor Wire/THHN, THWN

Assembly Tools

1. Ideal Industries
   http://www.idealindustries.com
   Review: View all products

Control Panel Layout Software

1. AutoDesk—Building Electrical
   http://www.autodesk.com
   Review: Products/AutoCAD Products/ AutoCAD Electrical

2. Microsoft Visio
   http://www.microsoft.com/office/visio/ default.asp
   Review: The feature or functions of this software.

Achievement Review

1. What is the importance of the use of wire numbers? Why would a common approach to a numbering scheme in a plant help troubleshoot problems? Explain your answer.

2. List a few items that would help to clean up an electrical circuit diagram and make it easier to read.

3. Sketch a typical control panel showing the location of components, terminal blocks, wiring duct, and the clear unused space. Use your own judgment as to what would be a good design.

4. List the advantages and disadvantages of bringing all conductors from limit switches, solenoids, and oil-tight units back to a control panel for checking purposes.

5. In using mechanically operated limit switches, what are the limitations on impact speed, overtravel, and contact rating?

6. Are there contact limitations when using vane limit switches? Would you recommend the use of a vane switch in a dusty atmosphere?

7. List five suggestions to follow when mounting and installing electrical components in a control cabinet or push-button station.

8. What are some advantages of indicating lights?

9. What factors would you consider in deciding on the use of plug-in components?

10. The generally accepted size of drawings is _____ and multiples thereof.

    a. 8½" × 11"
    b. 10" × 20"
    c. 9" × 12"

11. Why is the use of terminal blocks for electrical control on a machine important?

    a. To aid in checking circuits.
    b. They are always required when pilot lights are used.
    c. For changing voltage levels.

12. When mounting components on a panel in a control cabinet, they should be located within a range of _____ feet above the floor.

    a. 1–3
    b. 2–5½
    c. 1–6

13. For the circuit shown in Figure 20-6, sketch a panel layout showing all components (including optional devices) mounted to the sub-plate.

14. For the circuit shown in Figure 20-6, design a push-button layout.

15. For the circuit shown in Figure 20-6, sketch a panel build drawing.

Summary of Electrical Symbols

Common Electrical Symbols Used in This Textbook

Conductors Connected		Time-Delay Relay Coil	 1TR
Conductors Not Connected		Instantaneous Contact, NO	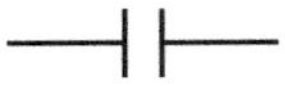
Battery		Instantaneous Contact, NC	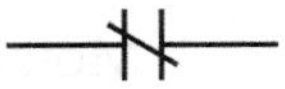
Thermocouple		NO Time Close Contact (ON Delay)	
Relay Coil	1CR	NC Time Open Contact (ON Delay)	
NO Relay Contact		NO Time Open Contact (OFF Delay)	
NC Relay Contact		NC Time Close Contact (OFF Delay)	

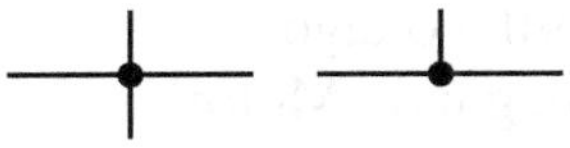

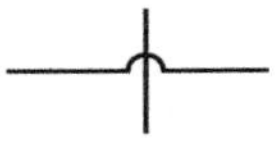

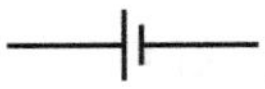

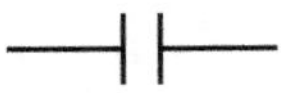

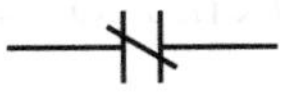

NO Push-button

NC Push-button

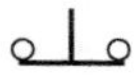

Voltmeter

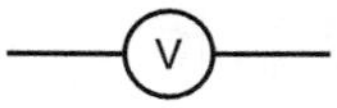

Ammeter

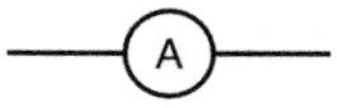

Solenoid Coil

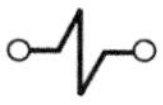

Timer

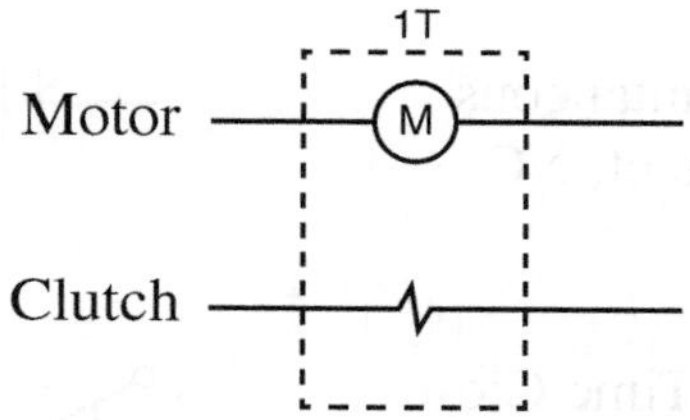

Motor

Clutch

1T

Contacts (operational sequence must be shown by contact: open o, closed ×)

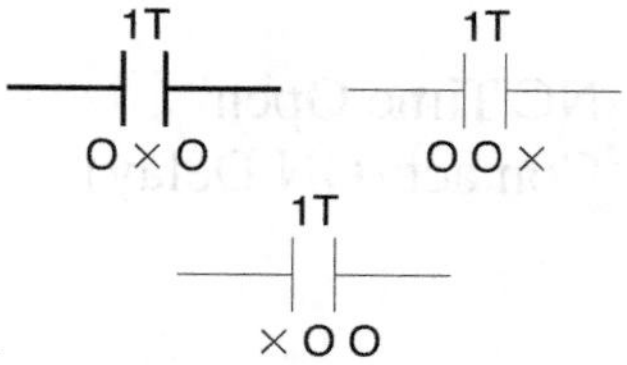

1T   1T

O × O   O O ×

1T

× O O

2 Position Selector Switch

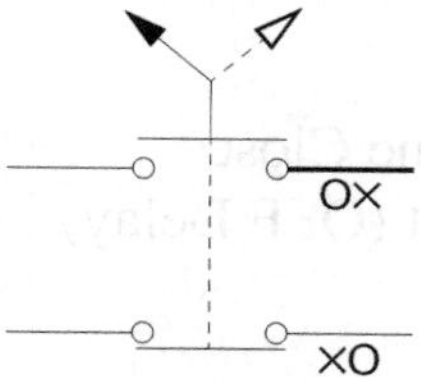

O×

×O

3 Position Selector Switch (Maintained)

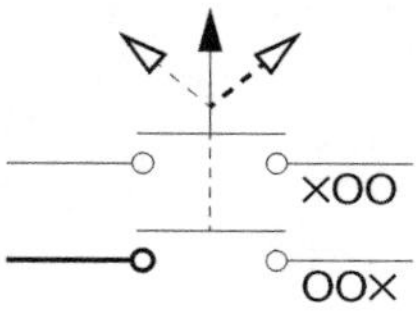

×OO

OO×

3 Position Selector Switch (Spring Return)

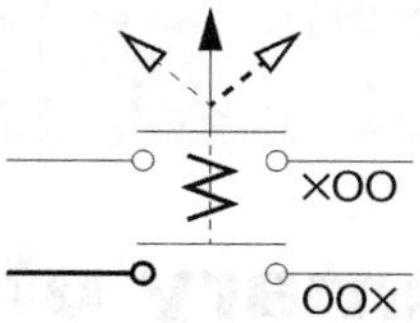

×OO

OO×

Dual-Primary, Single-Secondary Control Transformer

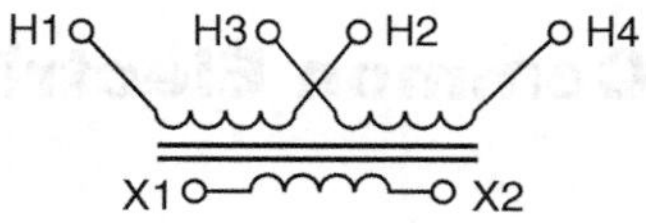

H1 H3 H2 H4

X1 X2

Full-Voltage Magnetic Motor Starter

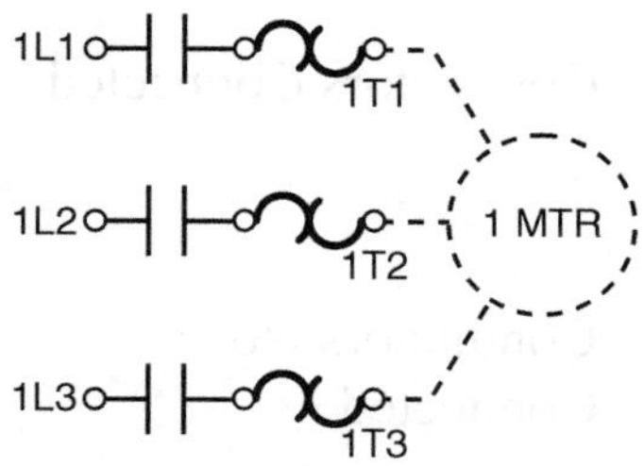

1L1 1T1

1L2 1T2 1 MTR

1L3 1T3

Limit Switch, NO

Limit Switch, NO, Held Closed

Limit Switch, NC

Limit Switch, NC, Held Open

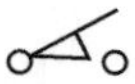

Pressure Switch, NO

Pressure Switch, NC

Temperature Switch, NO

Temperature Switch, NC

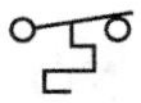

Proximity Limit Switch, NO

Proximity Limit Switch, NC

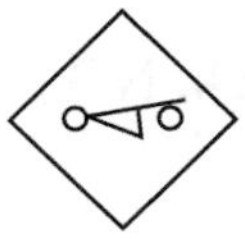

Float Switch, NO

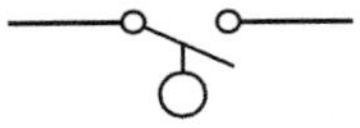

Float Switch, NC

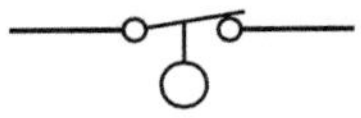

Ferrule-Type Fuse

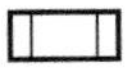

Contactor Coil

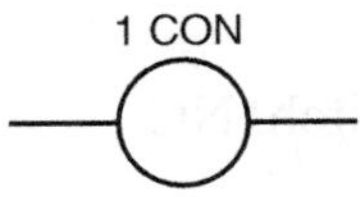

1 CON

Heating Element — HTR — OR — H —

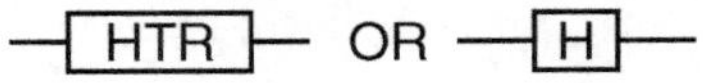

Thermal Overload Relay

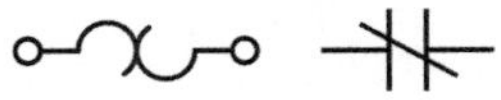

Ground Connection

Three-Pole Fused Disconnect Switch, Ganged with Handle

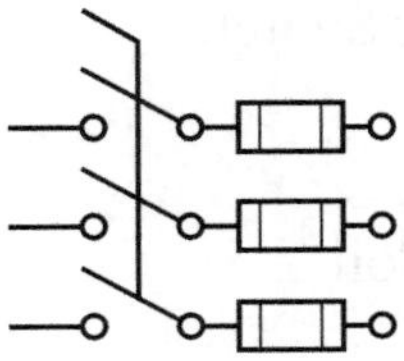

Three-Pole Circut Interrupter, Ganged with Handle

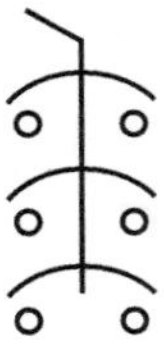

Three-Pole, Thermal-Magnetic Circuit Breaker, Ganged with Handle

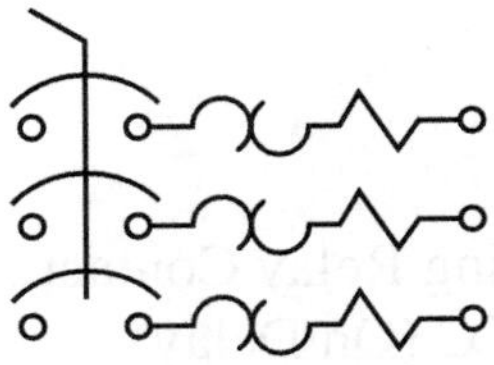

Circuit Breaker

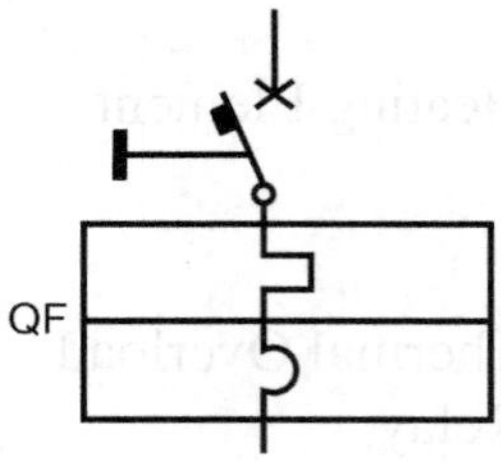

Pushbutton, NO

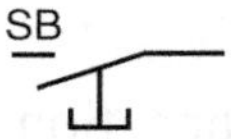

Pushbutton, NC

Selector Switch

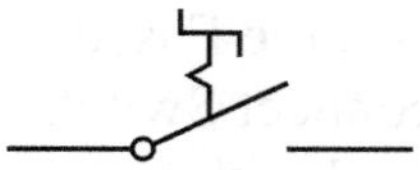

Relay Coil

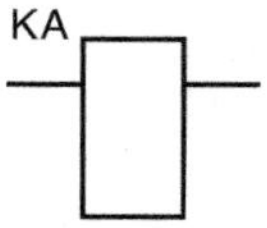

Relay Contact, NO

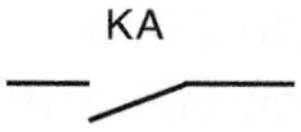

Relay Contact, NC

Timing Relay Coil

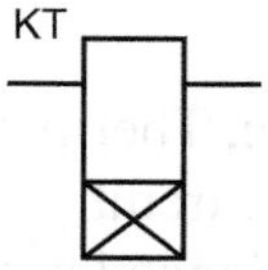

Timing Relay Contact,
NOTC (On Delay)

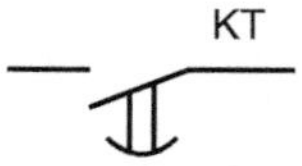

Timing Relay Contact,
NCTO (On Delay)

Timing Relay Contact,
NOTO (Off Delay)

Timing Relay Contact,
NCTC (Off Delay)

Solenoid, Electrical

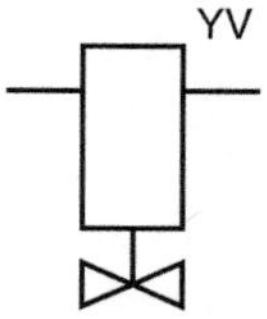

Limit Switch, NO

Limit Switch, NC

Proximity Switch, NO

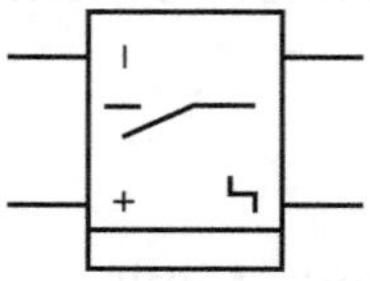

Proximity Switch, NC

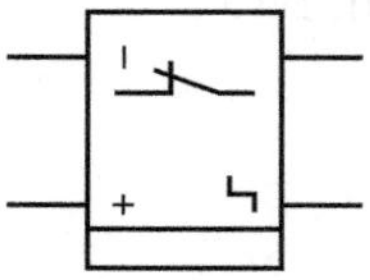

Pressure Switch, NO

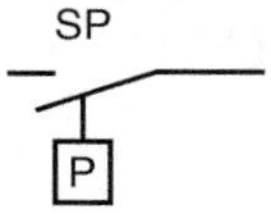

Pressure Switch, NC

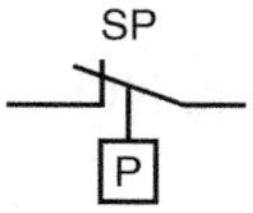

Flow Switch, NO

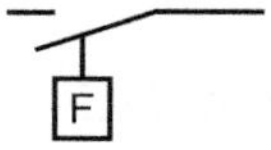

Flow Switch, NC

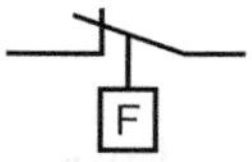

Level Switch, NO

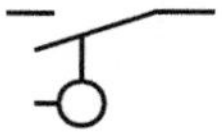

Level Switch, NC

Temperature Switch, NO

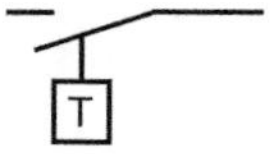

Temperature Switch, NC

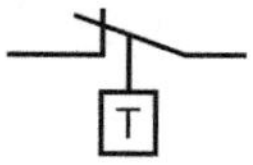

Pilot Light

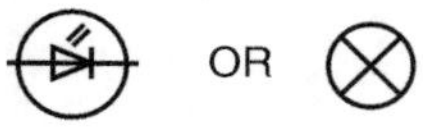

Units of Measurement

Metric System of Measurement
Principal unit for length: meter
Principal unit for capacity: liter
Principal unit for weight: gram

**Prefixes Used for Subdivisions
and Multiples**

pico-(p) =	10^{-12}	deka-(da) =	10^1
nano-(n) =	10^{-9}	hecto-(h) =	10^2
micro-(μ) =	10^{-6}	kilo-(k) =	10^3
milli-(m) =	10^{-3}	mega-(M) =	10^6
centi-(c) =	10^{-2}	giga-(G) =	10^9
deci-(d) =	10^{-1}	tera-(T) =	10^{12}

Measures of Length

10 millimeters (mm) = 1 centimeter (cm)
10 centimeters (cm) = 1 decimeter (dm)
10 decimeters (dm) = 1 meter (m)
1000 meters (m) = 1 kilometer (km)

Square Measure

100 square millimeters (mm^2) = 1 square centimeter (cm^2)
100 square centimeters (cm^2) = 1 square decimeter (dm^2)
100 square decimeters (dm^2) = 1 square meter (m^2)

Cubic Measure

1000 cubic millimeters (mm^3) = 1 cubic centimeter (cm^3)
1000 cubic centimeters (cm^3) = 1 cubic decimeter (dm^3)
1000 cubic decimeters (dm^3) = 1 cubic meter (m^3)

Dry and Liquid Measure

10 milliliters (mL) = 1 centiliter (cL)
10 centiliters (cL) = 1 deciliter (dL)
10 deciliters (dL) = 1 liter (L)
100 liters (L) = 1 hectoliter (hL)

1 liter = 1 dL^3 = the volume of 1 kilogram of pure
water at a temperature of 4°C (35.6°F).

Measure of Weight

10 milligrams (mg) = 1 centigram (cg)
10 centigrams (cg) = 1 decigram (dg)
10 decigrams (dg) = 1 gram (g)
10 grams (g) = 1 dekagram (dag)
10 dekagrams (dag) = 1 hectogram (hg)
10 hectograms (hg) = 1 kilogram (kg)
1000 kilograms (kg) = 1 (metric) ton (t)

Metric and U.S. Customary Conversions

Linear Measure

1 kilometer (km) = 0.6214 mile
1 meter (m) = 3.2808 feet
1 centimeter (cm) = 0.3937 inch
1 millimeter (mm) = 0.03937 inch

1 mile (mi) = 1.609 kilometers (km)
1 yard (yd) = 0.9144 meter (m)
1 foot (ft) = 0.3048 meter (m)
1 inch (in) = 2.54 centimeters (cm)

Square Measure

1 square kilometer (km^2) = 0.3861 square mile = 247.1 acres
1 hectare (ha^2) = 2.471 acres = 107.640 square feet
1 acre (a^2) = 0.0247 acre = 1076.4 square feet
1 square meter (m^2) = 10.764 square feet = 1.196 square yards
1 square centimeter (cm^2) = 0.155 square inch
1 square millimeter (mm^2) = 0.00155 square inch

1 square mile (mi^2) = 2.5899 square kilometers
1 acre (acre) = 0.4047 hectare = 40.47 acres
1 square yard (yd^2) = 0.836 square meter
1 square foot (ft^2) = 0.0929 square meter = 929 square centimeters
1 square inch (in^2) = 6.452 square centimeters = 645.2 square millimeters

Cubic Measure

1 cubic meter (m^3) = 35.314 cubic feet = 1.308 cubic yards
1 cubic meter (m^3) = 264.2 U.S. gallons
1 cubic centimeter (cm^3) = 0.061 cubic inch
1 liter (L) (cubic decimeter) = 0.0353 cubic foot = 61.023 cubic inches
1 liter (L) = 0.2642 U.S. gallon = 1.0567 U.S. quarts

1 cubic yard (yd^3) = 0.7645 cubic meter
1 cubic foot (ft^3) = 0.02832 cubic meter = 28.317 liters
1 cubic inch (in^3) = 16.38716 cubic centimeters
1 U.S. gallon (gal) = 3.785 liters
1 U.S. quart (qt) = 0.946 liter

Weight

1 metric ton (t) = 0.9842 ton (of 2240 pounds) = 2204.6 pounds

1 kilogram (kg) = 2.2046 pounds = 35.274 ounces avoirdupois

1 gram (g) = 0.03215 ounce troy = 0.3527 ounce avoirdupois

1 gram (g) = 15.432 grains

1 ton (t) (of 2240 pounds) = 1.016 metric ton = 1016 kilograms

1 pound (lb) = 0.4536 kilogram = 453.6 grams

1 ounce (oz) avoirdupois = 28.35 grams

1 ounce (oz) troy = 31.103 grams

1 grain (gr) = 0.0648 gram

1 kilogram per square millimeter (kg/mm^2) = 1422.32 pounds per square inch

1 kilogram per square centimeter (kg/cm^2) = 14.223 pounds per square inch

1 kilogram-meter (kg·m) = 7.233 foot-pounds ft.-lb.

1 pound per square inch (psi) = 0.0703 kilogram per square centimeter

1 kilocalorie (kcal) = 3.968 British thermal units (Btu)

Heat

Thermometer Scales

On the Fahrenheit (F) thermometer, the freezing point of water is marked at 32 degrees (°) on the scale, and the boiling point of water at atmospheric pressure is marked at 212°. The distance between these points is divided into 180°.

On the Celsius (C) thermometer, the freezing point of water is marked at 0° on the scale, and the boiling point of water is marked at 100°.

The following formulas are used for converting temperatures:

$$\text{Degrees Fahrenheit} = \frac{9 \times \text{degrees Celsius}}{5} + 32$$

$$\text{Degrees Celsius} = \frac{5 \times (\text{degrees Fahrenheit}) - 32}{9}$$

Circuits

Electrical Potential (in Volts)	$V = IR$ or $V = \mathcal{Z}$
Resistance (in Ω)	$R = \dfrac{V}{I}$
Capacitive Reactance (in Ω)	$X_C = \dfrac{1}{2\pi fC}$
Inductive Reactance (in Ω)	$X_L = 2\pi fL$

Series Impedance

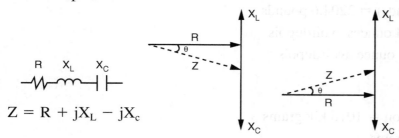

$$Z = R + jX_L - jX_c$$

Power

Real Power (in Watts)	$P = VI \cos \theta$
Apparent Power (in Volt Amperes)	$S = VI$
Reactive Power (in Vars)	$Q = VI \sin \theta$

Three-Phase Power

Wye Connection

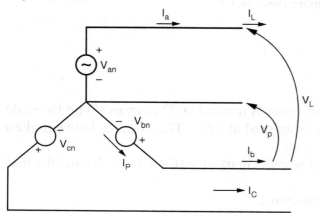

$$V_L = \sqrt{3}\, V_p$$

$$I_L = I_p$$

$$P_{3\phi} = \sqrt{3}\, V_L I_L \cos\theta$$

$$S_{3\phi} = \sqrt{3}\, V_L I_L$$

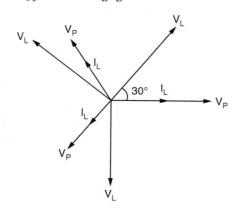

Delta Connection

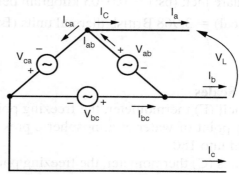

$$V_L = V_p$$

$$I_L = \sqrt{3}\, I_p$$

$$P_{3\phi} = \sqrt{3}\, V_L I_L \cos\theta$$

$$S_{3\phi} = \sqrt{3}\, V_L I_L$$

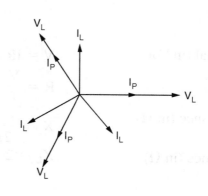

Rules of Thumb for Electric Motors

Here are some simple ways to take the mystery out of determining several important factors concerning electric motors.

C.1 Horsepower versus Amperes

For three-phase motors we find that approximately:

1 hp—575 V requires 1 A of current

1 hp—460 V requires 1.25 A of current

1 hp—230 V requires 2.50 A of current

1 hp—2300 V requires 0.25 A of current

The key to these relationships is 1 hp—1 A on 575 V. All others then become a direct ratio of this figure, e.g., 575/460 times 1.0 equals 1.25 A for 1 hp required on 460 V. Now, for example, with these facts at hand, you can say that a 150-hp motor on 230 V will require approximately 375 A. This fact is obtained as follows: 1 hp on 230 V requires 2.5 A per horsepower, so you simply multiply 2.5 times 150 hp.

C.2 Horsepower Revolutions per Minute—Torque

Torque is simply a twisting force that causes rotation around a fixed point. For example, torque is something we all experience every time we pass through a revolving door. *Horsepower* is what is required when we pass through the door because we are exercising torque at a certain rate of revolutions per minute (rpm). The faster we go through the revolving door, the more horsepower is required. In this case, the torque remains the same.

In relating this concept to motors, we use this rule of thumb:

1 hp at 1800 rpm delivers 3 ft.-lb. of torque

1 hp at 900 rpm delivers 6 ft.-lb. of torque

Using this simple rule, we can see that a 10-hp motor at 1800 rpm delivers 30 ft.-lb. of torque, a 20-hp motor at 1800 rpm delivers 60 ft.-lb. of torque, and a 1-hp motor at 1200 rpm delivers 4.5 ft.-lb. of torque. Here is how to figure torque at a different operating speed. Multiply the torque at 1800 rpm by the ratio 1800/1200. Torque, then, is the inverse ratio of the speed. In other words, for the same horsepower: SPEED DOWN—TORQUE UP.

A quick estimate for the torque of a 125-hp, 600-rpm motor can be figured by the following procedure: 125 hp times 3 ft.-lb. equal 375 ft.-lb. of torque for a 125-hp motor at 1800 rpm. Now, to convert to 600 rpm, multiply 375 times 1800/600 rpm or 1125 ft.-lb.

This rule will enable you to quickly determine the torque a motor is capable of delivering, down to 10 rpm. Below this speed, other factors must be taken into consideration.

C.3 Shaft Size—Horsepower— Revolutions per Minute

Remember the number 1150. It stands for the fact that a 1-inch diameter shaft can transmit 1 hp at 50 rpm. As the shaft speed goes up, so does the horsepower, and by the same ratio. Therefore, if you double the speed, you double the horsepower capacity of the shaft. However, when you double the shaft diameter, the capacity of the shaft to transmit horsepower is increased 8 times. Thus, whatever the shaft size is in inches, cube it and multiply the resulting figure by the proper speed ratio, and the horsepower-transmitting ability of the shaft is determined.

To express this relationship in a formula, use the following:

$$\text{Shaft horse power} = \frac{(\text{Shaft diameter in inches}) \times \text{rpm}}{50}$$

However, it is advisable to be conservative; so modify the results by 75%.

Electrical Formulas

1. Synchronous speed, frequency, and number of poles of AC motors and generators:

$$\text{rpm} = \frac{120 \times f}{P}, \quad f = \frac{P \times \text{rpm}}{120}, \quad P = \frac{120 \times f}{\text{rpm}}$$

where: rpm = revolutions per minute
f = frequency, in cycles per second
P = number of poles

2. Power in AC and DC circuits

Alternating current

To find	Three phase	Two phase *(Four wire)	Single phase	Direct current
Amperes when hp is known	$I = \dfrac{746 \times hp}{1.73 \times V \times Eff \times PF}$	$I = \dfrac{746 \times hp}{2 \times V \times Eff \times PF}$	$I = \dfrac{746 \times hp}{V \times Eff \times PF}$	$I = \dfrac{746 \times hp}{V \times Eff}$
Amperes when kW is known	$I = \dfrac{1000 \times kW}{1.73 \times V \times PF}$	$I = \dfrac{1000 \times kW}{2 \times V \times PF}$	$I = \dfrac{1000 \times kW}{V \times PF}$	$I = \dfrac{1000 \times kW}{V}$
Ampere when kVA is known	$I = \dfrac{1000 \times kVA}{1.73 \times V}$	$I = \dfrac{1000 \times kVA}{2 \times V}$	$I = \dfrac{1000 \times kVA}{V}$	-
Kilowatts input	$kW = \dfrac{1.73 \times V \times I \times PF}{1000}$	$kW = \dfrac{2 \times V \times I \times PF}{1000}$	$kW = \dfrac{V \times I \times PF}{1000}$	$kW = \dfrac{V \times I}{1000}$
Kilovolt-amperes	$kVA = \dfrac{1.73 \times V \times I}{1000}$	$kVA = \dfrac{2 \times V \times I}{1000}$	$kVA = \dfrac{V \times I}{1000}$	-
Horsepower output	$hp = \dfrac{1.73 \times V \times I \times Eff \times PF}{746}$	$hp = \dfrac{2 \times V \times I \times Eff \times PF}{746}$	$hp = \dfrac{V \times I \times Eff \times PF}{746}$	$hp = \dfrac{V \times I \times Eff}{746}$

I = Current, in Amps
V = Voltage, in Volts
Eff = Efficiency, in decimals
PF = Power factor, in decimals

kW = Real Power, in Kilowatts
kVA = Apparent Power, in Kilovolt-Amperes
hp = Horsepower output

*For two-phase, three-wire balanced circuits, the amperes in common conductor = 1.41 times that in either of the other two.

Use of Electrical Codes and Standards

E.1 Major Goals

All electrical codes and standards are written with at least two major goals in mind: maximum safety in operation and long, trouble-free operating life. These goals are best accomplished through good basic engineering design.

It is hard to write a single comprehensive code or standard because of the problems involved in applying electrical control. There are many different types of machines and user requirements. From the small machine shop to the large production-line manufacturer, conditions and requirements differ. Even within a given user's plant, conditions change from section to section and from time to time within the same section. Also, the needs or requirements of today will not necessarily be those of tomorrow.

During the past years, many groups have co-ordinated their efforts in writing codes and standards. In most cases they are written from the experience and knowledge of those groups. Therefore, codes and standards reflect the concepts of good design as held by the group that writes them. Differences, additions, or omissions among groups do not necessarily rank one group above another. Rather, it is an indication that they are written for a specific application in a certain locality or field. This specificity, of course, presents a problem to the manufacturer who supplies electrically controlled machines to every group.

One of the manufacturing problems for the builder is economics. (This statement is not meant to imply that good design is in direct proportion to cost. Often a poor design costs more.) The problems vary from one builder to another and arise in the fields of engineering, marketing, and production.

Problems are usually resolved by close cooperation between the builder and user on each specific machine or group of machines. Both the builder and user must be concerned with the need for codes and standards. From a practical sense they must weigh the relative merits of economy against complete adherence to safety requirements. Safety must be the top consideration, however.

The builder and user should be acquainted with the National Fire Protection Association (NFPA) 79, *Electrical Standard for Industrial Machinery*, which applies to industrial equipment for use in the United States. Machinery (and components) built for Europe must comply with the European Corformity or CE Marking directive. The manufacturer must certify that all applicable standards for the product (machine) have been complied with. For an industrial piece of equipment these standards would typically include safety, ergonomics, wiring, noise levels, and so on. Other countries may utilize other specific guidelines for the proper and safe use of machinery in their country. The NFPA electrical standard and others are reviewed, revised, and reissued on a regular basis. In addition, codes of local municipalities and states should be followed where applicable.

The point here is that the machine builder should be familiar with the latest version of the standards that apply to the equipment being built for the region where the machine will be shipped.

E.2 The History of the NFPA 79 Standard

In June 1981 the Joint Industrial Council (JIC) Board of Directors acknowledged the dated state of the electrical and electronic standards and requested that the NFPA 79 incorporate the material and topics covered by the JIC electrical (EMP-1-67, EGP-1-67) and electronic (EL-1-71) standards with the intention that the JIC standards would eventually be declared superceded. The NFPA Standards Council approved the request with the stipulation that the material and topics incorporated from the JIC standards be limited to areas related to electrical shock and fire hazards. The 1985 edition reflected the incorporation of the appropriate material from the JIC electrical (EMP-1-67, EGP-1-67) standards not previously covered. The 1991 edition includes additional references to international standards. The 1994 edition of NFPA 79 was reformatted in accordance with the ANSI style manual (8th edition).

The next publication of NFPA 79 was released in 2002/2003. In this release, the writing committee attempted to harmonize the NFPA 79 standard with HS1738, the SAE Electrical Standard for Industrial Machines and with EN60204, the European general requirements for electrical equipment of industrial machines. The ANSI style format was maintained in the release; however, the clauses were rearranged to match the European standard.

E.3 Tables

The following 9 tables are to be used only for general information and reference. In specific industrial applications, the current and applicable codes and standards should be consulted. The tables that follow are reproduced from the 1994 NFPA 79 *Electrical Standard for Industrial Machinery* through the courtesy of the National Fire Protection Association Inc.

Table 1—Fuse and circuit breaker selection: motor, motor branch circuit, and motor controller

Fuse class with non-time-delay	Full-load current (%)		
	AC-2	AC-3	AC-4
R	300	300	300
J	300	300	300
CC	300	300	300
T	300	300	300
Fuse class with time delay[1]	Type[2] of application		
	AC-2	AC-3	AC-4
RK-5[3]	150	175	175
RK-1	150	175	175
J	150	175	225
CC	150	300	300
Instantaneous trip circuit breaker[4]	800	800	800
Inverse trip circuit breaker[5]	150	250	250

Note: Where the values determined by Table 7.2.10.1 do not correspond to the standard sizes or ratings, the next higher standard size, rating, or possible setting shall be permitted.

[1] Where the rating of a time-delay fuse (other than CC type) specified by the table is not sufficient for the starting of the motor, it shall be permitted to be increased but shall in no case be permitted to exceed 225%. The rating of a time-delay Class CC fuse and non-time-delay Class CC, J, or T fuse shall be permitted to be increased but shall in no case exceed 400% of the full-load current.

[2] Types of starting duty are as follows:

(a) AC-2: All light-starting duty motors, including slip-ring motors; starting, switching off.

(b) AC-3: All medium starting duty motors including squirrel-cage motors; starting, switching off while running, occasional inching, jogging, or plugging but not to exceed 5 operations per minute or 10 operations per 10 minutes and all wye-delta and two-step autotransformer starting motors.

(c) AC-4: All heavy starting duty motors including squirrel-cage motors; starting, plugging, inching, jogging.

[3] Unless a motor controller is listed for use with RK-5 fuses, Class RK-5 fuses shall be used only with NEMA-rated motor controllers.

[4] Instantaneous trip circuit breakers shall be permitted to be used only if they comply with all of the following:

(a) They are adjustable.

(b) Part of a combination controller has motor-running protection and also short-circuit and ground-fault protection in each conductor.

(c) The combination is especially identified for use.

(d) It is installed per any instructions included in its listing or labeling.

(e) They are limited to single motor applications; circuit breakers with adjustable trip settings shall be set at the controller manufacturer's recommendation, but not greater than 1300% of the motor full-load current.

[5] Where the rating of an inverse time circuit breaker specified in Table 7.2.10.1 is not sufficient for the starting current of the motor, it shall be permitted to be increased but in no case exceed 400% for full-load currents of 100 amperes or less or 300% for full-load currents greater than 100 amperes.

Table 2—Relationship between conductor size and maximum rating or setting of short-circuit protective device for power circuits

Conductor size (AWG)	Maximum rating fuse or inverse time* circuit breaker (amperes)
18	See footnote
16	See footnote
14	60
12	80
10	100
8	150
6	200
4	250
3	300
2	350
1	400
0	500
2/0	600
3/0	700
4/0	800

* Maximum ratings and type of branch short-circuit and ground-fault protective devices for 16 AWG and 18 AWG shall be determined in accordance with 12.6.1.

Table 3—Running overcurrent limits

Kind of motor	Supply system	Number and location of overload units (such as trip coils, relays, or thermal cutouts)
1-phase, AC or DC	2-wire, 1-phase AC or DC ungrounded	One in either conductor
1-phase, AD or DC	2-wire, 1-phase AC or DC, one conductor grounded	One in ungrounded conductor
1-phase, AC or DC	3-wire, 1-phase AC or DC, one conductor grounded	One in either ungrounded conductor
3-phase, AC	Any 3-phase	Three, one in each phase

Note: For 2-phase power supply systems, see Section 430.37 of *NFPA 70*.

Table 4—Single conductor characteristics

Wire size (AWG or kcmil)	Cross-sectional area, nominal (cm/mm²)	DC resistance at 25°C (77°F) (ohms/1000 ft.)	Minimum number of strands		
			Nonflexing (ASTM Class)	Flexing (ASTM Class)	Constant flex (ASTM Class/ AWG Size)
22 AWG	640/0.324	17.2	7(')	7(')	19(M/34)
20	1020/0.519	10.7	10(K)	10(K)	26(M/34)
18	1620/0.823	6.77	16(K)	16(K)	41(M/34)
16	2580/1.31	4.26	19(C)	26(K)	65(M/34)
14	4110/2.08	2.68	19(C)	41(K)	41(K/30)
12	6530/3.31	1.68	19(C)	65(K)	65(K/30)
10	10380/5.261	1.060	19(C)	104(K)	104(K/30)
8	16510/8.367	0.6663	19(C)	(\)	(−)
6	26240/13.30	0.4192	19(C)	(\)	(−)
4	41740/21.15	0.2636	19(C)	(\)	(−)
3	52620/26.67	0.2091	19(C)	(\)	(−)
2	66360/33.62	0.1659	19(C)	(\)	(−)
1	83690/42.41	0.1315	19(B)	(\)	(−)
1/0	105600/53.49	0.1042	19(B)	(\)	(−)
2/0	133100/67.43	0.08267	19(B)	(\)	(−)
3/0	167800/85.01	0.06658	19(B)	(\)	(−)
4/0	211600/107.2	0.05200	19(B)	(\)	(−)
250 kcmil	− /127	0.04401	37(B)	(\)	(−)
300	− /152	0.03667	37(B)	(\)	(−)
350	− /177	0.03144	37(B)	(\)	(−)
400	− /203	0.02751	37(B)	(\)	(−)
450	− /228	0.02445	37(B)	(\)	(−)
500	− /253	0.02200	37(B)	(\)	(−)
550	− /279	0.02000	61(B)	(\)	(−)
600	− /304	0.01834	61(B)	(\)	(−)
650	− /329	0.01692	61(B)	(\)	(−)
700	− /355	0.01572	61(B)	(\)	(−)
750	− /380	0.01467	61(B)	(\)	(−)
800	− /405	0.01375	61(B)	(\)	(−)
900	− /456	0.01222	61(B)	(\)	(−)
1000	− /507	0.01101	61(B)	(\)	(−)

Notes:

(B), (C), (K): ASTM Class designation B and C per ASTM B 8; Class designation K per ASTM B 174.

('): A class designation has not been assigned to this conductor but it is designated as 22-7 AWG in ASTM B 286 and is composed of strands 10 mils in diameter (30 AWG).

(\): Nonflexing construction shall be permitted for flexing service per ASTM Class designation B 174, Table 3.

(−): Constant flexing cables are not constructed in these sizes.

Table 5—Single conductor insulation thickness of insulation in mils *[average/minimum (jacket)]

Wire size (AWG or kcmil)	A Average/ minimum	B Average/ minimum (jacket)
22 AWG	30/27	15/13(4)
20	30/27	15/13(4)
18	30/27	15/13(4)
16	30/27	15/13(4)
14	30/27	15/13(4)
12	30/27	15/13(4)
10	30/27	20/18(4)
8	45/40	30/27(5)
6	60/54	30/27(5)
4–2	60/54	40/36(6)
1–4/0	80/72	50/45(7)
250–500 kcmil	95/86	60/54(8)
550–1000	110/99	70/63(9)

*A: No outer covering. B: Nylon covering.

Source: ANSI/UL 1063, Table 1.1, *NEC* Construction.

Table 6—Conductor ampacity based on copper conductors with 60°C (140°F), 75°C (167°F), and 90°C (194°F) insulation in an ambient temperature of 30°C (86°F)

Conductor size (AWG)	Ampacity		
	60°C (140°F)	75°C (167°F)	90°C (194°F)
30	—	0.5	0.5
28	—	0.8	0.8
26	—	1	1
24	2	2	2
22	3	3	3
20	5	5	5
18	7	7	14
16	10	10	18
14	20	20	25
12	25	25	30
10	30	35	40
8	40	50	55
6	55	65	75
4	70	85	95
3	85	100	110
2	95	115	130
1	110	130	150
1/0	125	150	170
2/0	145	175	195
3/0	165	200	225
4/0	195	230	260
250	215	255	290
300	240	285	320
350	260	310	350
400	280	335	380
500	320	380	430
600	355	420	475
700	385	460	520
750	400	475	535
800	410	490	555
900	435	520	585
1000	455	545	615

Notes:

(1) Wire types listed in 12.3.1 shall be permitted to be used at the ampacities listed in this table.

(2) The sources for the ampacities in this table are Table 310.16 of NFPA 70.

Table 7—Maximum conductor size for given motor controller size

Motor controller size	Maximum conductor size (AWG or kcmil)
00	14 AWG
0	10
1	8
2	4
3	1/0
4	3/0
5	500 kcmil

Note: See NEMA ICS 2, Table 2, 110-1.

Table 8—Minimum radii of conduit bends

Conduit size		One-shot and full-shoe benders		Other bends	
Metric Designator	Trade Size	mm	in.	mm	in.
16	½	101.6	4	101.6	4
21	¾	114.3	4½	127	5
27	1	146.05	5¾	152.4	6
35	1¼	184.15	7¼	203.2	8
41	1½	209.55	8¼	254	10
53	2	241.3	9½	304.8	12
63	2½	266.7	10½	381	15
78	3	330.2	13	457.2	18
91	3½	381	15	533.4	21
103	4	406.4	16	609.6	24
129	5	609.6	24	762	30
155	6	762	30	914.4	36

Table 9—Minimum size of equipment grounding conductors and bonding jumpers

Rating or setting of automatic overcurrent device in circuit ahead of the equipment (not exceeding amperes)	Copper conductor size (AWG or kcmil)
10	16
15	14
20	12
30	10
40	10
60	10
100	8
200	6
300	4
400	3
500	2
600	1
800	1/0
1000	2/0
1200	3/0
1600	4/0
2000	250
2500	350
3000	400
4000	500
5000	700
6000	800

Application of Electric Heat

F.1 Calculating Heat Requirements

This appendix deals primarily with applications on and in the following:

- Relatively large metal objects, such as extruder barrels and platens for plastic molding machines
- Fluids used with hydraulic power and control systems

For the basic design of a heating element there are two important questions to answer:

1. How many watts (W) or kilowatts (kW; 1 kW = 1000 W) of electrical power will be required to bring the application to the required temperature in a given period of time?
2. How many watts or kilowatts of electrical power will be required to maintain the required temperature?

Under controlling conditions, the time required to bring the heated item from ambient temperature to the preset temperature, known as the *warm-up period,* varies depending on the amount of power applied. Unless the warm-up period of time is critical, it is generally more economical and practical to try to match the power applied to the load. The load requirements vary depending on the type of material to be heated, the percentage of time that the heat is required at the preset level, and heat losses. In practice, the power will generally be on 80% of the time to maintain the required temperature.

From a practical viewpoint, the required temperature is generally within a range. For example, in the plastic molding industry most materials are typically available within a 300°F to 700°F range. Therefore, if calculations are made for the top of the range (700°F), then the power will be on a lower percentage of the time and the warm-up period will be shorter for the bottom of the range (300°F).

It can be difficult to make the actual calculation of the watts or kilowatts of electrical power required to heat an object to a desired temperature in a definite time period. The main problem lies in determining heat losses, which include the following:

- Radiation from the heated object (affected by contour, condition of the surface exposed, and temperature difference between the heated item and the ambient)
- Material being processed (affected by type, amount, and rate of processing)
- Conduction to other parts of the machine (affected by temperature differential and type and placement of an insulating medium)

Tables and charts showing various types of heat losses are available from manufacturers of resistance heating elements. These items are a great help in approximating losses.

If the heating system can be assumed to be 100% efficient (no losses), and assuming that the object to be heated is steel, the information on kilowatt-hours (kW·h) required can be easily calculated:

$$kW \cdot h = \frac{\text{Weight of object in pounds} \times \text{specific heat of steel } (0.12) \times (\text{final temperature–initial temperature in degrees Fahrenheit})}{3412 \times \text{warm-up period in hours}}$$

For example, given a 300-pound steel object to raise its temperature from 70°F to 670°F in 1 hour the kilowatts required are:

$$\frac{300 \times 0.12 \times (670 - 70)}{3412 \times 1} = 6.3 \text{ kW}$$

If the warm-up period requirement is cut to one half-hour, then the kilowatt requirement doubles to 12.6 kW.

With cylindrical barrels such as are used on plastic injection molding machines, working in the 300°F to 700°F range, the designer will generally start with as many kilowatts for losses as were calculated for heating the barrel. The losses are based on the maximum operating temperature of 700°F. Operation is in an ambient temperature of 70°F.

F.2 Selection and Application of Heating Elements

Basically, a *heating element* is resistance wire in a magnesium-oxide filler with a sheath around it. The sheath material may be steel, stainless steel, or copper. The material may be tubular or flat rectangular in cross-sectional shape.

After the total kilowatt requirements are determined, the quantity and wattage of the individual heating elements must be determined. Heating elements with the exact wattage calculated are not usually commercially available. The practical approach is to select the next larger heating element. For example, if requirements are for 975 W per unit, and this size is not available, select a 1000-W heating element.

The number of heating elements is generally dictated by the available mounting space. Coverage of the area available should be as near to 100% as is practical. The most important factor in selecting and installing heating elements is the transfer of heat, which is affected by mating surfaces and how securely the heating elements are held in place. Whether the application is a flat plate, groove, hole, or curved surface, it must be clean and smooth. Efficiency of any heating system is reduced as oxides or foreign materials enter between the heating element and the heated surfaces, thus acting as an insulator.

In the case of heating elements with a high watt density inserted in a groove or a hole, the clearances should be held to 0.005 inch to 0.010 inch. (*Watt density* is the number of watts per unit area of surface, for example, watts per square inch [W/in²].) Two types of heating elements used in industry are shown in Figures F-1 and F-2.

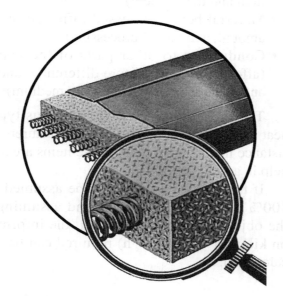

Figure F-1 Cutaway section through a strip heating element. *(Courtesy of Chromalox.)*

Figure F-2 Cutaway section through a cartridge heating element. *(Courtesy of Chromalox.)*

Power Factor Correction

G.1 Apparent Power and Actual Power

Consideration of power factor is essential in manufacturing plants that use relatively large amounts of alternating-current power, such as in plants using a large number of three-phase, squirrel-cage induction motors that are lightly loaded. In this appendix, the problem of power factor is explained.

Power factor describes the ratio between *apparent power* and *actual* or *useful power*. These two power components will be present when squirrel-cage induction motors are used.

To understand apparent power, assume that a coil is wound on a magnetic core and connected to a source of direct current. A switch is provided to close and open the circuit (Figure G-1). When switch A is closed, a current flows through the coil, setting up a magnetic field around the coil (Figure G-2). The magnetic field remains as long as the switch is closed and current flows in the coil. If the switch is opened, the magnetic field collapses, causing the lines of force to cut through the winding. This action causes a current to flow through the coil in the direction opposite to the current that flowed into the coil.

Now, assume that in place of using a direct-current source of power, the coil is connected to a source of alternating-current power (Figure G-3). With alternating current, the current flows in one direction, dies to zero, and then flows in the opposite direction (Figure G-4). Each time the current goes to zero, the lines of force collapse and cut through the winding. The current that flows out of the coil builds to a maximum value just as the supply current drops to zero. The current that flows out of the coil is said to be *out of phase* with the supply current.

If there is no loss in the coil because of heating, the current flowing into the coil (supply current) and the current flowing out of the coil because of the collapsing field will have the same value. If an ammeter is placed in the circuit, it will indicate a current. The amount of current will depend on the design of the coil.

It would appear that the coil (electromagnet) is consuming power, but that is not the case. The current is simply flowing back and forth. A rough mechanical analogy can be made by considering a coil spring being compressed and released. It is from this action of the current flowing back and forth that the term *apparent power* is obtained.

Actual, or *useful, power* is that power used in a heating element such as an electric iron or light bulb. Actual or useful power is also present in the induction motor when the load is applied on the shaft.

When a squirrel-cage induction motor is connected to a source of alternating current, a current flows in the windings of the stator. This current produces a revolving magnetic field. The lines of force from this magnetic field cut through the squirrel-cage rotor and set up a second magnetic field. The second magnetic field reacts against the field of the stator, forcing the rotor to rotate.

With no load connected to the motor, all the actual or useful power required is that amount

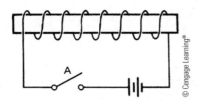

Figure G-1 Electromagnetic circuit with switch connected to DC power.

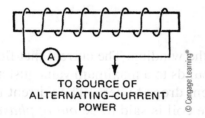

TO SOURCE OF
ALTERNATING–CURRENT
POWER

Figure G-3 Circuit connected to AC power.

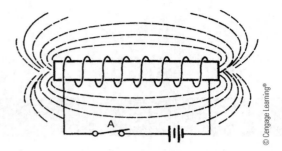

Figure G-2 With switch closed, current flows through the coil, producing magnetic field.

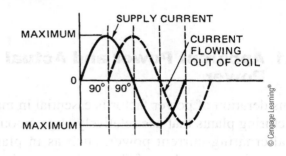

Figure G-4 Flow of AC current.

necessary to overcome the iron and copper losses. There are also friction and windage losses caused by the rotor turning. Therefore, most of the power used by a motor running idle is apparent power.

When a load is applied to the motor, such as connecting it to a pump, real power is supplied. This real power increases directly with the increase of load. Thus, there are two components of current flowing in the motor: *magnetizing current* and *power current* (Figure G-5).

The ratio between useful power and apparent power can be expressed as

$$\text{Power factor} = \frac{\text{Useful power}}{\text{Apparent power}}$$

A typical curve of a squirrel-cage induction motor (Figure G-6) shows that there is a low power factor when the motor is running at light loads. Therefore, the motor requires a large amount of apparent power compared to the amount of useful power.

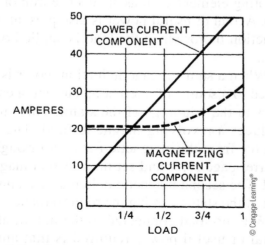

Figure G-5 Graph comparing power current and magnetizing current at various loads.

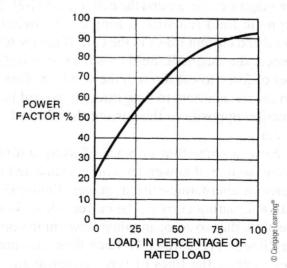

Figure G-6 Typical power factor curve of a squirrel-cage induction motor.

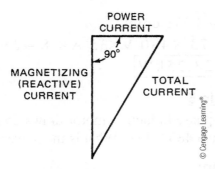

Figure G-7 Power triangle showing power current is 90° out of phase with magnetizing current.

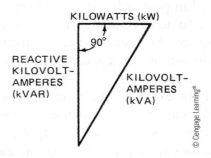

Figure G-8 Power triangle showing power (WATTS) is 90° out of phase with reacting power (kVAR).

G.2 Magnetizing Current and Power Current

Our explanation of apparent power began with describing current in a coil. The two current components, magnetizing and power, are present in the induction motor. However, they cannot be added directly because the magnetizing current is 90 degrees out of phase with the power current. The magnetizing current is said to *lag* the power current by 90 degrees.

Magnetizing current and power current can be added by using a triangle (Figure G-7). By knowing the amount of magnetizing current and power current, a triangle can be drawn to scale. By scaling the drawing, the total current can be determined.

When working with power, it is generally more convenient to work with the units *kilowatts* (kW), *kilovolt-amperes* (kVA), and *reactive kilovolt-amperes* (kVAR). Since the prefix *kilo* means 1000,

- 1 kilowatt-1000 watts (watts = volts × amperes)
- kilovolts-amperes

$$(\text{kVAC}) = \frac{\text{amperes} \times \text{volts}}{1000}$$

- reactive kilovolts-amperes

$$(\text{kVAC}) = \frac{\text{reactive amperes} \times \text{volts}}{1000}$$

The triangle can now be called a *power triangle,* as shown in Figure G-8.

Example 1

Using values of current in amperes and power in kilowatts, assume the following: A three-phase, 480-V squirrel-cage induction motor is carrying a line current of 121 A in each phase. The kilowatt load, measured with a wattmeter, is 80 kW. The kVA of a three-phase circuit is

Solution:
$$S_{3\phi} = \sqrt{3} \, VI$$

Therefore,

$$S_{3\phi} = 1.73 \times 480 \, V \times 121 \, A$$

$$S_{3\phi} = 100{,}000 \, VA = 100 \, kVA$$

The horizontal line of the power triangle is 80 kW. The power triangle is drawn to scale in Figure G-9.

As shown in Figure G-9, the extreme points of the horizontal line are marked A and B, and a vertical line is drawn down from A. Using point B, an arc is struck with a radius equal to the 100 kVA. The intersection of this arc with the vertical line is marked C. The reactive kVA can now be determined by measuring distance AC. The power factor may be expressed as the ratio of kilowatts, or working power, to the total kilovolt-amperes, or apparent power:

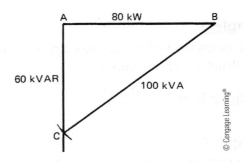

Figure G-9 Power triangle without power factor correction.

$$\text{Power factor} = \frac{AB}{BC} = \frac{kW}{kVA} = \frac{80}{100} = 80\%$$

The objective is to improve the power factor. Therefore, the magnetizing current, or kVAR, must be reduced in proportion to the power current, or kilowatts. To achieve this proportional reduction, a capacitor can be connected across the three lines feeding the motor. The capacitor will supply magnetizing current.

Using the example in Figure G-9, assume that a 15-kVAR capacitor is connected across the motor feed lines. As the capacitor supplies a leading power factor current, the kVAR of the capacitor is subtracted from the kVAR now being supplied to the motor (Figure G-10).

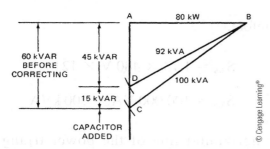

Figure G-10 **Power triangle with power factor correction.**

As represented by BD, the new kVA is 92. The power factor (PF) is now 87%, as follows:

$$PF = \frac{kW}{kVA} = \frac{80}{92} = 87\%$$

Example 2

A wye connected generator has a phase voltage of 69 V. What is the line voltage?

Solution: $V_L = \sqrt{3} \, V_P = 1.73 \times 69 \, V = \boxed{120 \, V}$

Example 3

A wye connected generator has a phase current of 12 A. What is the line current?

Solution: $I_L = I_P = \boxed{12 \, A}$

Example 4

A 480 VAC three-phase system delivers 35 A at 80% power factor. What is the real power delivered?

Solution:

Given $V_L = 480 \, V$
$\qquad I_L = 35 \, A$
$\qquad PF = 0.8$

$P_{3\phi} = \sqrt{3} \, V_L I_L \cos\theta$
$P_{3\phi} = 1.73 \times 480 \, V \times 35 \, A \times .8 = 23,251.2 \, W$
$\boxed{P_{3\phi} = 23.25 \, kW}$

Example 5

A three-phase induction motor draws 25 A and 18 kW from a 480 V line. What is the power factor?

Solution:

Given $V_L = 480 \, V$
$\qquad I_L = 25 \, A$
$\qquad P_{3\phi} = 18 \, kW$

$P_{3\phi} = \sqrt{3} \, V_L I_L \cos\theta$

$rearranging \ terms \ \cos\theta = \dfrac{P_{3\phi}}{\sqrt{3} \, V_L I_L}$

$$= \frac{18,000 \, W}{1.73 \times 480 \times 25} = 0.867$$

$pf = \cos\theta = \boxed{0.867 \, kW}$

Example 6

A three-phase delta connected generator is rated at 50 kVA and 240 V.

a) Find the real power output at a power factor of 90%.

Solution:

$P_{3\phi} = \sqrt{3} \, V_L I_L \cos\theta$

$P_{3\phi} = \sqrt{3} \, V_L I_L$

$P_{3\phi} = S_{3\phi} \cos\theta = 50,000 \, VA \times 0.9 = 45,000 \, W$

$\boxed{P_{3\phi} = 45 \, kW}$

b) Find the line current.

Solution:

$P_{3\phi} = \sqrt{3} \, V_L I_L \cos\theta$

rearranging terms

$I_L = \dfrac{P_{3\phi}}{\sqrt{3} \, V_L \cos\theta} = \dfrac{45,000 \, W}{1.73 \times 240 \, V \times .9} = 120.42 \, A$

$\boxed{I_P = 120.42 \, A}$

c) Find the coil current.

$I_p = \dfrac{I_L}{\sqrt{3}} = \dfrac{120.42}{1.73} = \boxed{69.6 \, A}$

G.3 Determining the Amount of Correction Required

A reasonable approximation of the capacitor size can be obtained from the following rule of thumb: For a 1200-rpm motor, use a figure of 1 kVAR for each 5 hp of induction motor load. This figure will vary slightly with motor speed and the amount of power factor correction required.

A more exact method is to use a table such as the one shown in Figure G-11, which gives the values of capacitors to correct power factor to approximately 95%.

There are many areas in an electrical system in which low power factor can present a problem. The concern can start with the alternator in the power company's station and continue through

Horsepower	Motor synchronous speed (rpm)	Capacitor rating correct power factor to approximately 95%* (kVAR)
	3600	3.0
20	1800	4.5
	1200	5.0
	900	8.0
	3600	4.0
30	1800	5.0
	1200	7.5
	900	8.5
	3600	5.0
40	1800	5.5
	1200	9.0
	900	14.0
	3600	6.5
50	1800	7.0
	1200	10.0
	900	14.5
	3600	7.5
60	1800	10.0
	1200	12.0
	900	17.5
	3600	9.5
75	1800	10.5
	1200	16.5
	900	18.0
	3600	13.0
100	1800	15.0
	1200	17.5
	900	22.0
	3600	20.0
150	1800	22.5
	1200	36.0
	900	45.0

*When these exact sizes are not commercially available, use the next smaller size.

Figure G-11   Capacitor sizes for power factor correction.

power distribution lines and transformers (both step up at the power station and step down at the user's plant). The distribution conductors in the user's plant are also involved.

Low power factor is of primary concern to the power company, as it must supply the magnetizing current that goes throughout the system. Therefore, it is to the power company's advantage if the power factor can be kept as high as practical (90–95%). As a result, most power companies include a penalty clause in their rate structure for low power factor and/or a discount for high power factor.

A high power factor also provides an advantage for the user. For any given distribution system in a user's plant, increased machine load can be added without increasing the size of the power plant feeders and associated equipment by improving the power factor.

In the example shown in Figure G-10, a capacitor is used to correct power factor. Using a capacitor is one of the most simple, direct, and relatively inexpensive methods for correcting power factor, and it can be made on the incoming lines or at individual motors. Remember, the correction from the capacitor always occurs toward the source of power.

If correction is desired only for the purpose of gaining rate advantage, one or more capacitors can be added on the incoming lines. A more satisfactory method for improving the user's load power factor is to connect individual capacitors on the load terminals of each motor starter. However, the optimum condition in installation economy for any given user's plant, may be the use of both methods. It is not unusual to find that savings in the user's power bill pay for the capacitors in one or two years.

Figures G-12 and G-13 show two installation circuits using capacitors.

Generally, it is not economical to apply capacitors directly to motors below 20 hp. If a plant has a predominance of small motors (below 20 hp), connecting capacitors on the line bus is practical (Figure G-12). However, this type of application can become difficult. It is suggested that the local power company be consulted on specific applications.

A check can be made to determine if too large a capacitor has been added on a motor. A voltmeter is attached across any two terminals of the motor (load side of starter). The load is disconnected by deenergizing the motor starter. The resultant rise in voltage as observed on the meter

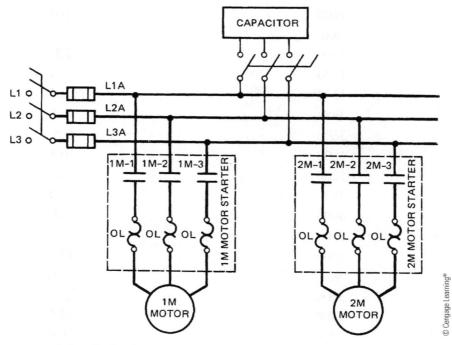

Figure G-12 Capacitors connected on the line bus.

© Cengage Learning®

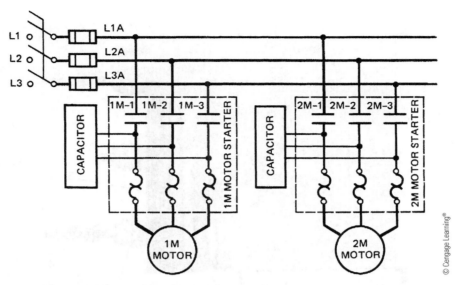

© Cengage Learning®

Figure G-13 Capacitors connected across motors for power factor correction. Conductors carrying load current at line voltage are denoted by heavy lines.

should not exceed 150% of the rated voltage on the motor nameplate.

The most practical location for the installation of a capacitor is on the outside of the motor starter enclosure. However, with long leads (more than 10 feet) from the starter to the motor, the capacitor should be mounted near the motor and the connections made in the motor terminal box.

Figure G-14 illustrates the use of capacitors mounted on a machine. The connections are made in the motor terminal box.

Figure G-14 Power factor correction capacitors mounted on a machine. *(Courtesy of HPM Corporation.)*

*Information courtesy of Westinghouse Electric Company.

G.4 Typical Capacitor Design Features*

Figure G-15 shows a typical capacitor as used for power factor correction. The following design features are listed in reference to the number designations on the photograph.

1. *Case*—The capacitor case is made from heavy-gauge mild steel with all joints welded and reinforced at point of wear for protection of working element during handling, shipment, and service. Capacitor to be either single phase or three phase internally delta connected.

2. *Gland-Sealed Bushing*—The gland-sealed bushing assembly is constructed such that the silicon rubber, thick-walled gland is to be compressed between the porcelain insulator, copper stud, and the top cover of the capacitor. This compression seal will prevent leakage of the dielectric fluid even if the bushing assembly is tilted 15 degrees, which occurs during installation of capacitor elements.

3. *Working Element*—The working element consists of individually wound sections of special capacitor-grade paper, low-loss capacitor-grade synthetic film, and soft aluminum foil. Multiple sections to be arranged to yield the specified KVAC at rated voltage. The grouped sections

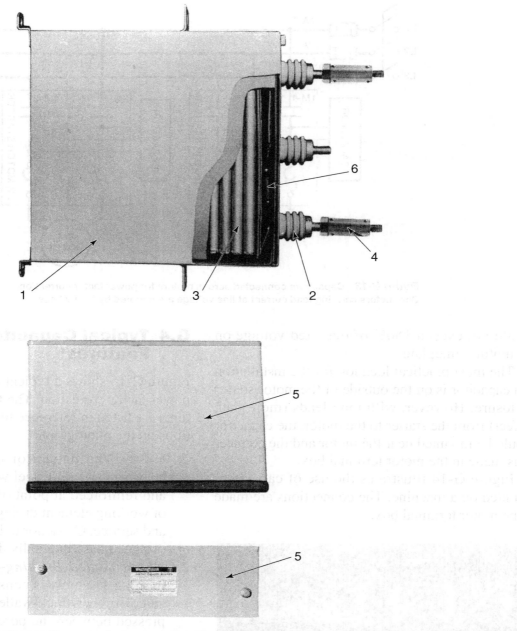

Figure G-15 **Typical capacitor used for power factor correction. See text for key.** *(Courtesy of TECO-Westinghouse Motor Company.)*

are insulated from the steel case by means of high purity Kraft insulation material.

4. *Fuses*—Each three-phase capacitor unit is protected by full range current-limiting fuses that have an interrupting capacity of 200,000 amps.

5. *Dust Cover*—A dust cover is provided consisting of a bushing enclosure and cover. The bushing enclosure is formed from 12-gauge mild steel with a neoprene gasket positioned between the bottom of the enclosure and the top of the capacitor unit. A bolt-on steel cover compresses a vinyl gasket to assure a dustproof and weatherproof seal between the top cover and the bushing enclosure. The dust cover is bolted to the top of the capacitor unit by means of weld nuts and slotted head studs enclosing the bushings, fuses, and connections.

6. *Discharge Resistors*—Individual internal discharge resistors are connected across the terminals to reduce the residual voltage to 50 V or less within one minute after removal from the energized circuit.

AC Input Interface (Module)—An input circuit that conditions various AC signals from connected devices to logic levels required by the processor.

AC Output Interface (Module)—An output circuit that switches the user-supplied control voltage required to control connected AC devices.

Accessible (as applied to equipment)—Admitting close approach because not guarded by locked doors, elevation, or other effective means.

Across-the-Line Starter—A motor starter that connects the motor to full voltage supply.

Actuate—The process of causing contacts to switch states from their normal condition.

Actuator—A cam, arm, or similar mechanical or magnetic device used to trip limit switches.

Address—A reference number assigned to a unique memory location. Each memory location has an address, and each address has a memory location.

Alphanumeric—A character set that contains both numeric and alphabetic characters.

Alternating Current (AC)—Electric current that periodically changes direction and magnitude.

Ambient Conditions—The condition of the atmosphere adjacent to the electrical apparatus; the specific reference may apply to temperature, contamination, humidity, and so on.

Ampacity—The current-carrying capacity, expressed in amperes.

Ampere (A)—The unit of current.

Analog Device—Apparatus that measures continuous information (e.g., current or voltage). The measured analog signal has an infinite number of possible values. The only limitation on resolution is the accuracy of the measuring device.

Analog Input Interface—An input circuit that employs an analog-to-digital converter to convert an analog value, measured by an analog measuring device, to a digital value that can be used by the processor.

Analog Output Interface—An output circuit that employs a digital-to-analog converter to convert a digital value sent from the processor to an analog value that will control a connected analog device.

Analog Signal—One having the characteristic of being continuous and changing smoothly over a given range, rather than switching suddenly between certain levels, as with discrete signals.

Analog-to-Digital Conversion—Hardware and/or software process that converts a scaled analog signal or quantity into a scaled digital signal or quantity.

AND—An operation that yields a logic "1" output if all the inputs are "1" and a logic "0" if any of the inputs are "0."

Apparatus—The set of control devices used to accomplish the intended control functions.

Arithmetic Capability—The ability to perform such math functions as addition, subtraction, multiplication, division, square roots, and so on. A given controller may have some or all of these functions.

Auxiliary Contacts—Contacts in addition to the main circuit contacts in a switching device. They function with the movement of the main circuit contacts.

Auxiliary Device—Any electrical device other than motors and motor starters necessary to fully operate the machine or equipment.

Available Short-Circuit Current—The maximum short-circuit current that can flow in an unprotected circuit.

Binary Coded Decimal (BCD)—A binary number system in which each decimal digit from 0 to 9 is represented by four binary digits (bits). The four positions have weighted values of 1, 2, 4, and 8, respectively, starting with the least significant bit. A thumbwheel switch is a BCD device; when connected to a programmable controller, each decade requires four wires. For example: decimal 9 = 1001 BCD.

Binary Number System—A number system that uses two numerals (binary digits), "0" and "1." Each digit position for a binary number has a place value of 1, 2, 4, 8, 16, 32, 64, 128, and so on, beginning with the least significant (right-hand) digit. Also called base 2.

Binary Word—A related grouping of 1s and 0s having coded meaning assigned by position or as a group; has some numerical value. 10101101 is an 8-bit binary word, in which each bit could have a coded significance, or as a group represents the number 173 in decimal.

Bit—One binary digit. The smallest unit of binary information. Bit is an abbreviation for BInary digiT. A bit can have a value of 1 or 0.

Block Diagram—A diagram that shows the relationship of separate subunits (blocks) in the control system.

Bonding—The permanent joining of metallic parts to form an electrical conductive path that will assure electrical continuity and the capacity to conduct safely any current likely to be imposed (NFPA 79-1984, *National Electrical Code*®).

Boolean Algebra—A mathematical shorthand notation that expresses logic functions, including AND, OR, EXCLUSIVE-OR, NAND, NOR, and NOT.

Branch Circuit—That portion of a wiring system extending beyond the final overcurrent device protecting the circuit. (A device not approved for branch circuit protection, such as a thermal cutout or motor overload protective device, is not considered an overcurrent device protecting the circuit.)

Breakdown Voltage—The voltage at which a disruptive discharge takes place, either through or over the surface of insulation.

Breaker—An abbreviated name for the circuit breaker.

Bus—Power distribution conductors.

Buss—A bank of storage capacitors whose function is to store energy and supply DC power to the transistors in the output bridge as it is required.

Byte—A group of adjacent bits usually operated on as a unit, such as when moving data to and from the memory. One byte consists of 8 bits.

Capacitor—Two conductors separated by an insulator.

Capacitor Sensor—Used in limit switches, these sensors contain an oscillator. When the object to be detected moves into the sensor's range, it creates enough capacitance to set the oscillator in motion.

Central Processing Unit (CPU)—That part of the programmable controller that governs systems activities, including interpretation and execution of programmed instructions. In general, the CPU consists of a logic unit, timing and control circuitry, program counter, address stack, and an instruction register.

Character—One symbol of a set of elementary symbols such as a letter of the alphabet or a decimal number. Characters can be expressed in many binary codes.

Chassis—A sheet metal box, frame, or simple plate on which electronic components and their associated circuitry can be mounted.

Chip—A very small piece of semiconductor material on which electronic components are formed. Chips are generally made of silicon and are generally less than one-quarter-inch square and one-thousandth-inch thick.

Circuit Breaker—A device that opens and closes a circuit by nonautomatic means or that opens the circuit automatically on a predetermined overload of current, without damage to itself when properly applied within its rating.

Circuit Interrupter—A nonautomatic manually operated device designed to open, under abnormal conditions, a current-carrying circuit without damage to itself.

CMOS—An abbreviation for complementary metal oxide semiconductor. A family of very low power, high-speed integrated circuits.

Coaxial Cable—A cable that is constructed with an outer conductor forming a cylinder around a central conductor. An insulating dielectric separates the inner and outer conductors. The complete assembly is enclosed in a protective outer sheath.

Combination Starter—A magnetic motor starter with a manually operated disconnecting device built into the same enclosure with the motor starter. Protection is always added. A control transformer may be added to provide 120 V for control. START-STOP pushbuttons and a pilot light may be added in the enclosure door.

Command—In data communication, an instruction represented in the control field or a frame and transmitted by the primary device. It causes the addressed secondary to execute a specific data link control function.

Communication Network—A communication system that consists of adapter modules and cabling. It is used to transfer system status and data between network PLCs and computers.

Compartment—A space within the base, frame, or column of the equipment.

Compatibility—The ability of various specified units to replace one another with little or no reduction in capability.

Component—*See* **Device.**

Computer Interface—An interface that consists of an electronic circuit designed to communicate data and instructions between a PLC and a computer. It also can be used to communicate with another computer-based intelligent device.

Conductor—A substance that easily carries an electrical current.

Conduit, Flexible Metal—A flexible raceway of circular cross section specially constructed for the purpose of pulling in or withdrawing wires or cables after the conduit and its fittings are in place.

Conduit, Flexible Nonmetallic—A flexible raceway of circular cross section specially constructed of nonmetallic material for the purpose of pulling in or withdrawing wires or cables after the conduit and its fittings are in place.

Conduit, Rigid Metal—A raceway specially constructed for the purpose of pulling in or withdrawing wires or cables after the conduit and its fittings are in place. It is made of metal pipes of standard weight and thickness, permitting the cutting of standard threads.

Contact—One of the conducting parts of a relay, switch, or connector that is engaged or disengaged to open or close an electrical circuit. When considering software, it is the junction point that provides a complete path when closed.

Contact Bounce—The continuing making and breaking of a contact after its initial engaging or disengaging.

Contact Symbology—A set of symbols used to express the control program using conventional relay symbols.

Contactor—A device for repeatedly establishing and interrupting an electrical power circuit.

Continuity—A complete conductive path for an electrical current from one point to another in an electrical circuit.

Continuous Rating—A rating that defines the substantially constant load that can be carried for an indefinite period of time.

Control Circuit—The circuit of a control apparatus or system that carries the electrical signals directing the performance of the controller but that does not carry the main power circuit.

Control Circuit Transformer—A voltage transformer used to supply a voltage suitable for the operation of control devices.

Control Circuit Voltage—The voltage provided for the operation of shunt coil magnetic devices.

Control Compartment—A space within a base, frame, or column of the machine used for mounting the control panel.

Control Logic—The program. Control plan for a given system.

Control Panel—*See* **Panel.**

Control Station—*See* **Push-Button Station.**

Controller, Electric—A device (or group of devices) that serves to govern, in some predetermined manner, the electric power delivered to the apparatus to which the device is connected.

Current-Carrying Capacity—The maximum amount of current that a conductor can carry without heating beyond a predetermined safe limit.

Current-Limiting Fuse—A fuse that limits both the magnitude and duration of current flow under short-circuit conditions.

Cycle—An alternating current waveform that begins from a zero reference point. It reaches a maximum value and then returns to zero. It then reaches a negative maximum value and returns to the zero reference point.

Data—Information encoded in digital form. It is stored in an assigned address of data memory for later use by the processor.

Data Bus—A system of wires (or conductors on a PC board) through which data is transmitted from one part of a computer to another.

Data Highway—A network of connections, which may be made up of any combination of twisted pair cables, coaxial cables, and fiber optic cables. It carries data allowing PLCs and other microcomputer-based devices to communicate.

Data Terminal—A peripheral device that can load, monitor, program, or save the contents of a PLC's memory.

De-energize—To remove electrical power.

Delta Connection—Connection of a three-phase system so that the individual phase elements are connected across pairs of the three-phase power leads A–B, B–C, C–A.

Derate—To reduce the current, voltage, or power rating of a device to improve its reliability or to permit operation at high ambient temperatures.

Device (Component)—An individual apparatus used to execute a control function.

Dielectric—The insulating material between metallic elements of any electrical or electronic component.

Digital—The representation of numerical quantities by means of discrete numbers. It is possible to express in binary form all information stored, transferred, or processed by dual state conditions; for example, ON-OFF, CLOSED-OPEN.

Disconnect Switch (Motor Circuit Switch)—A switch intended for use in a motor branch circuit. It is rated in horsepower and is capable of interrupting the maximum operating overload current of a motor of the same rating at the same rated voltage.

Disconnecting Means—A device that allows the current-carrying conductors of a circuit to be disconnected from their source of supply.

Documentation—An orderly collection of recorded hardware and software information covering the control system. These records provide valuable reference data for installation, debugging, and maintenance of the programmable controller.

Downtime—The time in which a system is not available for production due to required maintenance, either scheduled or unscheduled.

Drop-Out—The current, voltage, or power value that will cause energized relay contacts to return to their normal deenergized condition.

Dual-Element Fuse—Often confused with time delay, dual element is a manufacturer's term describing fuse element construction.

EEPROM—Electrically erasable programmable read-only memory (provides permanent storage of the program and can be easily changed with the use of a manual programming unit or standard CRT).

Electrical Equipment—The electromagnetic, electronic, and static apparatus as well as the more common electrical devices.

Electrical Optical Isolator—A device that couples input to output using a light source and detector in the same package. It is used to provide electrical isolation between input circuitry and output circuitry.

Electrical System—An organized arrangement of all electrical and electromechanical components and devices in a way that will properly control the machine or industrial equipment.

Electromechanical—A term applied to any device in which electrical energy is used to magnetically cause mechanical movement.

Electronic Control—Electronic, static, precision, and associated electrical control equipment.

Enclosure—A case, box, or structure surrounding the electrical equipment to protect it from contamination; the degree of tightness is usually specified (such as NEMA 12).

Encoder—An electronic, mechanical, or optical device used to monitor the circular motion of a device.

Energize—To apply electrical power.

EPROM—Erasable programmable read-only memory (PROM that can be erased with ultraviolet light and then reprogrammed).

Ethernet—A common system of switching devices and cables connected in a data highway.

Examine On—Refers to a normally open (NO) contact instruction in a relay ladder diagram.

Examine Off—Refers to a normally closed (NC) contact instruction in a relay ladder diagram.

Exposed (as applied to electrically live parts)—Capable of being inadvertently touched or approached nearer than a safe distance by a person. It is applied to parts not suitably guarded, isolated, or insulated (NFPA 70-1984, *National Electric Code*®).

External Control Devices—Any control device mounted external to the control panel.

Fail-Safe Operation—An electrical system designed so that the failure of any component in the system will prevent unsafe operation of the controlled equipment.

Fault—An accidental condition in which a current path that bypasses the connected load becomes available.

Feeder—The circuit conductors between the service equipment or the generator switchboard of an isolated plant and the branch circuit overcurrent device.

Ferroresonance (transformer)—Phenomenon usually characterized by overvoltages and very irregular wave shapes and associated with the excitation of one or more saturable inductors through capacitance in series with the inductor.

Fiber Optics—Transparent strands of glass or plastic used to conduct light energy.

Filter (electrical)—A network consisting of a resistor, capacitor, and inductor used to suppress electrical noise.

Frequency—The number of recurrences of an event within a specific time period, expressed in cycles per second or hertz.

Fuse—An overcurrent protective device containing a calibrated current-carrying member that melts and opens a circuit under specified overcurrent conditions.

Fuse Element—A calibrated conductor that melts when subjected to excessive current. The element is enclosed by the fuse body and may be surrounded by an arc-quenching medium such as silica sand. The element is sometimes referred to as a *link*.

Gate—A circuit having two or more input terminals and one output terminal, and in which an output is present when, and only when, the prescribed inputs are present.

Grounded—Connected to earth or to a conducting body that serves in place of the earth.

Grounded Circuit—A circuit in which one conductor or point (usually the neutral or neutral point of a transformer or generator windings) is intentionally grounded (earthed), either solidly or through a grounding device.

Grounded Conductor—A conductor that carries no current under normal conditions. It serves to connect exposed metal surfaces to an earth ground to prevent hazards in case of breakdown between current-carrying parts and exposed surfaces. If insulated, the conductor is colored green, with or without a yellow stripe.

Guarded—Covered, shielded, fenced, enclosed, or otherwise protected by suitable covers or easings, barriers, rails, screens, mats, or platforms to prevent contact or approach of persons or objects to a point of danger.

Hard Copy—A printed document of what is stored in the memory.

Hardware—Includes all the physical components of the programmable controller, including the peripherals.

Hardwired Logic—Logic control functions that are determined by the way devices are interconnected, as contrasted to programmable control in which logic control functions are programmable and easily changed.

Heat Rise—An increase in the temperature of a solenoid caused by resistance to current flow, eddy currents, and hysteresis.

Hermetic Seal—A mechanical or physical closure that is impervious to moisture or gas, including air.

Hertz (Hz)—Cycles per second; the unit measuring the frequency of alternating current.

Hexadecimal Numbering System (HEX)—A number system that uses the numerals 0 through 9 and the letters A through F. The system uses base 16.

Hysteresis (Magnetic)—Refers to a certain magnetic property of iron, which causes a power loss when iron is magnetized and demagnetized due to the change in magnetic flux in the iron lagging behind the change that causes it.

Image Table—An area in the PC memory dedicated to the I/O data. On and off are represented by 1s and 0s, conditions, respectively. During every I/O scan, each input controls a bit in the input image table. Each output is controlled by a bit in the output image table.

Impedance—Measure of opposition to flow of current, particularly alternating current. It is a vector quantity and is the vector sum of resistance and reactance. The unit of measurement is the ohm.

Inching—*See* **Jogging.**

Input—Information sent to the processor from connected devices through the input interface.

Input Device—Any connected equipment that will supply information to the central processing unit, such as switches, push-buttons, sensors, or peripheral devices. Each type of input device has a unique interface to the processor.

Inrush Current—In a solenoid or coil, the steady-state current taken from the line with the armature blocked in the rated maximum open position.

Instruction—A command or order that will cause a PC or a PLC to perform one prescribed operation.

Interconnecting Wire—A term referring to connections among subassemblies, panels, chassis, and remotely mounted devices; does not necessarily apply to internal connections of these units.

Interconnection Diagram—A drawing showing all terminal blocks in the system, with each terminal identified.

Interface—A circuit that permits communication between the central processing unit and a field input or output device.

Interlock—A device actuated by the operation of another device with which it is directly associated. The interlock prevents the operation of a device if the opposing operation is currently in process.

Interrupting Capacity—The highest current at rated voltage that a device can interrupt.

I/O—Abbreviation for input/output.

I/O Address—A unique number assigned to each input and output. The address number is used when programming, monitoring, or modifying a specific input or output.

I/O Module—A plug-in type assembly that contains more than one input or output circuit. A module usually contains two or more identical circuits, for example, 2, 4, 8, or 16 circuits.

I/O Update—The continuous process of revising each and every bit in the I/O tables, based on the latest results from reading the inputs and processing the outputs according to the control program.

Isolated I/O—Input and output circuits that are electrically separated from any and all other circuits of a module. They are designed to allow for connecting field devices that are powered from different sources to one module.

Isolation Transformer—A transformer used to separate one circuit from another.

Jogging (Inching)—A quickly repeated closure of the circuit to start a motor from rest to accomplish small movements of the driven machine.

Joint—A connection between two or more conductors.

Kilohertz (kHz)—1000 Hertz.

Kilowatts (kW)—1000 Watts.

Ladder Diagram—An industry standard for representing relay-logic control systems.

Ladder Element—Any one of the devices that can be used in a ladder diagram, including relays, switches, timers, counters, and so on.

Ladder Program—A type of control program that uses relay-equivalent contact symbols as instructions.

Language—A set of symbols and rules for representing and communicating information among people or between people and machines. The method used to instruct a programmable device to perform various operations.

Latch—A ladder program output instruction that retains its state even though the conditions that caused it to latch on may go off. A latched output must be unlatched. A latched output will retain its last state (on or off) if power is removed.

Latching Relay—A relay that can be mechanically latched in a given position manually, or when operated by one element, and released manually or by the operation of a second element.

LED—Abbreviation for light-emitting diode. A semiconductor diode, the junction of which emits light when passing a current in the forward direction.

Legend Plate—A plate that identifies the function of operating controls, indicating lights, and so on.

Limit Switch—A switch operated by some part or motion of a power-driven machine or equipment to alter the associated electric circuit.

Line Printer—A hard copy device that prints one line of information at a time.

Location—In reference to memory, a storage position or register identified by a unique address.

Locked Rotor Current—The steady-state current taken from the line with the rotor locked and with rated voltage (and rated frequency in the case of alternating current motors) applied to the motor.

Logic—A process of solving complex problems through the repeated use of simple functions that can be either true or false (on or off).

Logic Diagram—A drawing that shows the relationship of standard logic elements in a control system; it is not necessary to show internal detail of the logic elements.

Logic Level—The voltage magnitude associated with signal pulses that represent 1s and 0s in digital systems.

Machine Language—A program written in binary form.

Magnetic Device—A device operated by electromagnetic means.

Magnetostrictive Material—The phenomenon of magnetostriction relates to the stresses and changes in dimensions produced in a material by magnetization and the inverse effect of changes in the magnetic properties produced by mechanical stresses. Nickel, alloys of nickel and iron, invar, nichrome, and various other alloys of iron exhibit pronounced magnetostrictive effects.

Mean—A statistical value useful in analyzing quality, the mean is the true average of the sampled data measurements.

Median—The median is the middle value of all the sampled data collected. It is a statistic useful in quality analysis.

Memory—That part of the programmable controller in which data and instructions are stored either temporarily or semipermanently. The control program is stored in the memory.

Microprocessor—A digital electronics logic package capable of performing the program control, data processing functions, and execution of the central processing unit.

Microsecond—One millionth of a second.

Millisecond—One thousandth of a second.

Mode—Useful in quality analysis, the mode is the value that occurs most frequently in the sampled data.

Modem—Acronym for MODulator/DEmodulator. It is a device that employs frequency-shift keying for the transmission of data over a telephone line. It converts a two-level binary signal to a two-frequency audio signal and vice versa.

MOS—Metal-oxide semiconductor.

Motor Circuit Switch—*See* **Disconnect Switch.**

Motor Junction (Conduit) Box—An enclosure on a motor for the purpose of terminating a conduit run and joining the motor to power conductors.

Network—An organization of devices that are interconnected for the purpose of intercommunication.

Noise—Random, unwanted electrical signals normally caused by radio waves or electrical or magnetic fields generated by one conductor and picked up by another.

Nominal Voltage—The utilization voltage (see the appropriate NEMA standard for device voltage ratings).

Nonvolatile Memory—A type of memory whose contents are not lost or disturbed if operating power is lost.

Normally Closed, Normally Open—When applied to a magnetically operating switch device such as a contactor or relay or to the contacts thereof, these terms signify the position taken when the operating magnet is deenergized. When applied to a switching device such as a push-button or a limit switch, these terms signify the condition of the switch contact when no force is being applied to the device. The terms apply only to nonlatching types of devices.

Ohm (Ω)—Unit of electrical resistance.

Operating Floor—A floor or platform used by the operator under normal operating conditions.

Operating Overload—The overcurrent to which electrical apparatus is subjected under normal operating conditions; such overloads are currents that may persist for a very short time only, usually a matter of seconds.

Operator's Control Station—*See* **Push-Button Station.**

Optical Coupler—A device that couples signals from one circuit to another by means of electromagnetic radiation, usually visible or infrared.

Optical Isolation—Electrical separation of two circuits with the use of an optical coupler.

Oscilloscope—An instrument used to visually show voltage or current waveforms or other electrical phenomena, either repetitive or transient.

Outline Drawing—A diagram that shows approximate overall shape with no detail.

Output (PLC)—Information sent from the processor to a connected device through an interface.

Output Device (PLC)—Any connected equipment that will receive information or instructions from the central processing unit. It may consist of solenoids, motors, lights, and so on.

Overcurrent—Current in an electrical circuit that causes excessive or dangerous temperature in the conductor or conductor insulation.

Overcurrent Protective Device—A device that operates on excessive current and causes and maintains the interruption of power in the circuit.

Overlapping Contacts—Combinations of two sets of contacts actuated by a common means; each set closes in one of two positions and is arranged so that its contacts open after the contacts of the other set have been closed (NEMA IC-1).

Overload—Operation of equipment in excess of normal full-load rating or a conductor in excess of rated ampacity, which, if it were to persist for a sufficient length of time would cause damage or overheating.

Overload Relay—A device that provides overload protection for the electrical equipment.

Panel—A subplate on which the control devices are mounted.

Panel Layout—The physical position or arrangement of the components on a panel or chassis.

Parallel Circuit—A circuit in which two or more of the connected components or contact symbols in a ladder diagram are connected to the same set of terminals, so that current may flow through all the branches. This arrangement is in contrast to a series circuit where the parts are connected end to end so that current flow has only one path.

Pendant (Station)—A push-button station suspended from overhead, connected by means of flexible cord or conduit but supported by a separate cable.

Plugging—A control function that provides braking by reversing the motor line voltage polarity or phase sequence so that the motor develops a counter torque that exerts a retarding force (NEMA IC-1).

Plug-In Device—A plug arranged so that it may be inserted in its receptacle only in a predetermined position.

Potting—A method of securing a component or a group of components by encapsulation.

Power Factor—The ratio between apparent power (volt-amperes) and actual or true power (watts). The value is always unity or less than unity.

Power Supply—The unit that supplies the necessary voltage and current to the system circuitry.

Precision Device—A device that operates within prescribed limits and consistently repeats operations within those limits.

Pressure Connector—A conductor terminal applied with pressure to make the connection mechanically and electrically secure.

Processor—*See* **Central Processing Unit.**

Program—A planned set of instructions stored in memory and executed in an orderly fashion by the central processing unit.

Programming Device—A device for inserting the control program into the memory. The programming device is also used to make changes to the stored program.

PROM—Programmable read-only memory. It can be programmed once and cannot be altered after that.

Protocol—A structured format used to organize binary data. Protocols are necessary for clear communication between devices connected into a data highway because they provide a means of communication independent of the individual system languages used by these devices.

Push-Button Station—A unit assembly of one or more externally operable push-button switches. It sometimes includes other pilot devices, such as indicating lights and selector switches in a suitable enclosure.

Raceway—Any channel designed expressly for and used solely for the purpose of holding wires, cables, or bus bars.

RAM—Random access memory. Referred to as "read/write" as it can be written into as well as read from.

Read/Write Memory—A type of memory that can be read from or written to; can be altered quickly and easily by writing over the part to be changed or inserting a new part to be added.

Readily Accessible—Capable of being reached quickly for operation, renewal, or inspection without a worker having to climb over or to remove obstacles or use a portable ladder, and so on.

Rejection Fuse—A current-limiting fuse with high interrupting rating and with unique dimension or mounting provisions.

Relay—A device that is operative by a variation in the conditions of one electric circuit to affect the operation of other devices in the same or another electric circuit.

ROM—Read only memory. A type of memory that permanently stores information.

Rung—A ladder-program term that refers to the programmed instructions that drive one output.

Scan Time—The time required to read all the inputs, execute the control program, and update local and remote I/O.

Schematic Diagram—A diagram in which symbols and a plan of connections are used to illustrate in simple form the control scheme.

SCR—Silicon-controlled rectifier. A semiconductor device that functions as an electrically controlled switch for DC loads.

Seal-in circuit—A contact that is used to provide a path around an initiating condition so that the initiating condition may be removed while still maintaining continuity in the circuit.

Semiconductor—A device that can function either as a conductor or nonconductor, depending on the polarity of the applied voltage, such as a rectifier or transistor that has a variable conductance depending on the control signal applied.

Semiconductor Fuse—An extremely fast-acting fuse intended for the protection of power semiconductors. Sometimes referred to as a *rectifier fuse.*

Sequence of Operations—A detailed written description of the order in which electrical devices and other parts of the equipment should function.

Series Circuit—A circuit in which all the components or contact symbols are connected end to end. All must be closed to permit current flow.

Shielded Cable—A single-or multiple-conductor cable surrounded by a separate conductor (shield) to minimize the effects of other electrical circuits.

Short-Time Rating—A rating that defines the load that can be carried for a short, definitely specified time with the machine, apparatus, or device being at approximately room temperature at the time the load is applied.

Single Phasing—An occurrence in a three-phase system in which one phase is lost.

Software—Any written documents associated with the system hardware, such as the stored program or instructions.

Solenoid—An electromagnet with an energized coil, approximately cylindrical in form, and an armature whose motion is reciprocating within and along the axis of the coil.

Solid State—Circuitry designed using only integrated circuits, transistors, diodes, and so on.

Standard Deviation—A statistic used to determine the spread between all the data sampled. Its value is useful to analyze quality.

Starter—An electric controller that accelerates a motor from rest to normal speed. (A device for starting a motor in either direction of rotation includes the additional function of reversing and should be designated a controller.)

Static Device—As associated with electronic and other control or information-handling circuits, a device with switching functions that has no moving parts.

Status—The condition or state of a device; for example, on or off.

Stepping Relay (Switch)—A multiposition relay in which wiper contacts mate with successive sets of fixed contacts in a series of steps, moving from one step to the next in successive operations of the relay.

Subassembly—An assembly of electrical or electronic components, mounted on a panel or chassis, that forms a functional unit by itself.

Subplate—A rigid metal panel on which control devices can be mounted and wired.

Swingout Panel—A panel that is hinge mounted in such a way that the back of the panel may be made accessible from the front of the machine.

Symbol—A widely accepted sign, mark, or drawing that represents an electrical device or component thereof.

Temperature Controller—A control device responsive to temperature.

Terminal—A point of connection in an electrical circuit.

Terminal Block—An insulating base or slab equipped with one or more terminal connectors for the purpose of making electrical connections.

Termination—The load connected to the output end of a transmission line and provision for ending a transmission line and connecting to a bus bar or other terminating device.

Three Phase—Three different alternating currents or voltages, identical in magnitude, and 120 degrees out of phase with each other.

Tie Point—A distribution point in circuit wiring other than a terminal connection where junction of leads are made.

Time-Delay Fuse—A fuse that will carry an overcurrent of a specified magnitude for a minimum specified time without opening. The current/time requirements are defined in the UL 198 fuse standards.

Tolerance—A measurement that defines the acceptable range of values in meeting the specifications of quality.

Torque—The turning effect about an axis; it is measured in foot-pounds or inch-ounces and is equal to the product of the length of the arm and the force available at the end of the arm.

Transient (Transient Phenomena)—Rapidly changing action occurring in a circuit during the interval between closing of a switch and settling to steady-state condition or any other temporary actions occurring after some change in a circuit or its constants.

Triac—A semiconductor device that functions as an electrically controlled switch for AC loads.

TTL—Abbreviation for transistor-transistor logic. A semiconductor logic family in which the basic logic element is a multiple-emitter transistor. This family of devices is characterized by high speed and medium power dissipation.

Undervoltage Protection—The effect of a device operative upon the reduction or failure of voltage that causes and maintains the interruption of power to the main circuit.

Undervoltage Release—The effect of a device operative upon the reduction or failure of voltage that causes the interruption of power to the main circuit; voltage will return to the device when nominal voltage is reestablished.

User Memory—The memory where the application control program is stored.

Ventilated—Provided with a means to permit circulation of air sufficient to remove excess heat, fumes, or vapors (NFPA 70-1984, *National Electrical Code®*).

Viscosity—The property of a body that, when flow occurs inside, forces a rise in such a direction as to oppose the flow.

Volatile Memory—A memory whose contents are irretrievable when operating power is lost.

Volt (V)—The unit of electrical pressure or potential.

Watt (W)—The unit of electrical power.

Wheatstone Bridge—A circuit employing four arms in which the resistance of one unknown arm may be determined as a function of the remaining three arms that have known values.

Wireway—A sheet-metal trough with a hinged cover for housing and protecting electrical conductors and cable, in which conductors are laid in place after the wireway has been installed as a complete system.

Wobble Stick—A rod extended from a pendant station that operates the contacts; it functions when pushed in any direction.

Word—The unit number of binary digits (bits) operated on at any time by the central processing unit when it is performing an instruction or operating on data.

Write—The process of putting information into a storage location.

Wye Connections—A connection in a three-phase system in which one side of each of the three phases is connected to a common point or ground; the other side of each of the three phases is connected to the three-phase power line.